UNIVERSITY CONSORTIUM FOR
GEOGRAPHIC INFORMATION SCIENCE

A RESEARCH AGENDA FOR GEOGRAPHIC INFORMATION SCIENCE

UNIVERSITY CONSORTIUM FOR
GEOGRAPHIC INFORMATION SCIENCE

A RESEARCH AGENDA FOR GEOGRAPHIC INFORMATION SCIENCE

EDITED BY
**ROBERT B. MCMASTER
E. LYNN USERY**

CRC Press
Taylor & Francis Group
Boca Raton London New York

CRC Press is an imprint of the
Taylor & Francis Group, an **informa** business

Published 2005 by CRC Press
Taylor & Francis Group
6000 Broken Sound Parkway NW, Suite 300
Boca Raton, FL 33487-2742

ISBN-13: 978-0-367-45434-0 (pbk)
ISBN-13: 978-0-8493-2728-5 (hbk)

**Visit the Taylor & Francis Web site at
http://www.taylorandfrancis.com**

**and the CRC Press Web site at
http://www.crcpress.com**

Library of Congress Cataloging-in-Publication Data

A research agenda for geographic information science / edited by Robert B. McMaster, E. Lynn Usery.
   p. cm.
  Includes bibliographical references and index.
  ISBN 0-8493-2728-8 (alk. paper)
   1. Geographic information systems. 2. Geographic information systems—Research. I. McMaster, Robert Brainerd. II. Usery, E. Lynn (Eddy Lynn), 1951-

G70.212.R47 2004
910'.285--dc22   2004051870

Library of Congress Card Number 2004051870

Acknowledgements

A collective book project that spans over many years is inevitably indebted to many individuals. We would first like to thank the many authors who contributed to this project, and waited patiently for the long-awaited completion. The quality of this project reflects the very talented work of many of our discipline's leading researchers, and we appreciate their initial investment in, and support of, this volume. We also thank the UCGIS board for their continued support of this project throughout its inception, several twists and turns in scope and publishers, and final publication. In particular, the presidents of UCGIS during this period, including Greg Elmes, Harlan Onsrud, Art Getis, and Carolyn Merry, all supported the project in various ways. The project would not have seen the light of day without the talented work of Ms. Caryl J. Wipperfurth, cartographer for the U.S. Geological Survey, who copyedited and formatted the entire volume. Her careful reading improved the final version considerably. Additional thanks go to Michael Goodchild, who provided the Preface to the volume and suggested the final title, and David Mark, who contributed the concluding chapter. We also wish to thank the editors at CRC Press who helped in myriad ways, including Randi Cohen and Jessica Vakili.

In the end, it is the membership of UCGIS—the participants at the many assemblies—who enabled the completion of this book through the meticulous, thoughtful, and time-consuming process of generating and refining a research agenda for the UCGIS. For many of us, these research agenda events have reshaped our understanding of GIS. The project has taken countless hours of work at the assemblies, and after, and years of coordination and writing. We hope this effort, *A Research Agenda for Geographic Information Science*, will benefit our entire organization, and in particular our future students who wish to enter into this remarkable discipline of geographic information science.

Preface

Michael F. Goodchild, University of California, Santa Barbara

Inventiveness has always been a distinguishing characteristic of the human race, but over the past few centuries our rate of new inventions has been accelerating exponentially. Some inventions are so obviously beneficial as to be adopted without question, but others raise significant ethical and moral issues. Some, such as the telescope, have led to important discoveries and in some cases entirely new disciplines dedicated to exploiting their power. This book stems from one such invention, that of the geographic information system or GIS, which emerged in the 1960s, became commercially available in the late 1970s, and has since revolutionized virtually any activity that relies on information about the Earth's surface, from map-making and local government to resource management, wayfinding, warfare, and social and environmental research (for introductory texts on GIS see, for example, Burrough and McDonnell, 1998; Longley *et al.*, 2001; Chrisman, 2002; Clarke, 2003; or DeMers, 2003). Abler (1987, p. 322) has written that "GIS technology is to geographical analysis what the microscope, the telescope and computers have been to other sciences."

The invention and improvement of the telescope required significant advances in the theory of optics, and similarly there exists an intimate relationship between GIS and theory in relevant areas of numerous sciences: cartography, the science of map-making; photogrammetry, geodesy, and surveying, the sciences of Earth measurement; computer and information science; statistics, the science of error and uncertainty; and many more. In the early 1990s the term *geographic information science* (GIScience) was coined to describe this "science behind the systems", and this book is the most extensive effort to date to lay out exactly what it contains—the set of issues and fundamental scientific problems that must be solved if the use of GIS is to advance.

It is a great honor for me to be asked to write these introductory comments. I originally suggested the term GIScience in a keynote address to the 1990 International Symposium on Spatial Data Handling in Zurich (the "spatial" of the keynote title was later changed to "geographic" in the 1992 paper in part as a play on the "S" in "GIS", and in part to emphasize the focus of GIS on the Earth's surface and near-surface, rather than any other space), because it seemed to me that the research the participants were describing was far more profound and general than "data handling" would suggest—I argued that the discipline represented at that meeting was "more than the United Parcel Service of GIS" (Goodchild, 1992, p. 31). The publication of this book is a significant milestone in the development of GIScience, a term that now appears in the titles of numerous journals, in degree programs, and in the names of conferences.

The founding of the University Consortium for Geographic Information Science (UCGIS) was one of the more significant events in the development of GIScience in the United States. UCGIS rightly saw one of its first priorities as the identification of the field's research agenda, to provide some collective guidance on the topics that represented major challenges to the research community. What issues lay at the heart of GIScience, and how did they relate to the research agendas of cognate fields such as statistics or cognitive science? The chapters of this book present the results of a multi-year effort to answer these questions, using the process that is described in detail in Chapter 1. As such, the book will be essential reading for young scholars in GIScience—for the young faculty, graduate students, and senior undergraduates who will be pushing the frontiers of research and the development of GIS technology in the years to come.

REFERENCES

Alber, R.F., 1987, The NSF National Center for Geographic Information and Analysis. *International Journal of Geographical Information Systems,* 1(4): 303–326.

Burrough, P.A. and McDonnell, R.A., 1998, *Principles of Geographical Information Systems*, New York: Oxford University Press.

Chrisman, N.R., 2002. *Exploring Geographic Information Systems*, New York: Wiley.

Clarke, K.C., 2003, *Getting Started with Geographic Information Systems*, Upper Saddle River, NJ: Prentice Hall.

DeMers, M.N., 2003, *Fundamentals of Geographic Information Systems*, Hoboken, NJ: Wiley.

Goodchild, M.F., 1992, Geographical information science. *International Journal of Geographical Information Systems,* 6(1): 31–45.

Longley, P.A., Goodchild, M.F., Maguire, D.J. and Rhind, D.W., 2001, *Geographic Information Systems and Science.* New York: Wiley.

Contents

CHAPTER ONE

Introduction to the UCGIS
Research Agenda

E. Lynn Usery, University of Georgia and U.S. Geological Survey
Robert B. McMaster, University of Minnesota

1.1 INTRODUCTION

During 1987 the National Science Foundation, recognizing the rapid growth in geographic information systems (GIS), sent out a request for proposals for the creation of a national GIS center. This single action would result in major relocations of GIS personnel, as specialists were heavily recruited at key institutions, the creation of a national research agenda, and numerous inter and intra-campus professional relationships established across the country. The core activity of the newly established National Center for Geographic Information and Analysis (NCGIA), awarded to a consortium of the University of California at Santa Barbara, the State University of New York at Buffalo, and The University of Maine, was the creation of approximately 20 research initiatives and the NCGIA-sponsored "specialist meetings", which for a period of over a decade brought together researchers from around the world. As summarized in a 1995 *Geo Info Systems* paper:

> In the late fall of 1994, the University Consortium for Geographic Information Science was created at the founding meeting in Boulder, Colorado. Forty-two individuals representing 33 universities, research institutions, and the Association of American Geographers (AAG) met December 4–6, 1994, in Boulder, Colorado, to establish the University Consortium for Geographic Information Science (UCGIS). Representatives were invited by the presidents or chancellors of seven U.S. research universities with prominent programs in geographic information systems. The UCGIS is to be a non-profit organization of universities and other research institutions dedicated to advancing our understanding of geographic processes and spatial relationships through improved theory, methods, technology, and data.

At the founding meeting in Boulder, three major goals for UCGIS were identified, including:

- To serve as an effective, unified voice for the geographic information science research community;
- To foster multidisciplinary research and education; and
- To promote the informed and responsible use of geographic information science and geographic analysis for the benefit of society.

One major goal included the creation of a national research agenda, which was initiated in the summer of 1996 in Columbus, Ohio. In preparation for this significant event, UCGIS's Research Committee (with members John Bossler, William Craig, Jerome Dobson, Max, Egenhofer, George Hepner, and David Mark), and led by David Mark at SUNY Buffalo, organized a call for what would be labeled the "provisional research topics". This call led to 18 of the then UCGIS member institutions that contributed 81 research topics during April and May of 1996, which were then organized into 17 general themes and 3 "miscellaneous" categories by the Research Committee. The 18 institutions submitting research topics included:

- Boston University
- University of California, Berkeley
- University of California, Santa Barbara
- Clark University
- University of Colorado
- University of Maine
- Michigan State University
- University of Minnesota
- University of North Carolina, Chapel Hill
- State University of New York at Buffalo
- Oak Ridge National Laboratory
- Ohio State University
- Oregon State University
- San Diego State University
- University of South Carolina
- West Virginia University
- University of Wisconsin, Madison
- University of Wyoming

The above 18 institutions each sent representatives (along with other institutions as well) to the Columbus meeting. The 20 themes formed the basis for three rounds of breakout groups for in-depth discussion of the topics, as well as plenary sessions for the presentation of results and consensus building. The 20 categories and provisional themes included were: ([Note: Topics marked with an asterisk (*) appear under more than one theme.)

(1) Change Detection
BU-3* Methods of Change Detection at Various Scales (inc. global)
ORNL-3 Develop a Method for Measuring Global Change

(2) Conflation and Data Integration
OH-2 Integration of Spatial Data
SDSU-1 Integration of GIS Databases Across the United States–
     Mexico and United States–Canada Border
SC-1 Conflation of TIGER and Other Street Centerlines
UCSB-2 Conflation—Combining Geographic Information

(3) Dynamic Processes
BU-5 Development of Programming Environments Within GIS to
     Allow Modeling of Dynamic Spatial Processes
MSU-1 Improved Support for Multi-Temporal Data Management,
     Analysis, and Visualization

(4) Environmental Hazards
CLRK-2 Modeling Socioeconomic-Environmental Interactions
SC-5 Hazards Assessment and Mitigation
UB-4 Local Adjustment to Climate Change and Natural Hazards
UCB-3 Modeling Environmental Hazard and Risk: Emergency Response

(5) GIS and Society
COLO-4 Study Ethics of Access, Copyright, Personal Liability, and
     Protection of Intellectual Property for Spatial Data and Related
     Products Published, Delivered or Distributed Electronically
MN-1* Institutional Barriers to Use of GIS in Decision Making
OH-5 The Social Impacts of Spatial Information Technologies
WISC-5 The Social Impacts of Spatial Information Technologies
SDSU-4 Developing a Critical Social Theory of GIS
WVU-1 GIS and Society–Representing Multiple Realities, New
     Models of Space

(6) GIS Foundations
ME-5 Mobile Geographic Information Systems
UB-2 Ontological Foundations of Geographic Information Science
     Land, Law, & Property Rights
UCSB-3 Towards a Theoretical Foundation for GIS

UCSB-5 Distributed GIS Architectures and Semantic Interoperability for a Networked Information Society

(7) GIS in Decision-Making

CLRK-5 GIS and Decision-Making Technique

MN-1* Institutional Barriers to Use of GIS in Decision Making

ORNL-5 Research on Linkages Between GIS Research and Business/Government Decision Making

WY-1 Decision Support Systems Integrating Spatial Data with Models, Expert Knowledge and Graphical User Interfaces are Needed for Effective and Efficient Management of Natural Resources

(8) Global Change Modeling

ORE-1 Spatial Data Handling for Global Change Research

UCSB-4 GIS for Dynamic Environmental Modeling at Global Scales

(9) Miscellaneous Applications

UNC-1 Environmental Topics

UNC-2 Health Care Accessibility

UNC-3 Population-Environmental Topics

(10) Miscellaneous Data Modeling

ORE-2 Collection and Structuring of Spatial Data from the Deep

WISC-2 Object-Oriented Approach for Modeling Interacting Spatial Processes

(11) Model-GIS Interoperability

COLO-1 Integrative Links Between Spatial Statistics/Spatial Analytic Modeling Software and GIS Software

MSU-5 Better Integration of GIS with Other Computing Environments (*e.g.*, operations research, simulation modeling, *etc.*)

ORNL-1 Integrate Geographic Information Systems (GIS) with Process and Transport Models

SDSU-2 Spatial Modeling in a GIS Environment

UB-3 Unified Transportation, Location, and Land-Use Planning Models

UCB-5 Modeling the Metropolis

WY-2 Advancing the Interoperability Concept as a Process Model for Integrated Access and Utilization of Spatial Data Across Dissimilar Data Platforms

(12) Participatory GIS

CLRK-3 Linking Participatory Approaches in Institutional Environments

ME-4 Public Forum GIS

(13) Remote Sensing and Data Capture

BU-1 Improved Integration of Remote Sensing and GIS Data Structures

ME-2 GIS Data Capture

SDSU-3 Exploiting High Resolution Commercial Airborne and
Satellite Image Data
WISC-4 Thematic and Geometric Image Exploitation and Mapping
WVU-4 Natural Resource Data Development and Integration

(14) Scale
BU-3* Methods of Change Detection at Various Scales (inc. global)
CLRK-1 Spatial and Temporal Scaling
MN-4 Multiple Representations of Spatial Databases: Issues of Scale
and Generalization
MSU-2 Issues of Scale in Geographic Data and Analysis

(15) Spatial Cognition
ME-1 Spatial Metaphors
MSU-4 Environmental Psychology and Spatial Cognition
OH-3 The Effect of the Representational Schema of GIS on Spatial
Reasoning and Inference
ORNL-2 Create a Spatial Language
SC-4 Spatial Cognition: GIS and Humans
UCSB-1 Cognitive Issues in GIS

(16) Spatial Data Analysis
BU-2 Improved Use and Incorporation of Spatial Data Analysis in GIS
SC-3 Space by Time Statistical Models Which Complement GIS
UCB-1 New Metrics for Spatial Data Characterization and Analysis
UNC-4 Spatial Tools and Techniques
WVU 2 New Methods for the Exploratory Analysis of Geographical Data

(17) Spatial Data Infrastructure
COLO-5* Metadata Collection and Modeling
MN-2 Availability of Soil Survey Databases for GIS
UB-1 Spatial Data Infrastructure to Support Medical, Social,
and Environmental Research

(18) Statistical Aspects of Data Quality
CLRK-4 Error, Uncertainty, and Decision Risk
COLO-5* Metadata Collection and Modeling
MN-3 Spatial Data Quality and Responsible Use of GIS
MSU-3 Recognition and Representation of Inexactness Spatial Databases
OH-1 Improving Accuracy and Precision of Spatial Data in GIS Databases
SDSU-5 Methods for Handling Uncertainty and Error in Geographical
Information Used for Environmental Modeling Need to be
Developed and Tested
UCB-2 Integrated Analysis of Multisource and Multitemporal Spatial Data
WISC-1 Error and Uncertainty in Spatial/Temporal Data, Analysis,
and Modeling

(19) Three- and Four-dimensional Modeling
> COLO-2 Global Dimension GIS, Including Algorithm and Data Model
> Development that can Handle True Three- and Four-Dimensional
> Spatial Data
> OH-4 Four-Dimensional GIS

(20) Miscellaneous Topics not in any Existing Theme
> BU-4 Incorporation of Newer Technologies such as Neural Networks
> and Fuzzy Expert Systems into GIS
> COLO-3 Develop "Smart" Interfaces, that Employ Formalized
> Principles for Applying Appropriate Statistics, for Representing
> Reasonable Cartographic Displays, *etc.*
> ME-3 Spatial Data Mining
> ORNL-4 Solve the Problem of Reference Data for Accuracy Assessment
> SC-2 Environmental Modeling
> UB-5 Problems with Political Redistricting
> UCB-4 Electronic Libraries: Information, Selection and Uses
> WISC-3 Real-Time Processing of Large Volumes of Spatially Referenced
> Data from Multiple Sensors in Support of Advanced Applications
> WVU-3 The GIS Collaboratory—A Virtual Workplace for
> Collaborative Spatial Science

During the process, each voting delegate was invited to place one-third of the themes into a high priority category, one-third into a middle category, and one-third into a low category. This iterative process of simultaneous theme synthesis and "culling" eventually led to the UCGIS Assembly accepting 10 (what were called at the time) priorities. The final 10 included:

- Spatial Data Acquisition and Integration
- Distributed Computing
- Extensions to Geographic Representation
- Cognition of Geographic Information
- Spatial Analysis in a GIS
- Future of the Spatial Information Infrastructure
- Uncertainty in Spatial Data
- Interoperability
- Scale
- GIS and Society

It is clear that the process itself, the individuals and personalities, and the format of the UCGIS Assembly led to this particular outcome. Other models of "agenda" building would have likely led to other sets of research priorities, and there is a clear granularity to those selected, neither so broad as to seem as mega-topics, nor too narrow so that only a very few individuals could support them. It is also clear that topics of an applied nature did not make the list; the bias in the founding Assembly was to focus on basic research that would likely influence the research agendas of federal government funding agencies.

After the meeting in Columbus, teams of researchers crafted a series of white papers that were put on the UCGIS web site, and a summary article was published in the journal *Cartography and Geographic Information Systems* (1997, 115–127). Thus the results of this 6 month process had led to the creation of this nation's first truly comprehensive—in the sense of multiple individuals, multiple institutions, and multiple disciplines—research agenda in geographic information science.

Two years later the 10 priorities were revisited at the Third UCGIS Assembly in Park City, Utah, after the Second Assembly had worked at the development of an education agenda. The Park City Assembly utilized a different format, where expert panels were put together to update the thinking on each of the topics. In order to broaden participation, these panels contained individuals from academia, the private sector, and government and an attempt was made to achieve interdisciplinary diversity and representation from the previous 1996-produced white paper. The panel was to present, critique, and discuss each topic, with ample time for inter-action with the audience. The panel leaders for each of the 10 challenges included:

- Spatial Data Acquisition and Integration
  John Jensen, University of South Carolina

- Distributed Computing
  Michael Goodchild, University of California, Santa Barbara

- Extensions to Geographic Representation
  Donna Peuquet, Pennsylvania State University

- Cognition of Geographic Information
  Daniel Montello, University of California, Santa Barbara

- Spatial Analysis in a GIS
  Art Getis, San Diego State University

- Future of the Spatial Information Infrastructure
  Harlan Onsrud, University of Maine

- Uncertainty in Spatial Data
  Nicholas Chrisman, University of Washington

- Interoperability
  Joe Ferreira, Massachusetts Institute of Technology

- Scale
  Nina Lam, Louisiana State University

- GIS and Society
  Eric Sheppard, University of Minnesota

It is these 10 "challenges", or priorities, that have existed as the formal UCGIS Research Agenda since these events.

1.2 AN ANALYSIS OF THE UCGIS CHALLENGES

In an attempt to assess the interrelations of the original 10 challenges, an analysis of the concepts within each challenge was performed. A listing of concepts found in each challenge was independently prepared by each of six graduate students at the University of Georgia in a seminar on geographic information systems. The independent lists were then composited to form a final list of concepts in each topic that could be used for quantitative analysis. The composite list included the concept and an association of the challenges in which the concept occurs. For example, the concept "reliability and consistency of metadata" was found in Spatial Data Acquisition and Integration, Uncertainty in Spatial Data and GIS-Based Analysis, Scale, and Distributed Computing.

While the association of concepts with challenges attempts to capture the interrelations of the priorities by recording reappearing concepts, it must be understood that significant interpretation accompanies this process. One should not allow the association to imply a quantification precision which does not exist. In some challenges, concepts are very broadly expressed while in others a very narrow focus is applied. There is no attempt in the analysis to capture this containment information. For example, in Geographic Representation, the concepts of "temporal representation" and "new strategies for multiple representations" appear. In Cognition of Geographic Information, "alternative data models" appears as a concept. Are these the same or parts of the same concepts? In Uncertainty in Spatial Data and GIS-Based Analysis, "geographic representation" appears as a separate concept. Is this the same as "temporal representation" in Geographic Representation? Is it the same as "new strategies for multiple representation" also in Geographic Representation? Is it the same as the "alternative data models" concept of Cognition of Geographic Information? Given these types of problems of specificity of the research concepts and language of the white papers, any quantitative analysis will suffer from the initial requirement of interpretation of the exact meanings of the language in the white papers. However, following is one attempt to statistically analyze, through ordination and grouping mechanisms, the overlap and inter-linkages of the 10 research challenges.

1.2.1 Multidimensional Scaling and Cluster Analysis of the Research Challenges

In an attempt to quantify the relations among the 10 research challenges, the association of concepts with challenges was tabulated to indicate the number of common concepts between any 2 research challenges. Those numbers appear as an asymmetric data matrix in Table 1.1. The asymmetry results from the specificity of the concepts, discussed in the preceding section, which leads a mismatch in the numbers. Several specific concepts in one research priority will map to a

Table 1.1 Numbers of Common Topics Among Research Challenges

	DC	GR	SA	IO	C	U	GS	S	SI	SD
DC=Distributed Computing	0	4	4	3	2	0	2	2	1	3
GR=Geographic Representation	1	0	4	6	8	2	1	2	0	3
SA=Spatial Analysis	2	5	0	1	0	0	0	4	1	5
IO=Interoperability	3	7	2	0	3	2	2	3	1	5
C=Cognition	4	9	1	4	0	3	1	1	0	2
U=Uncertainty	0	3	1	4	3	0	0	1	0	6
GS=GIS and Society	2	1	0	2	1	1	0	0	4	1
S=Scale	2	1	5	5	1	0	0	0	0	1
SI=Spatial Information Infrastructure	4	0	1	1	0	0	4	0	0	1
SD=Spatial Data Acquisition & Integration	8	3	5	6	2	9	1	2	1	0

single general concept in another priority. For example, in 3) Spatial Analysis in a GIS Environment, research on "handling large datasets" is included. In 1) Distributed Computing, three concepts which are subsets of "handling large datasets" are included, "efficient use of bandwidth for large data volumes", "compression of raster data", and "parallel methods for vector". Thus the number 3 appears thrice in Distributed Computing while the number 1 appears only once in Spatial Analysis in a GIS Environment because of this containment relationship.

The numbers in Table 1.1 provide a basis for beginning an analysis of the interrelationships among the research challenges. Figure 1.1 displays the result of applying a multidimensional scaling algorithm to the data matrix in Table 1.1. The figure illustrates the proximity based simply on the absolute numbers of common topics between pairs of the research challenges. Note that the scaling actually required inverting the numbers from Table 1.1 to force larger numbers to reflect closer proximity. The result of this procedure shows several clusters in three-dimensional space. Specifically, cognition, geographic representation, and interoperability form a cluster which might be expected from a simple examination of the number concepts in common in Table 1.1. A more diffuse (uncertain) cluster of Uncertainty, GIS and Society, Spatial Data Acquisition and Integration, and Spatial Information Infrastructure is less easily explained from the listing of concepts. Figure 1.2 presents the two-dimensional scaling of the absolute numbers in Table 1.1 In this result Geographic Representation and Interoperability remain in close proximity and Cognition is in the same quadrant but further removed.

Since the actual numbers of topics in each priority vary, using the absolute counts of the recurring topics for scaling is somewhat suspect. Standardization of the numbers can be achieved by computing a percentage recurrence. For example, Cognition has 12 topics, 8 of which also appear in Geographic Representation.

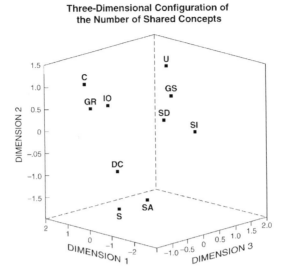

Figure 1.1 Three-dimensional result of multidimensional scaling of "distances" among the 10 research priorities. Numbers used for scaling represent absolute counts of common research topics. Abbreviations are the same as in Table 1.1

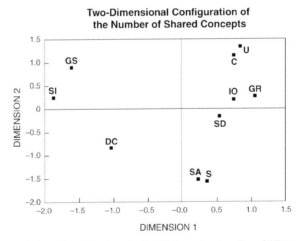

Figure 1.2 Two-dimensional result of multidimensional scaling of "distances" among the 10 research priorities. Numbers used for scaling represent absolute counts of common research topics. Abbreviations are the same as in Table 1.1

Thus, 66.67 percent of the Cognition topics are common to Geographic Representation. The relationship is not symmetric, however, since Geographic Representation has 15 topics, 9 of which are found in Cognition. Thus, this equates to a 60 percent overlap. The results of this standardization appear in Table 1.2 while the scaling of the percentages is shown in three- and two-dimensional configurations in Figures 1.3 and 1.4, respectively.

Table 1.2 Percentages of Common Topics Among Research Challenges

	DC	GR	SA	IO	C	U	GS	S	SI	SD
DC=Distributed Computing	0	27	31	25	17	0	14	20	17	6
GR=Geographic Representation	6	0	31	50	67	22	7	20	0	6
SA=Spatial Analysis	12	33	0	8	0	0	0	40	17	9
IO=Interoperability	18	47	15	0	25	22	14	30	17	9
C=Cognition	24	60	8	33	0	33	7	10	0	4
U=Uncertainty	0	20	8	33	25	0	0	10	0	11
GS=GIS and Society	12	7	0	17	8	11	0	0	67	2
S=Scale	12	7	38	42	8	0	0	0	0	2
SI=Spatial Information Infrastructure	24	0	8	8	0	0	29	0	0	2
SD=Spatial Data Acquisition and Integration	47	20	38	50	17	100	7	20	17	0

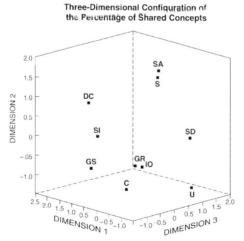

Figure 1.3 Three-dimensional result of multidimensional scaling of percentage of common research topics among the 10 research priorities. Abbreviations are the same as in Table 1.2

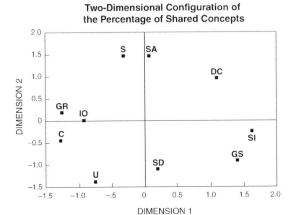

Figure 1.4 Two-dimensional result of multidimensional scaling of percentage of common research topics among the 10 research priorities. Abbreviations are the same as in Table 1.2

The percentages in Table 1.2 reveal much about the overlap among the research challenges. Higher percentages indicate significant overlap as in the case of Cognition and Geographic Representation (67 percent). Note that Interoperability and Geographic Representation also are highly interrelated (50 percent) but Cognition and Interoperability are less related (25 percent).

The graphics illustrate the closeness of Cognition, Geographic Representation, Interoperability, Uncertainty, and GIS and Society, especially along the vertical (Dimension 2) axis. The general concept of "geographic representation" is the common research theme among these challenges which accounts for much of the overlap and one could label the vertical axis of Figure 1.3 as a geographic representation axis. In Figure 1.4, the representation group appears again, but this time without GIS and Society. Spatial Data Acquisition and Integration is somewhat isolated in both Figures 1.3 and 1.4. The uniqueness of this research priority shows in the small percentages in Table 1.2 and results from the large number of research topics (53) compared to the other challenges.

A principal components analysis of the percentage overlap (Table 1.3) reveals that 82.7 percent of the variance in the matrix can be explained with only four orthogonal variables. This result along with the multidimensional analysis suggests that a smaller number of research challenges, perhaps 4 or 5, can be refined from the existing 10 and still account for the primary set of research topics.

A k-means cluster analysis reveals similar groupings. Creating 3 to 6 clusters always yields Spatial Data Acquisition as a single element cluster. Challenges commonly clustered are Spatial Analysis, Interoperability, and Cognition. Distributed Computing, Uncertainty, Scale, and Spatial Information Infrastructure form another cluster. Geographic Representation is a single element cluster in analysis with total numbers of clusters at 4, 5, or 6. These results are to be expected

Table 1.3 Principal Components Analysis of Research Challenges
Total Variance Explained

Component	Eigenvalues	% of Variance	Cumulative %
1	3.0	29.6	29.6
2	2.1	21.4	51.0
3	1.8	18.4	69.4
4	1.3	13.3	82.7
5	0.9	8.5	91.3
6	0.4	4.0	95.3
7	0.3	3.2	98.5
8	0.1	0.9	99.4
9	0.1	0.6	100.0
10	0.0	0.0	100.0

since the clustering techniques are based on distance and using the percentage values, Spatial Data Acquisition and Integration has the smallest percentages (largest distances) from the other challenges and Geographic Representation has the largest percentages (smallest distances). These logically form unique clusters.

1.2.2 Analysis Discussion

The 10 research challenges of UCGIS are intertwined and overlapping. Many of the topics appearing in one priority are reflected in other challenges; however, rarely is this relationship between challenges one-to-one or reciprocal. Varying levels of specificity of research topics and concepts provide for containment relations among challenges. Sorting the challenges into unique, orthogonal areas is probably not possible, but reduction in the number of challenges and stand-ardization of the level of detail in the presentation of the specific research concepts may be useful. A basic interpretation of the white papers reveals that the research challenges are divided into two broad areas, those topics dealing with theoretical, conceptual, and technical aspects of GIS, and those topics dealing with more human concerns such as social, legal, and management issues. In the first group, examples include Spatial Analysis and Spatial Data Acquisition and Integration. Examples in the second group are GIS and Society and Cognition. These two groups are not mutually exclusive and some challenges, such as Geographic Representation and Distributed Computing, include many elements of both groups.

Based on the multidimensional analysis, significant overlap occurs among Geographic Representation, Cognition, and Interoperability of Geographic Information. Spatial Data Acquisition and Integration is extremely broad encompassing more research topics and more specificity in the topics than any two of the other challenges combined. This suggests that this priority is too broad and could be divided. Principal components analysis of overlap percentages indicates that only four unique, orthogonal challenges are necessary to span the breadth of the topics in the current ten challenges. Cluster analysis corroborates these results.

A caveat to these statements is that the analysis is completely based on an original interpretation of research concepts within the 10 challenges. Alternative interpretations will certainly yield different results. Even with a completely valid interpretation of the research concepts, the varying level of detail from one priority to another could easily confuse the analysis results.

Notwithstanding the mitigating circumstances concerning the validity of the analysis discussed above, several recommendations for improvement of the research challenges can be made to UCGIS. First, consistency is needed in the level of detail at which research ideas are presented across the ten challenges. While this will be difficult to achieve, the current list is so variable that under-standing of the actual research intent is difficult. Second, a reduction in the number of existing challenges is warranted. Exactly how many are appropriate is not clear from the analysis because of the problem of varying detail level. While the analysis indicates that four or five appear to be optimum, the number may change with the standardization of the detail level of the topics. It is apparent that even with ten research challenges, some important topics are neglected or are not placed high enough in the lists. Thus, new topics need to be explored and included. Finally and perhaps most importantly, this analysis has indicated that the ten challenges are not linear and independent. Both hypotheses, that significant overlap exists among challenges and that the interrelations are nonlinear and non-Euclidean, have been demonstrated. Perhaps the organization of the challenges needs to be re-examined and a structure other than a linear list, such as a hierarchy or a graph theoretic structure like a semantic network, more appropriate to the content should be designed.

1.3 THE EMERGING THEMES

Recognizing the limitations of the original 10 challenges, some of which are discussed above, UCGIS embarked on a second initiative to define research themes. After the Winter Assembly in Washington in February of 2000, the UCGIS Research Committee issued a call for "Emerging Themes." To qualify, a topic was expected to be emerging as a research focus from one or more of the original 10 challenges, cross-cutting the 10, or a completely new topic not contained in the original set. A series of topics were submitted, most at a level of granularity similar to the original 10 challenges. The nine topics included:

- Public Participation GIS
- Geospatial Data Mining and Knowledge Discovery

- Simulation using Object Orientation
- Ontologies in Geographic Information Science
- The Social Construction of GIS and GIScience
- Geospatial Ontology
- Geographic Visualization
- Analytical Cartography
- Innovation and Integration Issues for Remotely Acquired Information

Over the next year and one-half, the UCGIS membership considered these themes in each of its Assemblies with refinements occurring to each topic that survived. Initially, all themes were required to be supported by at least three member institutions with multiple authorship of a white paper. A vote on the themes to be retained occurred at the Winter 2001 Assembly and discussion sessions were organized for the Summer 2001 Assembly. These discussions were to result in final changes to the white papers which would be placed on the UCGIS Web site to represent the final set of emerging themes. From the original list of nine topics, the four that survived the process with the final topical names for the white papers and became the UCGIS official list of emerging themes are:

- Geospatial Data Mining and Knowledge Discovery
- Geographic Visualization
- Ontological Foundations for Geographic Information Science
- Remotely Acquired Data and Information in GIScience

Of the four themes, one, Remotely Acquired Data and Information in GIScience, is a subset and update of Spatial Data Acquisition and Integration with increased priority. Two, Ontological Foundations for Geographic Information Science and Geographic Visualization, are priority areas that emerged from cross-cutting the 10 original challenges, and the theme Data Mining is a new area not covered in the 10.

UCGIS has continued to refine the research priorities and challenges and in 2003 adopted a bilevel approach with 14 long-term and short-term priorities documented as "Research Briefs" (*http://www.ucgis.org/research*). The current long-term challenges comprise the original 10 challenges and the 4 emerging themes documented in this book reorganized into 10 long-term challenges or problems about which GIScience will continually conduct research. The short-term priorities are viewed as topics for which research of 2–3 years will yield substantive results in resolving the problems in the area.

1.4 ORGANIZATION OF THIS BOOK

The remaining chapters in this book provide comprehensive documentation of the challenges and themes. Each chapter, written by a group of researchers involved in the conception of the challenge or theme, provides a comprehensive discussion of the basic research elements, the UCGIS approach, the need for the National research agenda, contributions to knowledge and society, and a complete set of references. Chapters 2–10 discuss nine of the original 10 challenges, with "Interoperability" not documented. Chapters 11–14 provide the written basis of the four "Emerging Themes." A final chapter draws some general conclusions about the UCGIS approach and the defined research challenges. UCGIS has used the definition of research challenges and priorities as a method to fulfill a part of its mission and goals discussed above. This book documents the developed challenges and provides sources of information for further exploration of research in GIScience.

CHAPTER TWO

Spatial Data Acquisition and Integration

John Jensen, University of South Carolina
Alan Saalfeld, Ohio State University
Fred Broome, Bureau of the Census
Dave Cowen, University of South Carolina
Kevin Price, University of Kansas
Doug Ramsey, Utah State University
Lewis Lapine, Chief South Carolina Geodetic Survey
E. Lynn Usery, University of Georgia and U.S. Geological Survey

2.1 OBJECTIVE

The objective of spatial data acquisition and integration is to improve the logic and technology for capturing and integrating spatial data in the form of *in situ* measurements, census enumeration data, thematic maps, and products derived from remotely sensed imagery. The priority also desires to identify where research should take place concerning: data collection standards, geoids and datums, positional accuracy, measurement sampling theory, classification systems (schemes), metadata, address matching, and privacy issues. The goal is to obtain accurate socioeconomic and biophysical spatial data that may be analyzed and modeled to solve problems.

2.1.1 Background

Geographic information provides the basis for many types of decisions ranging from simple wayfinding to management of complex networks of facilities, predicting complex socioeconomic and demographic characteristics (*e.g.*, population estimation), and the sustainable management of natural resources. Improved geographic data should lead to better conclusions and better decisions. According

0-8493-2728-8/05/$0.00+$1.50

to several "standards" and "user" groups, better data would include greater positional accuracy and logical consistency and completeness. But each new data set, each new data item that is collected can be fully utilized only if it can be placed correctly into the context of other available geospatial data and information.

To this end, the National Research Council (NRC) Mapping Science Committee (1995) made a strong case that the United States National Spatial Data Infrastructure (NSDI) consist of the following three foundation spatial databases (Figure 2.1a): 1) geodetic control, 2) digital terrain (including elevation and bathymetry), and 3) digital orthorectified imagery (Figure 2.1b,c). Foundation spatial data are the minimal directly observable or recordable data from which other spatial data are referenced and sometimes compiled. They used a metaphor from the construction industry wherein a building must have a solid foundation of concrete or other material. Then a framework of wood or steel beams is connected to the foundation to create a structure to support the remainder of the building. Examples of important thematic framework data might include hydrography and transportation. In fact, the NSDI framework incorporates the following three foundation and four framework data themes: geodetic control, orthoimagery, elevation, transportation, hydrography, governmental units (boundaries), and cadastral information (Federal Geographic Data Committee–FGDC, 2004).

Finally, there are numerous other themes of spatial information that may not be collected nationally, but may be collected on a regional or local basis. Examples include cultural and demographic data, vegetation (including wetland), soils, and geology and the myriad of data collected for the global climate change research initiative (Figure 2.1a). These thematic spatial data files must be rigorously registered to the foundation data, making it easy to use and share the spatial information.

It is clear that the human race has entered the information age. An unprecedented amount of spatial foundation and thematic framework information are being collected in a digital format. But do the current data collection and integration strategies fulfill our needs? Several important questions should continually be addressed by the UCGIS research community and others, including:

- Is it possible to integrate accurately the more abundant and precise spatial data with other current and historical datasets to solve complex problems? If it is difficult to integrate the data, what problems must be overcome to facilitate integration?

- Are there significant gaps between the *in situ* and remotely sensed data required by the public and scientific user communities and what data are collected? If the required data are not available, how can the data be obtained?

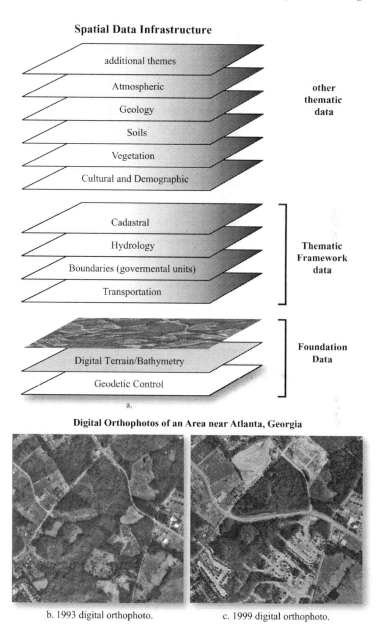

Figure 2.1 a) Foundation and thematic framework data found in most National Spatial Data Infrastructures (after NRC, 1995). b,c) Portions of digital orthophoto quarter quads (DOQQ) of an area near Atlanta, Georgia. These data reside in the Georgia Spatial Data Infrastructure

2.2 THE UCGIS APPROACH

The improved capture and integration of spatial data will require the collaboration of many disciplines including cartography, computer science, photogrammetry, geodesy, mathematics, remote sensing, statistics, geography, and various physical, social, and behavioral sciences with spatial analysis applications (Jensen *et al.*, 2002). We will solve key problems of capturing the right data and relating diverse data sources to each other by involving participants from all specialty areas, including the traditional data collectors, the applications users, and the computer scientists and statisticians who optimize data management and analysis for all types of data sets. We will develop mathematical and statistical models for integrating spatial data at different scales and different resolutions. We will especially focus on developing tools for identifying, quantifying, and dealing with imperfections and imprecision in the data throughout every phase of building a spatial database.

2.3 IMPORTANCE TO NATIONAL RESEARCH NEEDS

This paper identifies some of the major gaps or shortfalls in data integration and data collection strategies for more intensive investigation by UCGIS and other scientists. The paper first addresses important data integration issues that are generic to all data collection efforts. Then, a brief investigation of current and potential *in situ* and remote sensing socioeconomic and biophysical data collection requirements is presented.

2.4 GENERIC INTEGRATION (CONFLATION) ISSUES

Data integration strategies and methodologies have not kept pace with advances in data collection. It remains difficult to analyze even two spatial data sets acquired at different times, for different purposes, using different datums, positional accuracy (x,y,z), classification schemes, and levels of *in situ* sampling or enumeration precision. Scientists and the general public want to be able to conflate multiple sets of spatial data, *i.e.* integrate spatial data from different sources (Saalfeld, 1988). Conflation may be applied to transfer attributes from old versions of feature geometry to new, more accurate versions; to the detection of changes by comparing images of an area from *n* different dates; or to automatic Registration of one data set to another through the recognition of common features. In the past, however, methods of conflation (integration) have been *ad hoc*, designed for specific projects involving a specific pair of data sets and of no generic value. A general theoretical and conceptual framework is needed to be able to accommodate at a minimum five distinct forms of data integration:

- *In situ* measurement-to-*in situ* measurement (calibration, adjustment, variance, *etc.*);

- *In situ* measurement-to-foundation map (point-to-map, registration, verification);

- Vector-to-foundation map (map-to-map, vector segmentation scheme integration, different scales, different geographic coverage, *etc.*);

- Image-to-foundation map (image-to-map, for elevation mapping, map revision, *etc.*); and

- Image-to-foundation image (image-to-image fusion, involving different spatial, spectral, temporal, and radiometric resolutions).

An example of "*in situ* measurement-to-*in situ* measurement" calibration would be establishing the statistical relationship between vegetation canopy height and biomass (g/m^2) measured at a site. An example of "*in situ* measurement-to-foundation map" integration would be the conflation of all stream gauging, sediment load, and water quality data to a geodetically controlled foundation map. An example of "vector-to-foundation map" integration could involve the U.S. Census Bureau's Topographically Integrated Geographic Encoding and Referencing (TIGER) data. The Census Bureau is well aware that TIGER files lack accurate coordinates registered to the foundation, complete street addressing, and an ongoing maintenance program. Greater use of these data would be provided by improving coordinate accuracy using orthorectified imagery that is tied to the geodetic control network, completing the street and address coverage, and establishing an ongoing update mechanism. An example of "image-to-foundation map" integration would include the registration of a 1 x 1 meter (m) Ikonos panchromatic scene obtained at 20 degrees off-nadir to the digital road network of a 7.5-minute quadrangle. Examples of "image-to-foundation image" would include a) the rectification of the same Ikonos image to a digital orthophoto quarter-quad (DOQQ) foundation image, or b) the detection of change between two different Ikonos images obtained on different dates (one acquired at 20 degrees off-nadir and one at nadir) that were both registered to the foundation image DOQQ.

When developing the conceptual framework for spatial data integration it is important to remember that in a perfect, static world, feature-matching would be a one-to-one, always successful, nothing-left-over proposition. Each successful match would support previous choices and facilitate subsequent choices. Unfortunately, the real world is messy, and the real world problems involve dealing with and cleaning up the mess. A single common framework is needed that will integrate diverse types of spatial data. The single flexible framework would even allow some items to go unmatched or to be matched with limited confidence. Spatial data integration should include horizontal integration (merging adjacent data sets), vertical data integration (that allow map overlay operations to take place), and temporal data integration. Spatial data integration must handle differences in spatial data content, scales, data acquisition methods, standards, definitions, and practices; manage uncertainty and representation differences; and detect and deal with redundancy and ambiguity of representation.

The usual first step of a conflation system is feature-matching. Once the common components of two (or more) spatial data representations are identified, merging and situating feature information is an easier second step. Feature-matching tools differ with the types of data sets undergoing the match operation. Many ad hoc tools have been developed for specific data set pairs. One example is the plane-graph node-matching strategy used to conflate the TIGER files and U.S. Geological Survey's (USGS) digital line graph (DLG) files (Lynch *et al.*, 1985). Another example is an attribute-supported rule-based feature matching strategy applied to National Imagery and Mapping Agency (NIMA) Vector Product Format (VPF) products (Cobb *et al.*, 1998). Feature-matching that allows for uncertainty has been the focus of several research investigations (*e.g.*, Foley *et al.*, 1997). Tools for managing uncertainty in conflation systems currently under development include fuzzy logic, semantic constraints, expert systems, Dempster-Shafer theory, and Bayesian networks.

The following subsections briefly identify several additional generic spatial data integration issues (quality, consistency, and comparability) that should be addressed before data are collected, including: standards, geoid and datum, positional accuracy, classification system (scheme), *in situ* sampling logic, census enumeration logic, metadata collection, address matching, and privacy issues. Addressing these issues properly will facilitate subsequent data integration.

2.4.1 Standards: FGDC, Open GIS Consortium, and ISO

Many organizations and data users have developed and promoted standards for spatial data collection and representation (e.g., GETF, 1996; NAPA, 1998; NRC, 2001). In the U.S., the FGDC oversees the development of the NSDI. The UCGIS research community endorses the significant strides made by FGDC to establish and implement standards on data content, accuracy, and transfer (FGDC, 2004). FGDC's goal is to provide a consistent means to directly compare the content and positional accuracy of spatial data obtained by different methods for the same point and thereby facilitate interoperability of spatial data. Similarly, the Open GIS Consortium (OGC) is working with public, industry, and non-profit producers and consumers of geographic information system (GIS) technology and geospatial data to develop international standards for interoperability (OGC, 2004). The OGC sets standards so that the commercially available geographic information processing software and data produced by them are interoperable. UCGIS scientists should continue to be actively involved in the specification and adoption of FGDC and Open GIS Consortium standards.

UCGIS and other scientists should determine the impact on data collection if and when businesses and organizations implement international environmental standards as prescribed by the International Organization for Standardization (ISO). For example, the ISO 14000 series of environmental management standards (EMS) offers a consistent approach for managing a business or organization's environmental issues. The U.S. Department of Defense, U.S. Department of Energy, and U.S. Environmental Protection Agency conducted pilot projects to assess the effect of the ISO 14001 EMS on their facilities (FETC, 1998). The system is useful when placing data in environmental management systems, conducting environmental audits, performing environmental labeling, and evaluating the performance of an environment (*e.g.*, ISO 14031 provides guidance on the design and use of environmental performance evaluation and on the identification and selection of environmental performance indicators). Increasing environmental consciousness around the world is driving companies and agencies to consider environmental issues in their decisions. Therefore, companies and agencies are using the international standards to better manage their environmental affairs. Spatial environmental data collected and processed for these businesses and organizations may eventually have to meet a higher standard in order for the company or organization to maintain its ISO 14000 status.

2.4.2 Geodetic Control: Geoid and Datum

Scientists collect thematic framework data at $x,y,$ and z locations relative to the geodetically controlled foundation data. The FGDC compiled the Geospatial Positioning Accuracy Standard, Part 2, Geodetic Control Networks (FGDC-STD-007.2-1998) and the Geospatial Positioning Accuracy Standard, Part 3, National Standard for Spatial Data Accuracy (NSSDA), FGDC-STD-007.3-1998.

It is recommended that horizontal coordinate values be in North American Datum of 1983 (NAD 83) and that vertical coordinates be in the North American Vertical Datum of 1988 (NAVD 1988) or the National Geodetic Vertical Datum of 1929 (NGVD of 1929). While this is important for the creation of new data, what about all of the other spatial information compiled to other datums? How can these historical data be conflated (registered) to data compiled to the NAD 83 datum? For example, Welch and Homsey (1997) pointed out a classical data integration (conflation) problem involving the USGS 1:24,000-scale 7.5-minute topographic map sheets, DLG products, and Digital Elevation Models (DEMs) of the U.S. that were cast on the North American Datum of 1927 (NAD 27). These map products are a national treasure used for a variety of mapping and GIS database construction tasks. For some time there was no information readily available on the shifts in meters needed to convert NAD 27 Universal Transverse Mercator (UTM) Northing and Easting grid coordinates to NAD 83 values. Fortunately, we now have the NADCON federal standard for transforming latitude and longitude coordinate values in the NAD 27 to the NAD 83 (National Geodetic Survey, 2003).

Such translation is necessary if the historical topographic, DLG, DEM and other spatial information are to be registered to new data such as the USGS Digital

Orthophoto Quarter Quads (DOQQ) that are projected to NAD 83. It is important that the user be able to achieve registration between the data layers derived from these and other map products to accuracies commensurate with the U.S. National Map Accuracy standards. This means that all horizontal coordinates must be referenced to a single datum (Welch, 1995). UCGIS scientists should be actively involved in research that maximizes our ability to register a diverse array of spatial databases to a single, nationally approved datum.

2.4.3 Geodetic Control: Horizontal (*x,y*) and Vertical (*z*) Accuracy

The Federal Geodetic Control Network (FGCN) defines statistical methods for reporting the horizontal (*x,y*) circular error (radius of a circle of uncertainty) and vertical (*z*) linear error (linear uncertainty) of control (check) points in the National Spatial Reference System. The NSSDA standards define rigorous statistical methods for reporting the horizontal circular and vertical linear error of other well-defined points in spatial data derived from aerial photographs, satellite imagery, or maps. The NSSDA statistical reporting method replaces the traditional U.S. National Map Accuracy Standards and goes beyond the large-scale map accuracy specifications adopted by the American Society for Photogrammetry & Remote Sensing (ASPRS, 1990) to include scales smaller than 1:20,000.

While important advances have been made, there are still unresolved issues that need to be investigated, including: 1) the determination of error evaluation sample size based on map or image scale and other relevant criteria, 2) identification of the most unbiased method of allocating the test sample data throughout the study area (*e.g.*, by line, quadrant, stratified-systematic-unaligned sample, *etc.*), 3) development of improved methods for reporting the positional accuracy of maps or other spatial data that contain multiple geographic areas of different accuracy, 4) develop more rigorous criteria to identify coordinate "blunders", and 5) development of improved statistical methods for assessing horizontal and vertical positional accuracy.

2.4.4 Classification Standards: Logical Consistency and Completeness

Scientists collect biophysical and sociological data at *x,y,z* locations according to a classification system. Unfortunately, there may be several classification schemes that can be utilized for the same subject matter and their content may be logically inconsistent or incomplete. For example, until recently it was possible to map a large bed of cattail (*Typha latifolia*) on the edge of a freshwater lake using the following classification schemes: a) the "USGS Land Use and Land Cover Data Classification System for Use with Remote Sensor Data" (Anderson *et al.*, 1976), b) the "National Oceanic and Atmospheric Administration (NOAA) CoastWatch Landuse/ Land cover Classification System" (Klemas *et al.*, 1993), and c) the "U.S. Fish &

Wildlife Service Wetland Classification System" (Cowardin *et al.*, 1979; FGDC-STD-004). Using the three classification systems, the identical waterlily patch would be categorized as "non-forested wetland", "lacustrine aquatic bed-rooted vascular plant", and "lacustrine persistent emergent marsh" respectively. Wetland maps derived using these three classification systems are notoriously difficult to integrate.

There is also the issue of classification system attribute completeness (specificity). Some systems like the USGS (Figure 2.2) provide 1–2 levels of specificity and nomenclature and suggest that the user stipulate the classes associated with more detailed level 3–4 information. Conversely, the USDA Forest Service classification system provides specific level 4 and 5 classes that take into account plant characteristics, soils, and frequency of flooding. It is not surprising, therefore, that the Forest Service system titled "Classification of Wetlands and Deep Water Habitats" is the FGDC standard and should be utilized when conducting wetland studies. The Vegetation Classification Standard and Soil Geographic Data Standards have also been completed (Table 2.1).

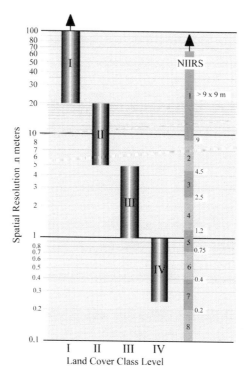

Figure 2.2 The general relationship between the USGS "Land Use and Land Cover Classification System for Use with Remote Sensor Data" land cover class level and the spatial resolution of the remote sensing system (ground resolved distance in meters). The National Image Interpretability Rating System (NIIRS) is also provided for comparison. A NIIRS "0" rating suggests that the interpretability of the imagery is precluded by obscuration, degradation, or very poor resolution

Table 2.1 Status of Federal Geographic Data Committee Standards (March, 2004; *http://www.fgdc.gov/standards/status/textstatus.html*)

FGDC Endorsed Standards
- Content Standard for Digital Geospatial Metadata (2.0) FGDC-STD-001-1998
- Content Standard for Digital Geospatial Metadata, Part 1: Biological Data Profile, FGDC-STD- 001.1-1999
- Metadata Profile for Shoreline Data, FGDC-STD-001.2-2001
- Spatial Data Transfer Standard (SDTS), FGDC-STD-002
- Spatial Data Transfer Standard (SDTS), Part 5: Raster Profile and Extensions, FGDC-STD-002.5
- Spatial Data Transfer Standard (SDTS), Part 6: Point Profile, FGDC-STD-002.6
- SDTS Part 7: Computer-Aided Design and Drafting (CADD) Profile, FGDC-STD-002.7-2000
- Cadastral Data Content Standard, FGDC-STD-003
- Classification of Wetlands and Deepwater Habitats of the United States, FGDC-STD-004
- Vegetation Classification Standard, FGDC-STD-005
- Soil Geographic Data Standard, FGDC-STD-006
- Geospatial Positioning Accuracy Standard, Part 1, Reporting Methodology, FGDC-STD-007.1-1998
- Geospatial Positioning Accuracy Standard, Part 2, Geodetic Control Networks, FGDC-STD-007.2-1998
- Geospatial Positioning Accuracy Standard, Part 3, National Standard for Spatial Data Accuracy, FGDC-STD-007.3-1998
- Geospatial Positioning Accuracy Standard, Part 4: Architecture, Engineering Construction and Facilities Management, FGDC-STD-007.4-2002
- Content Standard for Digital Orthoimagery, FGDC-STD-008-1999
- Content Standard for Remote Sensing Swath Data, FGDC-STD-009-1999
- Utilities Data Content Standard, FGDC-STD-010-2000
- U.S. National Grid, FGDC-STD-011-2001
- Content Standard for Digital Geospatial Metadata: Extensions for Remote Sensing Metadata, FGDC-STD-012-2002

Review Stage
- Content Standard for Framework Land Elevation Data
- Digital Cartographic Standard for Geologic Map Symbolization
- Facility ID Data Standard
- Geospatial Positioning Accuracy Standard, Part 5: Standard for Hydrographic Surveys and Nautical Charts
- Hydrographic Data Content Standard for Coastal and Inland Waterways
- NSDI Framework Transportation Identification Standard

Draft Stage
- Earth Cover Classification System
- Encoding Standard for Geospatial Metadata
- Geologic Data Model
- Governmental Unit Boundary Data Content Standard
- Biological Nomenclature and Taxonomy Data Standard

Proposal Stage
- Federal Standards for Delineation of Hydrologic Unit Boundaries
- National Hydrography Framework Geospatial Data Content Standard
- National Standards for the Floristic Levels of Vegetation Classification in the United States: Associations and Alliances
- Riparian Mapping Standard

Unfortunately, scientists are not as fortunate when dealing with urban land use. Research on urban classification systems is needed so that spatial data are collected using logical and complete nomenclature. The high spatial resolution remote sensor data (≤ 1 x 1 m) yield detailed level 4 urban/suburban land use and land cover information (Figure 2.2) and there is currently no standardized level 3–4 classification system for this information. At the present time the most complete urban land use classification system is the "Land-based Classification System" (LBSC) developed by the American Planning Association (APA, 2004).

Scientists should work closely with the FGDC to develop the Earth Cover Classification System. The Cultural and Demographic Content Standard should be implemented. Note that there are no standards associated with the collection of water quality, atmospherics, and snow/ice.

2.4.5 Single and Multiple Date Thematic Accuracy Assessment

Cartographers and photogrammetrists are adept at specifying the spatial positional accuracy (x,y,z) of a geographic observation in terms of root-mean-square-error (RMSE) statistics or circle of uncertainty. Scientists are also fairly adept at estimating the accuracy of an individual thematic map when compared with *in situ* "ground-reference" information using statistics such as the kappa coefficient-of-agreement (Stehman, 2000; Gahegan and Ehlers, 2000; Green and Congalton, 2003). Unfortunately, scientists have only begun to understand how to determine the statistical accuracy of map products derived from multiple dates of analysis. For example, only preliminary methods have been developed to measure the accuracy of a change detection map derived from the analysis of two dates (*e.g.*, Macleod and Congalton, 1998; Khorram *et al.,* 1998). Additional research is required to document a) the *in situ* sampling logic required, and b) the statistical analysis necessary to specify the accuracy of a change detection map or derivative products, especially when dealing with $n+2$ dates (Jensen, 2004).

2.4.6 Radiometric Correction of Remote Sensor Data

The FGDC Content Standard for Digital Orthoimagery (FGDC-STD-008-1999) is a thorough document that describes how digital orthophoto quad (DOQ) imagery should be prepared as one of the national foundation datasets. It is imperative that effective, easy to use algorithms be developed that radiometrically edge-match one quarter-quad to another. This is a serious, cumbersome problem that all scientists using DOQs must currently solve independently.

Similarly, it is difficult to compare the radiometric characteristics of two anniversary dates of almost any type of remotely sensed data due to atmospheric attenuation present in one or both images. The problem becomes even more acute when scientists desire to analyze $n+2$ images. Adequate atmospheric correction algorithms have only recently become available in commercial digital image processing programs (*e.g.*, ACORN, FLAASH). Improved easy-to-use atmospheric correction algorithms are required that can perform a) image-to-image scene normalization, b) absolute radiative transfer-based atmospheric correction of each date of imagery, and c) improved geometric and radiometric correction of remote sensor data for mountainous terrain (Jensen *et al.*, 1995; Bishop *et al.*, 1998; Jensen, 2004). Accurate absolute radiometric correction will allow biophysical measurements such as biomass or leaf-area-index (LAI) made on one date of imagery to be compared directly with those obtained on other dates. This is a serious data collection and processing problem.

2.4.7 Metadata

Data about data, *metadata*, are very important. Metadata allow us to understand the origin of the data, its geometric characteristics, attributes, and the type of modeling or digital image processing that has been applied to the data. The Content Standard for Digital Geospatial Metadata (CSDGM) is now in place and there are working groups focused on how to improve the standard (Table 2.1; FGDC, 2004). The "geodata.gov portal", also known as *Geospatial One-Stop*, serves as a public gateway for improving access to geospatial information and data (Geospatial One-Stop, 2004). It is one of 24 *e*-government initiatives sponsored by the Federal Office of Management and Budget (OMB) to enhance government efficiency and to improve citizen services. This portal makes it easier, faster, and less expensive for all levels of government and the public to access geospatial information. UCGIS should help design, implement, and refine the Geospatial One-Stop concept as much as possible. Additional research should continue on: a) how to organize, store, and serve metadata using Geospatial One-Stop; b) the development of improved web-based interfaces for efficiently browsing and downloading metadata; and c) documenting the genealogy (lineage) of all of the operations that have been performed or applied to a dataset (Lanter and Veregin, 1992; Jensen *et al.*, 2002). A user must have a complete understanding of the content and quality of a digital spatial dataset in order to make maximum use of its information potential.

Activities are also underway to harmonize the ISO metadata standard (ISO 19115) with FGDC's CSDGM (FGDC-STD-001-1998). ISO approval does not automatically present the U.S. with a new metadata standard. The American National Standards Institute (ANSI), which administers and coordinates voluntary standardization in the U.S. and represents U.S. interests with ISO, has to adopt ISO 19115 as an American National Standard (ANS) (FGDC, 2004).

2.4.8 Address-Matching Issues

The National Academy of Public Administration (NAPA, 1998) study evaluated the geographic information needs in the 21st century and found that 9 of the 12 public uses of spatial data required geocoded address files. Address information is important to assessors, appraisers, real estate agents, 911 emergency services, mortgage lenders, redistricting, and others. In fact, the billion dollar business geographics industry is founded on the concept that an address can be assigned to topologically correct geographic coordinates and that the address can be used to navigate to the correct location. Thus, there is great demand for an accurate street address data file for a myriad of business and public applications. The issue was raised by the original Mapping Science Committee (1990) and identified as an important aspect of the NSDI, *i.e.* a good place for local government, federal government and private sector cooperation. Unfortunately, the development of such a system on a nationwide basis is difficult for a number of reasons.

First, the address of a building or parcel of land may be the result of historical and administrative illogical decisions. This can result in addresses along a block face that are out of sequence, duplicated, or missing (Figure 2.3). It is very difficult accurately to locate addresses using any form of spatial interpolation along the block face. For example when a set of business addresses are geocoded with TIGER street centerlines, they typically are lumped towards the beginning of the address range for a street segment (Figure 2.4). The Postal Service Zip + 4 system is now widely used for geocoding purposes because it contains a more current set of streets than available from the Census Bureau or a commercial provider. However, the nine digit zip code is usually only able to assign an address to a midpoint of the street centerline for a block. Significant problems can also arise when building locations and their addresses were derived from source materials that were not at the same scale or date. For example, in Figure 2.5, many of the parcel centroids could not be properly referenced from the TIGER street centerlines and would be assigned to the incorrect Census Block based on a point-in-polygon search.

The solution to this problem is to develop a comprehensive set of street centerlines at a scale that ensures that the location of lots, houses, and other buildings will be topologically correct. In the United Kingdom, the Ordinance Survey solved the problem by digitizing buildings and roads from large-scale map sources. In the United States, this problem will most likely be solved using TIGER data. The Census Bureau understands the need to improve the positional accuracy of the TIGER representation of street center lines to conform with the accuracy of points collected with global positioning system (GPS) technology. Improvement of the address/street location accuracy is one of the five stated objectives of the Census Bureau's MAF/TIGER Enhancement program (Broome and Godwin, 2003). According to the Census Bureau, a combined positional accuracy of 7.6 m for linear features and structures will ensure that 99.6 percent of the points used to represent residential structures will fall within the proper census block (Broome and Godwin, 2003). Assembling the source material in the

3,232 counties and equivalent areas to realign the TIGER line files to meet this specification is a major technical, financial, and institutional challenge. In fact, there is no assurance that the $200 million contract to acquire or develop these sources will meet the need. The program relies heavily on the voluntary contribution by local government entities of high-resolution street centerlines. In many cases, these data have licenses that govern their distribution and prohibit them from becoming part of the TIGER system that must remain in the public domain.

The TIGER street centerlines will likely become the road transportation features of the *National Map* (Cowen and Craig, 2004), a consistent online, seamless geospatial database (Clarke *et al.,* 2003; Kelmelis *et al.,* 2003). It will provide the user community with an excellent resource for geocoding that is continuously updated.

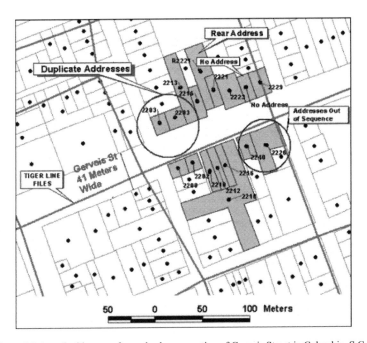

Figure 2.3 Actual addresses of parcels along a section of Gervais Street in Columbia, S.C. highlighting the type of inconsistencies that are typical with urban addresses

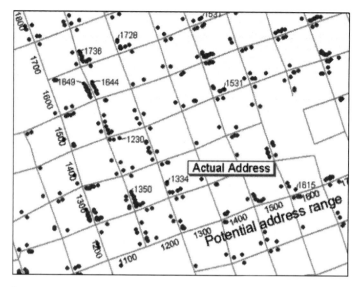

Figure 2.4 Geocoded locations of business addresses that demonstrate the problem of clustering of addresses at the lower end of the potential address range

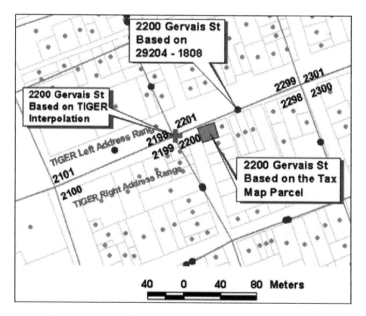

Figure 2.5 Three different locations for the same address based on different sources of geocoding. Also note that the TIGER block boundaries would not capture the correct set of parcel centroids

Accurate street centerlines with address ranges will improve the accuracy of geocoded addresses; however, it will not overcome the problems associated with interpolation of addresses along a street segment. The ultimate solution to the problem is to replace the interpolation process with geocoding based on a one to one correspondence to the actual location of the building. The Census Bureau's goal for the 2010 decennial census is to build such a file using GPS acquired coordinates. This method may be supplemented using information extracted from high spatial resolution imagery. Therefore, the solution to the geocoding problem becomes a feature extraction problem coupled with the integration of official records of street addresses. Structures must be identified, converted into a vector presentation of building footprints and centroids. This level of representation will constitute the building block for a national multipurpose cadastre (Cowen and Craig, 2004). UCGIS scientists need to be involved in many aspects of this research.

2.4.9 Privacy

Geographic information systems and the technological family associated with them—global positioning systems and the high spatial resolution remote surveillance systems—raise important personal privacy questions (Onsrud *et al.*, 1994; Curry, 1997; Slonecker *et al.*, 1998). Of immediate significance is the fact that the systems store and represent data in ways that render ineffective the most popular safeguards against privacy abuse. UCGIS scientists and others should delve deeply into the ethical and moral issues associated with technological change, the impact of improvements in the specificity and resolution of the data collected, and the changing "right-to-privacy" for countries, communities, businesses, and the "digital individual".

2.5 *IN SITU* DATA COLLECTION

Much of the data collected about people, flora, fauna, soils/rocks, the atmosphere, and water in its various forms are obtained by *in situ* measurement. These data hopefully are collected using a well thought-out sampling scheme or by a complete census of the population. In order to integrate spatial information derived from diverse *in situ* measurements, several issues must be investigated.

2.5.1 *In Situ* Instrument Calibration

Instruments such as thermometers, radiometers, and questionnaires must be calibrated. The logic and methods used to calibrate the instrument at the beginning, at intermediate stages throughout the data collection process, and at the end must be rigorously defined and reported as part of the metadata. Also, there is the ever-present problem of how to calibrate the human operator of the instrument.

Research is required to document the impact of integrating spatial information derived from perhaps multiple studies with instruments that were poorly or even improperly calibrated. The situation becomes more complex when poorly calibrated point observations are subjected to an interpolation algorithm that creates a geographically extensive continuous statistical surface.

A monograph on *in situ* instrument calibration and data collection covering most of the relevant issues associated with population (people) questionnaires, traditional surveying, GPS, atmospheric sampling, soil/rock sampling, water sampling, vegetation sampling, and spectroradiometer instrumentation would be heavily used. At the present time, one must obtain such information from very diverse sources, often with conflicting opinions about instrument calibration procedures.

2.5.2 *In situ* Sampling Logic

The world is a geographically extensive, complex environment that generally does not lend itself well to a complete wall-to-wall enumeration (census). Consequently, it is usually necessary to sample the environment with a calibrated instrument while hoping to capture the essence of the attributes under investigation. Sampling may save both time and money, but may not be as accurate as a complete census. Nevertheless, it may be acceptable within certain statistically defined confidence limits. Research is required to identify effective sampling logic and more robust statistical analysis techniques to analyze the sampled data (*e.g.*, Stehman, 2000).

Sampling often requires determining attributes at specific locations defined as single points, *e.g.*, soil samples in an agricultural field; lines, such as transects through geological strata; areas, as in zones for urban population estimation; or volumes, as in ore body analysis. The complexity of sampling approaches to geographical data has increased with the variety of applications of GIS. For example, in agricultural applications, soil samples have traditionally been based on a rectangular grid sampling pattern. With the advent of precision farming methods, grid sampling at sufficient resolution for the application of variable nutrient inputs across a field is cost prohibitive. Research is needed to determine appropriate stratified sampling methods or new technology such as remote sensors for sampling soil characteristics. Other examples of sampling methods which do not fit the traditional spatial processing models include weed and insect scouting in agriculture. Often the results of these infield scouting efforts are single point locations or isolated polygon locations with random positioning and spacing. Use of these types of data for geographical analysis requires development of new and different methods for interpolating and combining these data with more traditional sampling schemes and data types (Whelan *et al.*, 1996; Stehman, 2001).

In addition, research is necessary to identify the optimum method of interpolating between point observations to derive a continuous statistical surface in one of several data structures, including: raster, triangular-irregular-network (TIN), quad-tree, *etc.* Research should determine the wisdom of comparing multiple continuous surfaces that were created using different methods of interpolation, *e.g.*, inverse distance-weighting versus geostatistical kriging (Atkinson and Lewis, 2000; Davidson and Csillag, 2003).

2.6 CENSUS ENUMERATION LOGIC

A census is not a sample, but a complete enumeration of the population. There are many ways to conduct a census, including: direct enumeration, self enumeration, and administrative enumeration. If appropriate census design and operations methods are not followed, serious error can enter the database such as overcount, undercount, and misallocation. Several of the most important census issues to be resolved are a) the impact of the geographic database used during field enumeration, b) avoiding incomplete coverage, c) minimizing response errors due to measurement instrument problems, d) selecting the most appropriate data transformation(s), and e) assessing the quality or accuracy of a census.

2.61 Census of Agricultural Yield Example

The development of yield monitoring equipment for grains such as wheat and corn and for peanuts and cotton, in conjunction with the use of global positioning technology, has created new types of data which are not appropriately handled with current GIS processing methods. These data most closely approximate the raster structure of spatial data; however, each raster cell is based on a fixed time interval and movement of harvester equipment through the interval. Since the time interval is used to capture geographic position using a GPS receiver and the harvester equipment speed is not constant, each raster cell is potentially a different size. Added to the confusion is the time delay between the actual GPS signal receipt and the volume of the harvested crop which moves through the equipment. Deconvolution of the signal and creation of the exact yield as a specific point or area in the field is a non-trivial spatial data processing problem (Boydell, 1997). Effective representation of the total yield across the field as a raster image of varying cell sizes or appropriate interpolation to a fixed grid is an important research question. Additional research is needed to determine how best to use these yield data in conjunction with infield sampling and with variable rate application equipment which are driven by digital maps to allow site specific management of farms. The potential benefits are 1) a reduction in the volume of agrochemicals applied, thereby reducing the degradation to the environment, and 2) the application of the chemicals where they are needed most and in the correct amounts thereby increasing crop yields. The successful use of these technologies depends on significant advances in spatial data representation, processing, and analysis methods (Usery *et al.*, 1995).

2.7 REMOTE SENSING DATA COLLECTION

Remote sensor data may not provide the level of completeness (*i.e.* specificity) nor the rigorous spatial position information that can be obtained when the data are collected in the field by a scientist armed with an appropriate *in situ* measurement device and a GPS unit. Fortunately, calibrated remote sensor data can, in certain instances, provide geographically extensive information about human occupancy and biophysical characteristics (*e.g.*, biomass, temperature, moisture content) in much greater detail than extremely costly point *in situ* investigations. The key is knowing when it is appropriate to use each technology alone or in conjunction with the other.

Several important observations are in order concerning remote sensor data. First, remote sensor data may be used to collect information for many of the Spatial Data Themes of the FGDC Subcommittees summarized in Figure 2.1 and Table 2.2 and (FGDC, 2004). In fact, it is difficult to collect the required spatial information for many of the themes without using remote sensor data. The standards being developed by each of the FGDC subcommittees (*e.g.*, the Vegetation Classification Standard) recognize that remote sensor data calibrated with *in situ* observation are the only way to collect much of the data that must populate the database.

Table 2.2 Spatial Data Themes of FGDC Subcommittees (FGDC, 2004)

Data Theme/Subcommittee	Agency Chairing Subcommittee
Basic cartographic	Dept. of Interior (DOI): USGS
Cadastral	DOI: Bureau of Land Management
Cultural and demographic	Dept. of Commerce (DOC): U.S. Census Bureau
Federal Geodetic Control	DOC: National Geodetic Survey
Geologic	DOI: USGS
Ground transportation	Bureau of Transportation Statistics
Marine and Coastal Spatial Data	National Oceanic and Atmospheric Administration Coastal Services Center
Spatial Climate	U.S. Dept. of Agriculture (USDA)–NRCS
Spatial Water Data	DOI: USGS
Soils	USDA
Vegetation	USDA
Wetlands	DOI: U.S. Fish and Wildlife Service

Unfortunately, there is a growing conception that a) the historical declassified imagery; b) the commercial high spatial resolution sensor systems such as Ikonos, QuickBird, and EROS 1A; and c) the suite of National Aeronautics and Space Administration (NASA) Earth Observing System (EOS) sensors will solve most of our remote sensing data collection requirements. This is not the case (Pace *et al.,* 1997; Cowen and Jensen, 1998; Stoney, 1998; Townshend and Justice, 2002). In fact, the data may create entirely new problems (Miller *et al.,* 2000). For example, the cost of commercially available imagery may be prohibitive and there may be impractical copyright restrictions placed on the data that limit its utility (Miller *et al.,* 2003). Only research will determine if the remote sensor data can solve old and perhaps entirely new problems. The following sections briefly document the state-of-the-art of: a) urban/suburban socioeconomic data requirements, and b) biophysical attribute data requirements compared with the current and near-future proposed sensor systems to document where significant gaps in data collection capability and utility exist. Important research topics are identified within each section.

2.7.1 Remote Sensing of Urban/Suburban Socioeconomic Characteristics

The relationship between the temporal and spatial data requirements for selected urban/suburban attributes and the temporal and spatial characteristics of selected remote sensing systems is presented in Table 2.3 and Figure 2.6. The attributes summarized in the table were synthesized from practical experience reported in journal articles, symposia, chapters in books, and government and society manuals (specific references are reported in Cowen and Jensen, 1998; Jensen and Cowen, 1999; Jensen, 2000; Bossler *et al.,* 2002; Jensen and Hodgson, 2004). Sensors operating in the visible and near-infrared portions of the spectrum are usually sufficient for collecting urban information, unless the area is shrouded in clouds in which case active microwave (radar) is more appropriate (Leberl, 1990). Hyperspectral data is not normally required for urban applications. Therefore, this discussion focuses on whether the urban spatial and temporal resolution data collection requirements are satisfied.

2.7.1.1 Land Use/Land Cover

The relationship between USGS land cover classification system levels (I–IV) and the spatial resolution of the sensor system (ground resolved distance in meters) was presented in Figure 2.2. The National Image Interpretability Rating System (NIIRS) guidelines were provided for comparative purposes. Level I classes may be inventoried using sensors such as the Landsat Multispectral Scanner (MSS) with a nominal spatial resolution of 79 x 79 m, the Thematic Mapper (TM) at 30 x 30 m, and SPOT HRV (XS) at 20 x 20 m. Sensors with a minimum spatial resolution of 5–20 m are required to obtain Level II information. The SPOT HRV and the Russian SPIN-2 TK-350 are the only operational satellite sensor systems providing 10 x 10 m panchromatic data. Radarsat provides 11 x 9 m spatial resolution data for Level I and II land cover inventories even in cloud-shrouded tropical landscapes. Landsat 7 with its 15 x 15 m panchromatic band was launched in 1999. More detailed Level III classes may be inventoried using a sensor with a spatial resolution of approximately 1–5 m (Welch, 1982; Forester, 1985) such as IRS-1CD pan (resampled to 5 x 5 m), large scale aerial photography,

Table 2.3 Urban/suburban attributes and the minimum remote sensing
resolutions required to provide such information
(updated from Cowen and Jensen, 1998; Jensen and Cowen, 1999; Jensen, 2004)

	MINIMUM RESOLUTION REQUIREMENTS		
	Temporal	Spatial	Spectral
Land Use/Land Cover			
L1–USGS Level I	5–10 years	20–100 m	V-NIR-MIR-Radar
L2–USGS Level II	5–10 years	5–20 m	V-NIR-MIR-Radar
L3–USGS Level III	3–5 years	1–5 m	V-NIR-MIR-Pan
L4–USGS Level IV	1–3 years	0.25–1 m	Panchromatic
Building and Property Line Infrastructure			
B1–building perimeter area, volume, height	1–2 years	0.25–0.5 m	Panchromatic
B2–cadastral mapping	1–6 months	0.25–0.5 m	Panchromatic
Transportation Infrastructure			
T1–general road centerline	1–5 years	1–30 m	Panchromatic
T2–precise road width	1–2 years	0.25–0.5	Panchromatic
T3–traffic count studies	5–10 min	0.25–0.5 m	Panchromatic
T4–parking studies	10–60 min	0.25–0.5 m	Panchromatic
Utility Infrastructure			
U1–general utility line mapping and routing years	1–5 years	1–30 m	Panchromatic
U2–precise utility line width, right-of-way years	1–2 years	0.25–0.6 m	Panchromatic
U3–location of poles, manholes, substations	1–2 years	0.25–0.6 m	Panchromatic
Digital Elevation Model (DEM) Creation			
D1–large scale DEM	5–10 years	0.25–0.5 m	Panchromatic
D2–large scale slope map	5–10 years	0.25–0.5 m	Panchromatic
Socioeconomic Characteristics			
S1–local population estimation	5–7 years	0.25–5 m	Panchromatic
S2–regional/national population estimation	5–15 years	5–20 m	V-NIR
S3–quality of life indicators	5–10 years	0.25–30 m	Pan-NIR
Energy Demand and Conservation			
E1–energy demand and production potential	1–5 years	0.25–1 m	Pan-NIR
E2–building insulation surveys	1–5 years	1–5 m	Thermal infrared
Meteorological Data			
M1–daily weather prediction	30 min–12 hrs	1–8 km	V-NIR-TIR
M2–current temperature	30 min–1 hr	1–8 km	TIR
M3–current precipitation	10–30 min	4 km	Doppler Radar
M4–immediate severe storm warning	5–10 min	4 km	Doppler Radar
Critical Environmental Area Assessment			
C1–stable sensitive environments	1–2 years	1–10 m	V-NIR-MIR
C2–dynamic sensitive environments	1–6 months	0.3–2 m	V-NIR-MIR-TIR
Disaster Emergency Response			
DE1–pre-emergency imagery	1–5 years	1–5 m	V-NIR
DE2–post-emergency imagery	12 hr–2 days	0.3–2 m	V-Pan-NIR-Radar
DE3–damaged housing stock	1–2 days	0.3–1 m	V-Pan-NIR
DE4–damaged transportation	1–2 days	0.3–1 m	V-Pan-NIR
DE5–damaged utilities, services	1–2 days	0.3–1 m	V-Pan-NIR

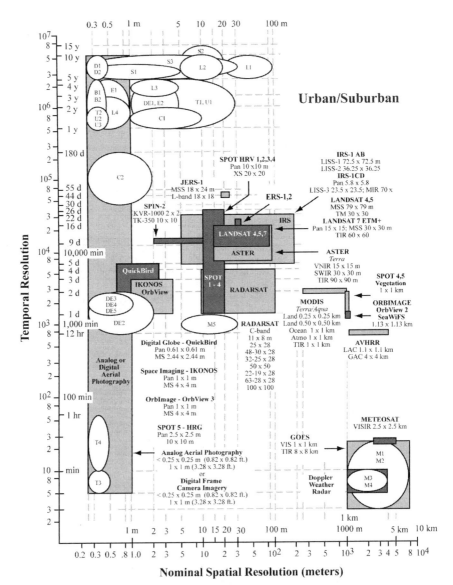

Figure 2.6 Spatial and temporal resolution requirements for urban/suburban attributes (ellipses) overlaid on the spatial and temporal capabilities of selected remote sensing systems (rectangles)

Space Imaging's Ikonos (1 x 1 m pan and 4 x 4 m multispectral), DigitalGlobe's QuickBird (0.61 x 0.61 cm pan and 2.24 x 2.24 m multispectral), and Indian Remote Sensing (IRS) P5 (2.5 x 2.5 m). The synergistic use of high spatial resolution panchromatic data (e.g., 1 x 1 m) and merged, lower spatial resolution multispectral data (*e.g.*, 4 x 4 m) is providing an image interpretation environment that is superior to using panchromatic data alone (Jensen, 2000). Level IV classes and cadastral (property line) information is best monitored using high spatial resolution panchromatic sensors including aerial photography (< 0.25–1 m) or QuickBird panchromatic (0.61 x 0.61 cm) data. Urban land use/cover classes in Levels I through IV have temporal requirements ranging from 1 to 10 years (Table 2.3 and Figure 2.6). All the sensors mentioned have temporal resolutions of < 55 days so the temporal resolution of the land use/land cover attributes is satisfied by the current and proposed sensor systems.

Additional research is required to automatically extract landuse/cover information from the high spatial resolution (< 1 x 1 m) panchromatic remote sensor data. This will require new approaches including 1) the use of neural networks (Jensen and Qiu, 1998; Qui and Jensen, 2004), 2) inductive logic decision-tree algorithms that make use of machine learning (Huang and Jensen, 1997), and 3) classification algorithms that take into account spatial and spectral characteristics such as object-oriented image segmentation algorithms (Benz, 2001). For example, Figure 2.7 depicts polygons of homogeneous material identified using object-oriented image segmentation classification logic applied to high spatial resolution hyperspectral data.

2.7.1.2 Building and Cadastral Infrastructure

Architects, real estate firms, planners, utility companies, and tax assessors often require information on building footprint perimeter, area, volume and height, and property line dimensions (Cullingworth, 1997). Such information is of significant value when creating a multi-purpose cadastre associated with land ownership (Warner, 1996; Cowen and Craig, 2004). Detailed building height and volume data can be extracted from stereoscopic high spatial resolution (0.25 – 0.5 m) photography or other similar stereoscopic remote sensor data (Figure 2.8). The digital building DEM is finding great value for virtual reality walk-throughs. Ikonos and QuickBird provide stereoscopic images with approximately 0.61–1 m spatial resolution. However, such imagery may still not obtain the detailed planimetric (perimeter, area) and topographic detail and accuracy (terrain contours and building height and volume) that can be extracted from high spatial resolution stereoscopic aerial photography (0.25 – 0.5 m).

**Image Segmentation Based on Spectral (Green, Red, Near-infrared)
and Spatial (Smoothness and Compactness) Criteria**

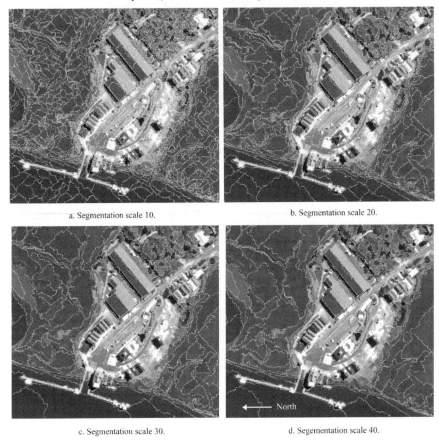

a. Segmentation scale 10.

b. Segmentation scale 20.

c. Segmentation scale 30.

d. Segementation scale 40.

Figure 2.7 Image segmentation of HyMap hyperspectral data of an area near Pritchard's Island, South Carolina. The polygons encompass areas of relatively homogeneous terrain extracted using spectral and spatial criteria (adapted from Jensen, 2004).

Research is required to develop improved hardware and software to inexpensively extract building infrastructure information using soft-copy photogrammetric techniques (NRC, 1995; Jensen, 2000). Photogrammetric studies should document the building footprint perimeter and height information that can be extracted using the high spatial resolution (<1 x 1 m) satellite data and what *in situ* ground control is required to obtain the desired *x,y,z*-coordinate precision.

Figure 2.8 Building wire-frame perimeter, area, and volume information may be extracted from high spatial resolution stereoscopic remote sensor data using soft-copy photogrammetric techniques (courtesy of OrbImage, Inc.)

2.7.1.3 Transportation Infrastructure

Tremendous resources are being spent on revitalizing our nation's transportation infrastructure. Transportation planners use remote sensor data to 1) update transportation network maps, 2) evaluate road condition, 3) study urban traffic patterns at choke points such as tunnels, bridges, shopping malls, and airports, and 4) conduct parking studies (Haack *et al.*, 1997). One of the more prevalent forms of transportation data is the street centerline spatial data (SCSD). Three decades of practice have proven the value of differentiating between the left and right sides of each street segment and encoding attributes to them such as street names, address ranges, ZIP codes, census and political boundaries, and congressional districts. SCSD provide a good example of a framework spatial data theme by virtue of their extensive current use in facility site selection, census operations, socioeconomic planning studies, and legislative redistricting (NRC, 1995). However, additional research should determine when it is necessary to extract one to many centerlines. Is it when it is more than two lanes? What about turn and on-and off-ramp lanes? When is a divided highway divided? These are significant issues that are important when creating the transportation infrastructure so central to many geographic information systems.

Road network centerline updating is done once every 1–5 years and in areas with minimum tree density (or leaf-off) can be accomplished using imagery with a spatial resolution of 1–30 m (Lacy, 1992). If more precise road dimensions are required such as the exact center of the road, the width of the road and sidewalks, then a spatial resolution of 0.25–0.5 m is required (Jensen *et al.*, 1994). Currently, only aerial photography can provide such planimetric information. Road, railroad, and bridge condition (cracks, potholes, *etc.*) may be monitored both *in situ* and using high spatial resolution (< 0.25 x 0.25 m) remote sensor data (Swerdlow, 1998).

Traffic count studies of automobiles, airplanes, boats, pedestrians, and people in groups require very high temporal resolution data ranging from 5 to 10 minutes. It is often difficult to resolve a car or boat using even 1 x 1 m data. This requires high spatial resolution imagery from 0.25 to 0.5 m. Such information can only be acquired using aerial photography or video sensors that are a) located on the top edges of buildings looking obliquely at the terrain, or b) placed in aircraft or helicopters and flown repetitively over the study areas. When such information is collected at an optimum time of day, future parking and traffic movement decisions can be made. Parking studies require the same high spatial resolution (0.3–0.5 m) but slightly lower temporal resolution (10–60 minutes). Doppler radar has demonstrated some potential for monitoring traffic flow and volume. High spatial resolution imagery obtained from stable satellite platforms make it possible to geometrically mosaic multiple flightlines of data together without the radiometric effects of radial/relief displacement or vignetting away from the principal point of each photograph. Improved edge detection algorithms are required to extract street (centerline) information automatically from the imagery.

In 2000, the U.S. Department of Tranportation (USDOT) Research and Special Programs Administration and NASA created the National Consortia on Remote Sensing in Transportation (NCRST), with funding for 4 years under the Transportation Equity Act for the 21st century (see Chapter 13). Four university consortia conducted remote sensing and spatial information research in four focus areas of transportation: environmental assessment (NCRST-E), infrastructure management NCRST-I), traffic flow (NCRST-F), and disaster assessment, safety and hazards ((NCRST-DASH). Many of these scientists are UCGIS associates.

2.7.1.4 Utility Infrastructure

Urban/suburban environments create great quantities of refuse, waste water, and sewage and require electrical power, natural gas, telephone service, and potable water (Schultz, 1988; Haack *et al.*, 1997). Automated mapping/facilities management (AM/FM) and geographic information systems have been developed to manage extensive right-of-way corridors for various utilities, especially pipelines (Jadkowski *et al.*, 1994; Jensen, 2000). The most fundamental task is to update maps to show a general centerline of the utility of interest such as a powerline right-of-way. This is relatively straightforward if the utility is not buried and 1–30 m spatial resolution remote sensor data are available. It is also often necessary to identify prototype utility (*e.g.*, pipeline) routes (Feldman *et al.*, 1995). Such studies require more geographically extensive imagery such as Landsat TM data (30 x 30 m). Therefore, the majority of the actual and proposed rights-of-way may be observed well on imagery with 1–30 m spatial resolution obtained once every 1–5 years. When it is necessary to inventory the exact location of the footpads or transmission towers, utility poles, manhole covers, the true centerline of the utility, the width of the utility right-of-way, and the dimensions of buildings, pumphouses, and substations then it is necessary to have a spatial resolution of from 0.25 to 0.6 m (Jadkowski *et al.*, 1994). The nation is spending billions on improving transportation and utility infrastructure. It would be wise to provide funds for mapping (inventorying) the improvements.

2.7.1.5 Digital Elevation Model Creation

It is possible to extract relatively coarse *z*-elevation information using SPOT 10 x 10 m data, SPIN-2 data (Lavrov, 1997), and even Landsat TM 30 x 30 m data (Gugan and Dowman, 1988). However, any DEM to be used in an urban/ suburban application should have a *z*-elevation and *x,y* coordinates that meet draft Geospatial Positioning Accuracy Standards. The only sensors that can provide such information at the present time are stereoscopic large scale metric aerial photography with a spatial resolution of 0.25–0.5 m and light detection and ranging (LIDAR) sensors (Raber *et al.*, 2002; Hodgson *et al.*, 2003ab). A LIDAR-derived DEM of Manhattan obtained on September 27, 2001, is shown in Figure 2.9.

A DEM of an urbanized area need only be acquired once every 5–10 years unless there is significant development and the analyst desires to compare two different date DEMs to determine change in terrain elevation, identify unpermitted additions to buildings, or changes in building heights. The DEM data can be modeled to compute slope and aspect statistical surfaces for a variety of applications. Digital desktop soft-copy photogrammetry is revolutionizing the creation and availability of special purpose DEMs (Petrie and Kennie, 1990; Jensen, 2000). However, additional research is required that extracts detailed DEMs from the imagery using inexpensive hardware and software. Too many of the systems are costly and very cumbersome, making it difficult for the scientist to develop a local DEM on demand. DEMs can also be produced using interferometric synthetic aperture radar (IFSAR) techniques.

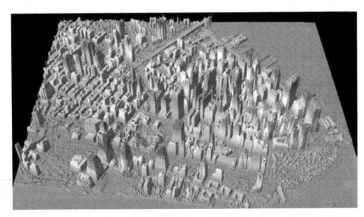

Figure 2.9 Digital elevation model derived from LIDAR 1 x 1 m posting data obtained on September 27, 2001 by EarthData International, Inc. for NOAA (courtesy of EarthData International, NOAA, and the American Society for Photogrammetry & Remote Sensing)

2.7.1.6 Socioeconomic Characteristics

Selected socioeconomic characteristics may be extracted directly from remote sensor data. Two of the most important attributes are population estimation and quality-of-life indicators. Population estimation can be performed at the local, regional, and national level based on (Sutton *et al.*, 1997; Jensen *et al.*, 2002): a) counts of individual dwelling units, b) measurement of urbanized land areas (often referred to as settlement size), and c) estimates derived from land use/land cover classification. Remote sensing of population using the individual dwelling unit method is based on the following assumptions (Lo, 1995; Haack *et al.*, 1997; Jensen, 2000; Lo and Yeung, 2002):

- the imagery must be of sufficient spatial resolution to identify individual structures even through tree cover and whether they are residential, commercial, or industrial buildings,

- some estimation of the average number of persons per dwelling unit must be available,

- some estimate of the number of homeless, seasonal, and migratory workers is required, and

- it is assumed all dwelling units are occupied, and only *n* families live in each unit.

This is usually performed every 5–7 years and requires high spatial resolution remotely sensed data (0.25–5 m). Broome (1998) suggested that this method requires so much *in situ* data to calibrate the remote sensor data that it can be impractical. Research is required to document the utility of the method in a variety of cultures and population densities.

There is a relationship between the simple urbanized built-up area (settlement size) extracted from a remotely sensed image and settlement population (Olorunfemi, 1984), where $r = a \times P^b$ and r is the radius of the populated area circle, a is an empirically derived constant of proportionality, P is the population, and b is an empirically derived exponent. Sutton *et al.* (1997) used Defense Meteorological Satellite Program Operational Linescan System (DMSP-OLS) visible near-infrared nighttime 1 x 1 kilometer (km) imagery to inventory urban extent for the entire U.S. When the data were aggregated to the state or county level, spatial analysis of the clusters of the saturated pixels predicted population with an $R^2 = 0.81$. Unfortunately, DMSP imagery underestimates the population density of urban centers and overestimates the population density of suburban areas (Sutton *et al.*, 1997). Research is required to calibrate this population estimation technique in diverse cultures and population densities.

Most quality-of-life studies make use of census data to extract socioeconomic indicators. Only recently have factor analytic studies documented how quality-of-life indicators (such as house value, median family income, average number of rooms, average rent, education, and income) can be estimated by extracting the urban attributes from relatively high spatial resolution (0.25–30 m) imagery (Jensen, 1983; Avery and Berlin, 1993; Haack *et al.*, 1997; Lo and Faber, 1998). Sensitivity analysis of these methods should take place to see if the quality-of-life indicators are transferable across time and space among various cultures.

2.7.1.7 Energy Demand and Production Potential

Local urban/suburban energy demand may be estimated using remotely sensed data. First, the square footage (or m²) of individual buildings is determined. Local ground reference information about energy consumption is then obtained for a representative sample of homes in the area. Regression relationships are derived to predict the energy consumption anticipated for the region. This requires imagery with a spatial resolution of from 0.25 to 1 m. Regional and national energy consumption may be predicted using DMSP imagery (Welch, 1980; Elvidge, 1997; Sutton *et al.*, 1997).

It is also possible to predict how much solar photovoltaic energy potential a geographic region has by modeling the individual rooftop square footage and orientation with known photovoltaic generation constraints. This requires very high spatial resolution imagery (0.25–0.5 m) (Angelici *et al.*, 1980). The creation of local and regional energy demand and production potential should be a high priority UCGIS research topic as the results could have significant national energy policy implications, especially if energy conservation becomes important once again.

2.7.1.8 Disaster Emergency Response

The Federal Emergency Management Agency (now part of the U.S. Department of Homeland Security) is utilizing remote sensing data as it conducts the Multi-Hazard Flood Map Modernization program. This includes mapping and analyzing data for all types of hazards. The program requires geodetic control, aerial imagery, elevation, surface-water extent, and other thematic data which are used to produce digital flood maps and other hazard-related products. All data will be served via the *Geospatial One-Stop* portal and *The National Map* (Lowe, 2003).

Floods (Mississippi River in 1993; Albany, Georgia in 1994), hurricanes (Hugo in 1989; Andrew in 1992; Fran in 1996), tornadoes (every year), fires, tanker spills, and earthquakes (Northridge, CA in 1994) demonstrated that a rectified, pre-disaster remote sensing image database is indispensable. The pre-disaster data only needs to be updated every 1–5 years; however, it should have high spatial resolution (1–5 m) multispectral data if possible. When disaster strikes, high resolution (0.25–2 m) panchromatic and/or near-infrared data should be acquired within 12 hours to 2 days (Schweitzer and McLeod, 1997). If the terrain is shrouded in clouds, imaging radar might provide the most useful information. Post-disaster images are registered to the pre-disaster images and manual and digital change detection takes place (Jensen, 2004). If precise, quantitative information about damaged housing stock, disrupted transportation arteries, the flow of spilled materials, and damage to above ground utilities are required, it is advisable to acquire post-disaster 0.25–1 m panchromatic and near-infrared data within 1–2 days. Such information was indispensable in assessing damages and allocating scarce clean-up resources during Hurricane Hugo, Hurricane Andrew, Hurricane Fran (Wagman, 1997), and the Northridge earthquake. The role of remote sensing data and GIS modeling in disaster and risk management is an important area of UCGIS research.

2.7.2 Remote Sensing of Biophysical Characteristics

Scientists and natural resource managers of wetland, forests, grassland, rangeland, *etc.* recognize that spatially-distributed biophysical information (e.g., vegetation biomass, leaf-area-index, temperature) are essential for ecological modeling and planning (Johannsen *et al.*, 2003; Jensen and Hodgson, 2004). It is very difficult to obtain such information using *in situ* measurement. Therefore, public agencies and scientists have expended significant resources developing methods to obtain the required information using remote sensing science (Nemani *et al.*, 2003). The UCGIS community of scientists and scholars should be at the forefront of conducting research to extract biophysical information from remote sensor data. The following sections identify the ability of sensor systems to provide some of the required biophysical data. Emphasis is given to the spatial and spectral characteristics of the vegetation requirements in this brief summary.

2.7.2.1 Vegetation: Type, Biomass, Stress, Moisture Content, Landscape Ecology Metrics, Surface Roughness, and Canopy Structure

Vegetation type and biomass may be collected for continental, regional, and local applications, each requiring a different spatial resolution generally ranging from 250 m to 8 km, 20 m to 1 km, and 0.5 to 10 m, respectively (Table 2.4; Figure 2.10). The general rule of thumb is to utilize one band in the visible (preferably a chlorophyll absorption band centered on 0.675 mm), one in the near-infrared (0.7–1.2 mm), and one in the middle-infrared region (1.55–1.75 or 2.08–2.35 mm). Biomass (productivity) prediction algorithms such as the normalized difference vegetation index (NDVI) and the enhanced vegetation index (EVI) applied to EOS Moderate Resolution Imaging Spectrometer (MODIS) data make use of these spectral regions (Running *et al.*, 1994; Townshend and Justice, 2002).

Studies by Carter (1993; 1996) and others suggest that plant stress is best monitored using the 0.535–0.640 and 0.685–0.7 μm visible light wavelength ranges. The optimum spatial resolution is 0.5–10 m to identify very specific regions of interest. Atmospherically corrected hyperspectral data are likely to provide the most informative stress information. Unfortunately, there are no orbital hyperspectral sensors that will obtain data at such a high spatial resolution. Vegetation moisture content best is measured using either thermal infrared (10.4–12.5 μm) and/or L-band (24 cm) radar data. The ideal would be 0.5–10 m spatial resolution. Unfortunately, there currently are no satellite thermal infrared or L-band sensors that function at this spatial resolution.

Landscape ecology metrics derived from remote sensor data are becoming the de facto standard indicators of local and regional ecosystem health (Ritters *et al.*, 1995; Frohn, 1998; Jones *et al.*, 1998; Jensen, 2004). The metrics may be obtained using the same spatial and spectral resolution criteria as vegetation type and biomass. Only a few studies have used high spatial resolution data with instantaneous-field-of-view (IFOV) < 20 x 20 m. Research should document the scale dependency of the metrics.

The surface roughness of vegetated surfaces is ideally computed using C-, X-, and L-band radars with spatial resolutions of 10–30 m. The actual selection of the optimum wavelength (frequency) to use is a function of the dominant local micro-relief of the local terrain components (*e.g.*, grass, shrubs, or trees) and needs further research.

Canopy structure information is best extracted using long wavelength radar data (L-band) at 5–30 m spatial resolution. The longer the wavelength, the greater the penetration into the canopy and the greater the volume scattering among the trunk, branches, and stems. Significant research is required to document the relationship between canopy parameters and the backscattering coefficient.

Table 2.4 Relationship between biophysical attributes and the minimum
remote sensing resolutions required to provide such data

	MINIMUM RESOLUTION REQUIREMENTS		
	Temporal	**Spatial**	**Spectral**
Vegetation			
V1—Type and biomass—Level I (continental)	Daily	250 m–8 km	0.5–1.2 μm
V2—Level II (regional)	1–5 years	20 m–1 km	0.5–1.2; 1.55–1.75 μm
V3—Species (local)	1–10 years	0.5 m–10 m	0.4–1.2; 1.55–1.75 μm
V4—Stress	1–2 weeks	0.5 m–10 m	0.4–0.675; 0.7–1.2; 1.55–1.75 um
V5—Moisture content	1–2 weeks	0.5 m–10 m	0.4–1.2; 1.55–1.75; 10.4–12.5 μm; L-band
V6—Landscape ecology metrics (patch)	1–2 years	5–30 m	0.5–1.2 μm
V7—Surface roughness	1–2 years	10–30	C-, X-, and L-band
V8—Canopy structure (stems, branches)	1–2 months	5–30 m	L-band
Water			
W1—Land surface water	1–2 years	10 m–8 km	0.725–1.10 μm
W2—Ocean water extent	Daily	1–8 km	0.725–1.10 μm
W3—Depth (bathymetry)	1–10 years	1–10 m	0.44–0.54 μm
W4—Inorganic matter - Suspended sediment	1–10 days	10 m–4 km	0.4–1.2 μm
W5—Organic matter - Phytoplankton, Chl *a*	1–10 days	10 m–4 km	0.4–0.675 μm
W6—Dissolved organics	1–10 days	10 m–4 km	0.4–1.2 μm
W7—Temperature	1–2 days	10 m–4 km	10.5–12.5 μm
Soils and Rocks			
SR1—Inorganic matter - mineral content	1–10 years	10–100 m	0.4–1.2 μm; L-band
SR2—Organic matter - humus	1–10 years	20–30 m	0.4–1.2 μm
SR3—Hydrothermal alteration (clay, mica)	1–10 years	20–30 m	1.55–1.75; 2.08–2.35 μm
SR4—Soil moisture	monthly	20–30 m	1.55–1.75 μm; X-, C-band
Snow and Ice			
SI1—Snow extent	daily	1–8 km	0.55–0.7 μm
SI2—Ice extent	daily	1–8 km	0.55–0.7 μm
SI3—Snow versus clouds	daily	1–8 km	1.55–1.75 μm
Atmosphere			
A1—Cloud extent day	hourly	1–8 km	0.55–0.7; 10.5–12.5 μm
A2—Cloud extent night	hourly	1–8 km	3.5–3.93; 10.5–12.5 μm
A3—Cloud temperature	hourly	1–8 km	10.5–12.5 μm
A4—Water vapor	hourly	1–8 km	6.7 μm
A5—Ozone	monthly	1–8 km	9.58–9.88 μm

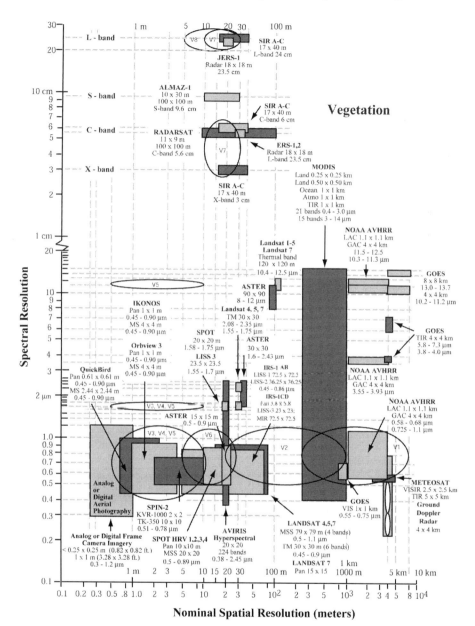

Figure 2.10 Spatial and spectral resolution requirements for extracting vegetation biophysical variables (ellipses) from selected remote sensor data (rectangles)

Notice the lack of a high resolution middle-infrared band for vegetation stress and moisture studies, the lack of a thermal channel for moisture studies, and high resolution radar data for surface roughness and canopy structure information (Figure 2.9). Improved algorithms are also required that perform on-board processing of the spectral data and then telemeter the biophysical vegetation information to the ground receiving station. Improved soil and atmospherically-resistant vegetation index algorithms and on-board absolute atmospheric correction of the data are required. MODIS hyperspectral data are providing valuable land cover, leaf-area-index (LAI), and absorbed photosynthetically active radiation (APAR) information on a repetitive basis (Townshend and Justice, 2002).

2.7.2.2 Water: Land and Ocean Extent, Bathymetry (depth), Organic and Inorganic Matter, Temperature, Snow and Ice Extent

Remote sensing in the near-infrared region from 0.725 to 1.10 μm provides good discrimination between land and water. Oceanic studies require a spatial resolution from 1 to 8 km while land water surface extent studies may be from 10 m to 8 km. However, improved algorithms are required when the water column contains significant quantities of organic and/or inorganic matter.

The optimum spectral region for obtaining bathymetric information in clear water is from 0.44 to 0.54 μm with the best water penetration at 0.48 μm. Bathymetric charting normally requires a spatial resolution of from 1 to 10 m. Research is required to remove the effects of a) suspended organic and/or inorganic matter in the water column, and b) bottom type on the depth estimate.

Water contains clear water, inorganic suspended materials (*e.g.*, suspended sediment), organic constituents (especially phytoplankton and associated chlorophyll *a*), and dissolved organic matter. Obtaining information in the chlorophyll *a* (0.4–0.5 μm) and *b* (centered on 0.675 μm) absorption bands provides very useful information about phytoplankton distribution both in oceanic and land surface water. The SeaWiFS sensor was designed to be sensitive to these spectral regions. Visible and near-infrared bands (0.4–1.2 μm) provide information on suspended sediment distribution. The spatial resolution requirements may range from 10 m to 4 km when conducting local to regional studies. The visible region from 0.4 to 0.7 μm has been shown to be effective in identifying the dissolved organic matter gelbstoff (yellow stuff) in water. Disentangling the organic and inorganic constituents from the spectral response of clear water remains one of the most serious problems. Significant water quality research is required following the logic suggested by Bukata *et al.* (1995).

Water temperature is routinely collected using thermal infrared sensors operating in the region from 10.5 to 12.5 μm and at spatial resolutions ranging from 10 to 4 km. The spectral region from 0.55 to 0.7 μm is sufficient for identifying the surface extent of snow and ice in daytime images. However, to discriminate between snow/ice and clouds it may be necessary to use the middle-infrared bands from 1.55 to 1.75 and 2.08 to 2.35 μm. Spatial resolution should range from 1 to 8 km.

2.7.2.3 Soils and Rocks: Inorganic Matter, Organic Matter, and Soil Moisture

Rocks are composed of specific minerals. Soils contain inorganic matter (soil texture is the proportion of sand, silt, and clay size particles), organic matter (humus), and moisture (Vincent, 1997). One of the most important remote sensing data collection problems is disentangling the contribution of these constituents to the remote sensing spectra. For example, it is still difficult to determine the proportion of sand, silt, and clay in soils using traditional visible and near-infrared bands (0.4–1.2 μm). When conducting such studies it is best to use relatively high spatial resolution imagery (20–30 m). The mid-infrared band (2.08–2.35 μm) coincides with an important absorption band caused by hydrous minerals (*e.g.*, clay, mica, and some oxides and sulfates) making it valuable for lithologic mapping and for detecting clay alteration zones associated with mineral deposits, such as copper (Avery and Berlin, 1993). Longer wavelength radar imagery (L-band) has shown some usefulness for penetrating beneath dry alluvium to detect subsurface inorganic constituents.

It is still difficult to determine the amount of organic matter (humus) in a soil. Some information may be obtained in the region from 0.4 to 1.2 μm at relatively high spatial resolutions of 20–30 m. High spatial resolution hyperspectral imagery may be used to identify the proportion of the various land cover and mineral end-members found within each pixel. Mixture-tuned matched-filtering and spectral angle mapper algorithms may be used to extract the information (Jensen, 2004).

If vegetation is present on the soil then it is difficult to disentangle the contribution from soil moisture and vegetation moisture. Nevertheless, on relatively unvegetated soil it is possible to obtain relatively accurate soil moisture estimates using active microwave X- and C-band radar imagery. Spatial resolutions of from 20 to 30 m are useful. Remote sensing of soil moisture must become an operational reality if we are ever to have farmers embrace geospatial technology.

2.7.2.4 Atmosphere: Meteorological Data, Clouds, and Water Vapor

Great expense has gone into the development of near real-time monitoring of frontal systems, temperature, precipitation, and especially severe storm warning. The Geostationary Operational Environmental Satellites (GOES) West obtains information about the western U.S. and is parked at 135° W while GOES East obtains information about the Caribbean and eastern United States from 75° W. Every day millions of people watch the progress of frontal systems that sometimes generate deadly tornadoes and hurricanes. The visible (0.55–0.70 μm) and near-infrared (10.5–12.5 μm) data are obtained at a temporal resolution of 30 minutes. Some of the images are aggregated to create 1- and 12-hour animation. The spatial resolution of GOES East and West is 0.9 x 0.9 km for the visible bands and 8 x 8 km for the thermal infrared band. The public also relies on ground-based Doppler radar for near real-time precipitation and severe storm warning. Doppler radar obtains 4 x 4 km data every 10–30 minutes when monitoring precipitation and every 5–10 minutes in severe storm warning mode.

Clouds are best discriminated in the daytime using the spectral region from 0.55 to 0.7 μm at spatial resolutions ranging from 1 to 8 km. At night, a thermal infrared sensor operating in the region from 10.5 to 12.5 μm is required. Water vapor in the atmosphere is mapped using the spectral region centered on 6.7 μm at spatial resolutions ranging from 1 to 8 km. Dual frequency GPS data may also provide information about precipitable water.

2.8 SUMMARY

This chapter identified some of the major gaps or shortfalls in data integration and data collection strategies for more intensive investigation by UCGIS and other scientists. The paper first addressed important data integration issues that are generic to all data collection efforts. Then, a brief investigation of current and potential *in situ* and remote sensing socioeconomic and biophysical data collection requirements was presented.

2.9 REFERENCES

Anderson, J.R., Hardy, E., Roach, J. and Witmer, R., 1976, *A Land Use and Land Cover Classification System for Use with Remote Sensor Data.* Washington: USGS Professional Paper 964, 28 p.

Angelici, G.L., Bryant, N. A., Fretz, R.K. and Friedman, S.Z., 1980, *Urban Solar Photovoltaics Potential: An Inventory and Modeling Study Applied to the San Fernando Valley of Los Angeles*, Pasadena: JPL, Report #80-43, 55 p.

APA, 2004, *Land-based Classification System*, American Planning Association, *http://www.planning.org/LBCS/index.html.*

ASPRS, 1990, ASPRS Accuracy Standards for Large Scale Maps, *Photogrammetric Engineering & Remote Sensing*, 56(7): 1,068–1,070.

Atkinson, P. M. and P. Lewis, 2000, Geostatistical Classification for Remote Sensing: An Introduction. *Computers & Geosciences*, 2000: 361–371.

Benz, U., 2001, Definiens Imaging GmbH: Object-Oriented Classification and Feature Detection. *IEEE Geoscience and Remote Sensing Society Newsletter*, (September), pp. 16–20.

Bishop, M.P., Shroder, J.F., Hickman, B.L. and Copland, L., 1998, Scale-dependent Analysis of Satellite Imagery for Characterization of Glaciers in the Karakoram Himalaya. *Geomorphology*, 21: 217–232.

Bossler, J.D., Jensen, J.R., McMaster, R.B. and Rizos, C., 2002, *Manual of Geospatial Science and Technology*, London: Taylor & Francis, 623 p.

Boydell, B.C.J., 1997, *Yield Mapping of Peanut: A First Stage in the Development of Precision Farming for Peanut*, Unpublished Master's Thesis, The University of Georgia, Athens, Georgia, 86 p.

Broome, F., 1998, Correspondence, Washington: U.S. Census Bureau.

Broome, F.R. and Godwin, L.S., 2003, Partnering for the People: Improving the U.S. Census Bureau's MAF/TIGER Database. *Photogrammetric Engineering & Remote Sensing*, 69(10): 1,119–1,126.

Bukata, R.P., Jerome, J.H., Kondratyev, K.Y. and Pozdynyakov, D.V., 1995, *Optical Properties and Remote Sensing of Inland and Coastal Waters*, Boca Raton, FL: CRC Press, 362 p.

Carter, G.A., 1993, Responses of Leaf Spectral Reflectance to Plant Stress. *American Journal of Botany*, 80(3): 239–243.

Carter, G.A., Cibula, W.G. and Miller, R.L., 1996, Narrow-band Reflectance Imagery Compared with Thermal Imagery for Early Detection of Plant Stress. *Journal of Plant Physiology*, 148: 515–522.

Clarke, K.C., Armstrong, M.R., Cowen, J.D., Koepp, D.P., Lopez, X., Miller, R.D., Teselle, G.W., Tobler, W.R. and von Meyer, N., 2003, *Weaving a National Map: Review of the U.S. Geological Survey Concept of The National Map*. Washington: The National Academies Press, 128 p.

Cobb, M.A., Chung, M.J., Foley, H., Petry, F.E., Shaw, K.B. and Miller, H.V., 1998, A Rule-based Approach for the Conflation of Attributed Vector Data. *GeoInformatica*, 2(1): 7–36.

Congalton, R.G. and Green, K., 1999, *Assessing the Accuracy of Remotely Sensed Data: Principles and Practices*. Boca Raton, FL: Lewis Publishers, 137 p.

Cowardin, L.M., Carter, V., Golet, F.C. and LaRoe, E.T., 1979, *Classification of Wetlands and Deepwater Habitats of the U.S.*, Washington: U.S. Fish & Wildlife Service, FWS/OBS-79/31, 103 p.

Cowen, D.J. and Craig, W., 2004, A Retrospective Look at the Need for a Multi-purpose Cadastre. *Surveying and Land Information Science*, 63(4): 205–214.

Cowen, D.J. and Jensen, J.R., 1998, *Extraction and Modeling of Urban Attributes Using Remote Sensing Technology, in People and Pixels: Linking Remote Sensing and Social Science*, Washington: National Research Council, National Academy Press, 164–188.

Cowen, D., Jensen, J.R., Bresnahan, G., Ehler, D., Traves, D., Huang, X., Weisner, C. and Mackey, H. E., 1995, The Design and Implementation of an Integrated GIS for Environmental Applications. *Photogrammetric Engineering & Remote Sensing*, 61: 1,393–1,404.

Cullingworth, B., 1997, *Planning in the USA: Policies, Issues and Processes*, London: Routledge, 280 p.

Curry, M.R., 1997, The Digital Individual and the Private Realm. *Annals of the Association of American Geographers*, 87(4): 681–699.

Davidson, A. and Csillag, F., 2003, A Comparison of Analysis of Variance (ANOVA) and Variograms for Characterizing Landscape Structure under a Limited Sampling Budget. *Canadian Journal of Remote Sensing*, 29(1): 43–56.

Elvidge, C.D., Baugh, K.E., Kihn, E.A., Kroehl, H.W. and Davis, E.R., 1997, Mapping City Lights with Nighttime Data from the DMSP Operational Linescan System. *Photogrammetric Engineering & Remote Sensing*, 63: 727–734.

Feldman, S.C., Pelletier, R.E., Walser, E., Smoot, J.R. and Ahl, D., 1995, A Prototype for Pipeline Routing Using Remotely Sensed Data and Geographic Information System Analysis. *Remote Sensing of Environment*, 53: 123–131.

FETC, 1998, *ISO 14000 Update*, Washington: Federal Energy Technology Center, 12 p.

FGDC, 2004, Federal Geographic Data Committee, Washington: Federal Geographic Data Committee, *http://www.fgdc.org*.

Foley, H., Petry, F., Cobb, M. and Shaw, K., 1997, Using Semantics Constraints for Improved Conflation in Spatial Databases. *Proceedings, 7th International Fuzzy Systems Association World Congress*, Prague, 193–197.

Forester, B.C., 1985, An Examination of Some Problems and Solutions in Monitoring Urban Areas from Satellite Platforms. *International Journal of Remote Sensing*, 6: 139–151.

Frohn, R.C., 1998, *Remote Sensing for Landscape Ecology*, Boca Raton, FL: Lewis, 99 p.

Gahegan, M. and Ehlers, M., 2000, A Framework for the Modeling of Uncertainty Between Remote Sensing and Geographic Information Systems. *ISPRS Journal of Photogrammetry & Remote Sensing*, 55:176–188.

Geospatial One-Stop, 2004, *Geospatial One-Stop, http://www.geo-one-stop.gov*

GETF, 1996, *EARTHMAP: Design Study and Implementation Plan*, Annandale: Global Environment & Technology Foundation, 57 p.

Green, K. and Congalton, R.G., 2003, An Error Matrix Approach to Fuzzy Accuracy Assessment: The NIMA Geocover Project. In Lunetta, R.L. and Lyons, J.G. (eds.), 2003, *Geospatial Data Accuracy Assessment,* Las Vegas: U.S. Environmental Protection Agency, EPA/600/R-03/064, 335 p.

Gugan, D.J. and Dowman, I.J., 1988, Topographic Mapping from SPOT Imagery. *Photogrammetric Engineering & Remote Sensing,* 54: 1,409–1,404.

Haack, B., Guptill, S., Holz, R., Jampoler, S., Jensen, J. and Welch, R., 1997, Chapter 15: Urban Analysis and Planning. *Manual of Photographic Interpretation,* Bethesda: American Society for Photogrammetry & Remote Sensing, pp. 517–553.

Hickman, B.L., Bishop, M.P. and Rescigno, M.V., 1995, Advanced Computational Methods for Spatial Information Extraction. *Computers & Geosciences,* 21(1): 153–173.

Hodgson, M.E., Jensen, J.R., Schmidt, L., Schill, S. and Davis, B.A., 2003a, An Evaluation of LIDAR- and IFSAR-Derived Digital Elevation Models in Leaf-on Conditions with USGS Level 1 and Level 2 DEMS. *Remote Sensing of Environment,* 84(2003): 295–308.

Hodgson, M.E., Jensen, J.R., Tullis, J.A. Riordan, K.D., and Archer, C.M., 2003b, Synergistic Use of Lidar and Color Aerial Photography for Mapping Urban Parcel Imperviousness. *Photogrammetric Engineering & Remote Sensing,* 69(9): 973–980.

Huang, X. and Jensen, J.R., 1997, A Machine Learning Approach to Automated Construction of Knowledge Bases for Image Analysis Expert Systems That Incorporate GIS Data. *Photogrammetric Engineering & Remote Sensing,* 63(10): 1,185–1,194.

Jadkowski, M.A., Convery, P., Birk, R.J. and Kuo, S., 1994, Aerial Image Databases for Pipeline Rights-of-Way Management. *Photogrammetric Engineering & Remote Sensing,* 60: 347–353.

Jensen, J.R., 1983, Urban/Suburban Land Use Analysis. In Colwell, R.N. (ed.), *Manual of Remote Sensing,* 2nd edition, Falls Church, American Society of Photogrammetry, pp. 1,571–1,666.

Jensen, J.R., 1995, Issues Involving the Creation of Digital Elevation Models and Terrain Corrected Orthoimagery Using Soft-Copy Photogrammetry. *Geocarto International: A Multidisciplinary Journal of Remote Sensing,* 10: 1–17.

Jensen, J.R., 2004, *Introductory Digital Image Processing: A Remote Sensing Perspective,* 3rd edition, Upper Saddle River, NJ: Prentice-Hall, 545 p.

Jensen, J.R., 2000, *Remote Sensing of the Environment: An Earth Resource Perspective,* Upper Saddle River, NJ: Prentice-Hall, 544 p.

Jensen, J.R., Botchway, K., Brennan-Galvin, E., Johannsen, C., Juma, C., Mabogunje, A., Miller, R., Price, K., Reining, P., Skole, D., Stancioff, A. and Taylor, D.R.F., 2002, *Down to Earth: Geographic Information for Sustainable Development in Africa*, Washington: National Academy Press, 155 p.

Jensen, J.R. and Qiu, F., 1998, A Neural Network Based System for Visual Landscape Interpretation Using High Resolution Remotely Sensed Imagery. *Proceedings, Annual Meeting of the American Society for Photogrammetry & Remote Sensing*, Tampa, FL, CD–ROM: 15 p.

Jensen, J.R. and Cowen, D.C., 1999, Remote Sensing of Urban/Suburban Infrastructure and Socioeconomic Attributes. *Photogrammetric Engineering & Remote Sensing*.

Jensen, J.R. and Hodgson, M.E., 2004, Remote Sensing of Biophysical Variables and Urban/Suburban Phenomena. In Brunn, S.D., Cutter, S.L., and Harrington, Jr., J.W. (eds.), *Geography and Technology*, Kluwer Academic Publishers, Dodrecht, Netherlands, pp. 109–154.

Jensen, J.R., Cowen, D.C., Halls, J., Narumalani, S., Schmidt, N., Davis, B.A. and Burgess, B., 1994, Improved Urban Infrastructure Mapping and Forecasting for BellSouth Using Remote Sensing and GIS Technology. *Photogrammetric Engineering & Remote Sensing*, 60: 339–346.

Johannsen, C.J., Petersen, G.W., Carter, P.G. and Morgan, M.T., 2003, Remote Sensing: Changing Natural Resource Management. *Journal of Soil and Water Conservation*, 58(2): 42–45.

Jones, K.B., Ritters, K.H., Wickham, J.D., Tankersley, R.G., O'Neill, R.B., Chaloud, D.J., Smith, E.R. and Neale, A.C., 1998, *An Ecological Assessment of the United States Mid-Atlantic Region*, Washington: Environmental Protection Agency.

Kelmelis, J.A., DeMulder, M.L., Ogrosky, C.E., Van Driel, N.J. and Ryan, B.J., 2003, The National Map: From Geography to Mapping and Back Again. *Photogrammetric Engineering & Remote Sensing*, 69(10): 1,109–1,118.

Khorram, S., Biging, G, Chrisman, N., Colby, D., Congalton, R., Dobson, J., Ferguson, R., Goodchild, M., Jensen, J. and Mace, T., 1998, Accuracy Assessment of Land Cover Change Detection. *ASPRS Monograph Series*, Bethesda: American Society for Photogrammetry & Remote Sensing, 78 p.

Klemas, V., Dobson, J.E., Ferguson, R.L. and Haddad, K.D., 1993, A Coastal Land Cover Classification System for the NOAA CoastWatch Change Analysis Project. *Journal of Coastal Research*, 9(3): 862–872.

Lacy, R., 1992, South Carolina Finds Economical Way to Update Digital Road Data. *GIS World*, 5: 58–60.

Lanter, D.P. and Veregin, H., 1992, A Research Paradigm for Propagating Error in Layer-based GIS. *Photogrammetric Engineering & Remote Sensing*, 58(6): 825–835.

Lapine, L., 1989, Correspondence, Columbia: South Carolina Geodetic Survey.

Lavrov, V.N., 1997, Space Survey Photocameras for Cartographic Purposes. *Proceedings of the Fourth International Conference on Remote Sensing for Marine and Coastal Environments*, Michigan: ERIM, 7 p.

Leberl, F.W., 1990, *Radargrammetric Image Processing*, Norwood, Artech House.

Lo, C.P., 1986, The Human Population. *Applied Remote Sensing*, New York: Longman, pp. 40–70.

Lo, C.P., 1995, Automated Population and Dwelling Unit Estimation from High-Resolution Satellite Images: A GIS Approach. *International Journal of Remote Sensing*, 16: 17–34.

Lo, C.P. and Faber, B.J., 1998, Interpretation of Landsat Thematic Mapper and Census Data for Quality of Life Assessment. *Remote Sensing of Environment*.

Lo, C.P., Quattrochi, D.A. and Luvall, J.C., 1997, Application of High-resolution Thermal Infrared Remote Sensing and GIS to Assess the Urban Heat Island Effect. *International Journal of Remote Sensing*, 18(2): 287–304.

Lo, C.P. and Yeung, A.K., 2002, *Concepts and Techniques of Geographic Information Systems*. Upper Saddle River: Prentice-Hall, 492 p.

Lowe, A.S., 2003, The Federal Emergency Management Agency's Multi-Hazard Flood Map Modernization and The National Map. *Photogrammetric Engineering & Remote Sensing*, 69(10): 1,133–1,135.

Lynch, M. and Saalfeld, A., 1985, Conflation: Automated Map Compilation—A Video Game Approach. *Proceedings, AutoCarto 7*, Washington: ACSM, pp. 343–352.

Mapping Science Committee, 1990, *Spatial Data Needs: The Future of the National Mapping Program*, Washington: National Academy Press.

Mcleod, R.D. and Congalton, R.G., 1998, A Quantitative Comparison of Change-Detection Algorithms for Monitoring Eelgrass from Remotely Sensed Data. *Photogrammetric Engineering & Remote Sensing*, 64(3): 207–216.

Miller, R.B., Abbott, M.R., Harding, L.W., Jensen, J.R., Johannsen, C.J., Macauley, M., MacDonald, J.S. and Pearlman, J.S., 2001, *Transforming Remote Sensing Data into Information and Applications*, Washington: National Research Council, 75 p.

Miller, R.B., Abbott, M.R., Harding, L.W., Jensen, J.R., Johannsen, C.J., Macauley, M., MacDonald, J.S. and Pearlman, J.S., 2003, *Using Remote Sensing in State and Local Government: Information for Management and Decision Making*, Washington: National Research Council, 97 p.

NAPA, 1998, *Geographic Information for the 21st Century: Building a Strategy for the Nation*, Washington: National Academy of Public Administration, 358 p.

National Geodetic Survey, 2003, *NADCON*, Version 2.10, *http://www.ngs.noaa.gov/TOOLS/Nadcon/Nadcon.html*

Nemani, R.R., Keeling, C.D., Hashimoto, H., Jolly, W.M., Piper, S.C., Tucker, C.J., Myneni, R.B. and Running, S.W., 2003, Climate-Driven Increases in Global Terrestrial Net Primary Production from 1982 to 1999. *Science*, 300(6): 1,560–1,563.

NRC, 1993, *Towards a Spatial Data Infrastructure for the Nation*, Washington: National Research Council: National Academy Press.

NRC, 1995, *A Data Foundation for the National Spatial Data Infrastructure*, Washington: Mapping Science Committee, National Research Council, 55 p.

NRC, 2001, *National Spatial Data Infrastructure Partnership Programs: Rethinking the Focus*, Washington: National Academy Press.

NSTC, 1996, *Our Changing Planet: The FY 1996 U.S. Global Change Research Program*, Washington: National Science and Technology Council, Subcommittee on Global Change Research, 152 p.

OGC, 2004, *Open GIS Consortium, http://www.opengis.org*.

Onsrud, H.J., Johnson, J.P. and Lopez, X.R., 1994, Protecting Personal Privacy in Using Geographic Information Systems. *Photogrammetric Engineering & Remote Sensing*, 60(9): 1,083–1,095.

Pace, S., O'Connell, K.M. and Lachman, B.E., 1997, *Using Intelligence Data for Environmental Needs: Balancing National Interests*, Washington: Rand, 75 p.

Petrie, G. and Kennie, T.J.M., 1990, *Terrain Modeling in Surveying and Civil Engineering*, London: Whittles Publishing, 351 p.

Qui, F. and Jensen, J.R., 2004, Opening the Neural Network Black Box and Breaking the Knowledge Acquisition Bottleneck of Fuzzy Systems for Remote Sensing Image Classification. *International Journal of Remote Sensing*, in press.

Raber, G.T., Jensen, J.R., Schill, S.R. and Schuckman, K., 2002, Creation of Digital Terrain Models Using An Adaptive Lidar Vegetation Point Removal Process, *Photogrammetric Engineering & Remote Sensing*, 68(12): 1,307–1,315.

Ritters, K.H., O'Neill, R.V., Hunsaker, C.T., Wickham, J.D., Yankee, D.H., Timmins, S.P., Jones, K.B. and Jackson, B.L., 1995, A Factor Analysis of Landscape Pattern and Structure Metrics. *Landscape Ecology*, 10(1): 23–39.

Running, S.W., Justice, C.O., Salomonson, V., Hall, D., Barker, J., Kaufmann, Y.J., Strahler, A.H., Huete, A.R., Muller, J.P., Vanderbilt, V., Wan, Z.M., Teillet, P. and Carneggie, D., 1994, Terrestrial Remote Sensing Science and Algorithms Planned for EOS/MODIS. *International Journal of Remote Sensing*, 15(17): 3,587–3,620.

Saalfeld, A., 1988, Conflation: Automated Map Compilation. *International Journal of Geographical Information Systems*, 2(3): 217–228.

Schultz, G.A., 1988, Remote Sensing in Hydrology. *Journal of Hydrology*, 100: 239–265.

Schweitzer, B. and McLeod, B., 1997, Marketing Technology that is Changing at the Speed of Light. *Earth Observation Magazine*, 6: 22–24.

Slonecker, E.T., Shaw, D.M. and Lillesand, T.M., 1998, Emerging Legal and Ethical Issues in Advanced Remote Sensing Technology. *Photogrammetric Engineering & Remote Sensing*, 64(6): 589–595.

Stehman, S.V., 2000, Practical Implications of Design-based Sampling for Thematic Map Accuracy Assessment. *Remote Sensing of Environment*, 72: 35–45.

Stehman, S.V., 2001, Statistical Rigor and Practical Utility in Thematic Map Accuracy Assessment. *Photogrammetric Engineering & Remote Sensing*, 67: 727–734.

Sutton, P., Roberts, D., Elvidge, C. and Meij, H., 1997, A Comparison of Nighttime Satellite Imagery and Population Density for the Continental United States. *Photogrammetric Engineering & Remote Sensing*, 63: 1,303–1,313.

Swerdlow, J.L., 1998, Making Sense of the Millennium. *National Geographic*, 193: 2–33.

Townshend, J.R.G. and Justice, C.O., 2002, Towards Operational Monitoring of Terrestrial Systems by Moderate-resolution Remote Sensing. *Remote Sensing of Environment*, 83: 351–359.

Usery, E.L. and Pape, D., 1995, Extracting Geographic Features from Raster Data. *Proceedings, American Society for Photogrammetry and Remote Sensing Annual Convention*, Charlotte, North Carolina, pp. 733–740.

Usery, E.L., Pocknee, S. and Boydell, B., 1995, Precision Farming Data Management Using Geographic Information Systems. *Photogrammetric Engineering and Remote Sensing*, 51(11): 1,383–1,391.

Warner, W.S., Graham, R.W. and Read, R.E., 1996, Chapter 15: Urban Survey. *Small Format Aerial Photography*, Scotland: Whittles Publishing, pp. 253–256.

Welch, R., 1980, Monitoring Urban Population and Energy Utilization Patterns from Satellite Data. *Remote Sensing of Environment*, 9: 1–9.

Welch, R., 1982, Spatial Resolution Requirements for Urban Studies. *International Journal of Remote Sensing*, 3: 139–146.

Welch, R., 1995, Emerging Technologies for Low Cost, Integrated GPS, Remote Sensing and GIS Applications. *Proceedings, Cambridge Conference for National Mapping Organizations*, Cambridge, England, 6 p.

Welch, R. and Homsey, A., 1997, Datum Shifts for UTM Coordinates. *Photogrammetric Engineering & Remote Sensing*, 63(4): 371–375.

Whelan, B.M., McBratney, A.B. and Viscarra Rossel, R.A., 1996, Spatial Prediction for Precision Agriculture. *Proceedings of the 3rd International Conference on Precision Agriculture*, Minneapolis, Minnesota, pp. 331–342.

CHAPTER THREE

Cognition of Geographic Information

Daniel R. Montello, University of California at Santa Barbara
Scott Freundschuh, University of Minnesota at Duluth

3.1 INTRODUCTION

"Geographic information science" has newly emerged as the study of basic and applied research issues involving geospatial information. This multi-disciplinary field is concerned with the collection, storage, processing, analysis, and depiction and communication of digital information about spatiotemporal and thematic attributes of the earth, and the objects and events found there. One area of research within geographic information science involves the cognition of geographic information. Cognition of geographic information deals with human perception, memory, reasoning, problem-solving, and communication involving earth phenomena and their representation as geospatial information. Research in cognition is relevant to many issues involving geographic information: data collection and storage, graphic representation and interface design, spatial analysis, interoperability, decision-making, the societal context of geographic information systems (GIS), and more. We believe that many aspects of GIS usability, efficiency, and profitability can be improved by greater attention to cognitive research.

Research on geographic cognition is important to many areas of high priority within the national research and development agenda. An understanding of how humans conceptualize geographic features and information will help promote interoperability of systems, including distributed information systems. Good examples of this include attempts to develop national and international data standards, and attempts to create digital geographic libraries. Research on geographic cognition will improve the functionality and dissemination of many information technologies, including data capture technologies, GIS, and intelligent transportation systems. It will also help provide ways to externalize the divergent belief and value systems of different stakeholders in land use debates. Finally, the study of geographic information cognition will play a major role in improving the effectiveness of geographic education at all levels.

Inadequate attention to cognitive issues impedes fulfillment of the potential of geographic information technologies to benefit society. Cognitive research will lead to improved systems that take advantage of an understanding of human geographic perception and conception, including that of spatial and geographic "experts". It will aid in the design of improved user interfaces and query languages. The possibility that it might lead to improvements in representations, operations, or data models is very real and should be investigated as well. In any case, a geographic information technology that is more responsive to human factors in its design will greatly improve the effectiveness and efficiency of GIS. In addition, cognitive research holds great promise for the advance of education in geography and geographic information at all levels. This includes both traditional general concerns about the poor state of geographic knowledge in the populace, and more specific concerns, such as education about the critical issues of global and environmental change, or extracting the concepts and approaches of geographic information experts.

To provide more equitable and effective access to GIS, it must be recognized that consumers of geographic information are not all the same. Some of these variations among individuals include differences in perceptual and cognitive styles, abilities, and preferences. Cognitive research will therefore allow us to respond to differences among users. Relatively inexperienced or disadvantaged users will gain access to geographic information technologies, and experienced or expert users will gain power and efficiency in their use of the technologies. Information access will be afforded to those with sensory disabilities, the young and the old, people from different cultures who speak different languages, the poor as well as the rich. Intelligent defaults and effective training programs will make systems accessible to the largest possible segment of the population. Alternatively, systems that are flexible may be customized to the particular needs of the individual.

A good example of the potential importance of cognitive research to geographic information science and technology is the development of the *Digital Earth*. Vice President Gore's speech introducing the concept of the Digital Earth was subtitled "Understanding Our Planet in the 21st Century." Understanding is a cognitive act. In the context of Digital Earth, it encompasses the knowledge we can acquire about the earth and its people with the help of new technologies. As such, a project like Digital Earth would only reach its optimal effectiveness with research on the cognition of geographic information. It may very well be an expensive and massive failure without this research. In addition to technology research on hardware and software development, we will need research on human cognition in order to improve the technology, making it help us understand the earth better, including ongoing natural and human processes. Cognitive research, as broadly construed in this chapter, will tell us what and how much information people want and can comprehend, and in what formats it should be presented. Research on the display and visualization of complex geographic information will be of crucial importance. The perception of patterns in space and time is a research issue of ongoing interest in the cognitive sciences. How do people integrate multiple sources of information presented in different sensory and representational modalities? In particular, how does this occur in immersive virtual environments, during a "magic carpet ride"? Digital Earth will allow rapid panning and zooming of displays to view places and landscapes at multiple resolutions, from the very large to the very small. It will also allow simultaneous views at multiple scales. Research on the comprehension and communication of scale and scale changes, in both space and time, will be needed in order to make this a reality. The development of an effective natural language interface for Digital Earth will require cognitive research on spatial and geographic language. Furthermore, it will be essential to understand ways that individuals and groups differ in their cognition of geographic information. Of particular importance, research on education, experience, and age differences will make it possible to build a system that can be used by the young and the old, the expert and the novice. Cognitive research will also help us develop the artificial intelligence components of Digital Earth, such as those involved in automatic imagery interpretation and intelligent data agents. In Mr. Gore's words: "The hard part of taking advantage of this flood of geospatial information will be making sense of it—turning raw data into understandable information". Research on the cognition of geographic information will play a central role in solving this difficult problem.

3.1.1 Background

A growing number of researchers are addressing cognitive questions about geographic information. Such work stems from a research tradition begun primarily in the 1950s and 1960s (with just a few pieces of work earlier) by behavioral geographers, cartographers, urban planners, and environmental psychologists. Behavioral geographers began developing theories and models of the human reasoning and decision-making involved in spatial behavior, such as migration, vacationing, and daily travel (Cox & Golledge, 1969; Golledge & Stimson, 1997). Geographers working in the area of "environmental perception" investigated questions about human responses to natural hazards (White, 1945; Saarinen, 1966), including cognitive responses. Cartographers initiated research on how maps and map symbols are perceived and understood by map users, both expert and novice (Robinson, 1952). Finally, environmental psychologists joined planners and environmental perception researchers in refocusing traditional questions about psychological processes and structures to understand how they operate in built and natural environments, such as public buildings, neighborhoods, cities, and wilderness areas (Lynch, 1960; Appleyard, 1969).

During the decades since the 1960s, several additional disciplines within the behavioral and cognitive sciences have contributed their own research questions and methodologies to this topic. Within research psychology, the subfields of perceptual, cognitive, developmental, educational, industrial/organizational, and social psychology have all conducted research on questions relating to how humans acquire and use spatial and nonspatial information about the world.

Architects have joined planners in attempting to improve the design of built environments through an understanding of human cognition in and of those environments. Both linguists and anthropologists have conducted research on human language and conceptualization about space and place. Artificial intelligence (AI) researchers within computer science and other disciplines have developed simulations of spatial intelligence, in some cases as part of the design of mobile robots. Fundamental theoretical questions about alternative conceptualizations of space and place, and their representations in formal systems, have been investigated by mathematicians, computer scientists, and philosophers.

More recently, within the past 10 years, an interest in geographic cognition has developed within the geographic information science community, a community that now includes many of the disciplines described above. Several specialty groups of The Association of American Geographers are populated by researchers who concern themselves with questions at the intersection of cognition and geographic information, including Environmental Perception & Behavioral Geography, Cartography, GIS, Geography Education, Hazards, Disability, and Urban Geography Specialty Groups. GIS research labs are increasingly focusing on questions about the human comprehension of geographic information and the human factors of GIS (Medyckyj-Scott & Hearnshaw, 1993; Davies & Medyckyj-Scott, 1994, 1996; Nyerges, Mark, Laurini, & Egenhofer, 1995; Egenhofer & Golledge, 1998). The Conference on Spatial Information Theory (COSIT) has taken place every 2 years since 1993, bringing together researchers from several different countries and disciplines to discuss cognitive aspects of spatial information. The National Center for Geographic Information and Analysis (NCGIA) sponsored several workshops and research initiatives dealing with questions of human cognition; examples include I-2 on "Languages of Spatial Relations", I-10 on "Spatio-temporal Reasoning", and I-21 on "Formal Models of Common Sense Geographic Worlds". In its recent incarnation as Project Varenius, the NCGIA's research agenda was composed of three research panels. One of the panels was "Cognitive Models of Geographic Space", comprised of three specialist topics: "Scale and Detail in the Cognition of Geographic Information", "Cognition of Dynamic Phenomena and Their Representation", and "Multiple Modes and Multiple Frames of Reference for Spatial Knowledge". These meetings took place during 1998 and 1999; a summary may be found in Mark, Freksa, Hirtle, Lloyd, & Tversky (1999).

3.2 THEORETICAL PERSPECTIVES ON COGNITION

During the 20th century, several theoretical perspectives or frameworks have been developed in the study of cognition. These perspectives organize research, and provide competing and cooperating explanations for cognitive phenomena. One of the earliest was *constructivism*, emerging from the work of the experimental psychologist Bartlett (1932) and the child psychologist Piaget

(Piaget, 1926/1930; Piaget & Inhelder, 1948/1967). According to this perspective, knowledge of the earth and features on the earth is stored in the mind in the form of cognitive representations that are constructed from perceptual information combined with existing knowledge schemata that serve to organize the perceptual information. Earth knowledge is not simply a perceptual copy of the world but a construction that represents some properties accurately, and distorts or omits other properties. This perspective has been subsequently expressed in research on the structure, acquisition, and use of *cognitive maps*, reviewed below.

A clear alternative to the constructivist framework is the *ecological* perspective of J.J. Gibson (1950, 1979). Contrary to the dualist (according to Gibson) idea of constructivism, the ecological perspective asserts that knowledge exists in a mutual fit between organism and environment. Knowledge need not be constructed from perceptual input but is "directly" available in perceptual arrays encountered by moving organisms. These perceptual arrays are not collections of atomistic sensory properties (lights, tones, *etc.*), but meaningful higher-level units such as openings and support surfaces that provide information for the organism about functional properties of the environment, called *affordances*. More recently, the ecological approach has been mathematically developed by researchers working with "dynamic systems" theory (Thelen & Smith, 1994).

An *information-processing* perspective emerged in the late 1960s and 1970s. It agrees with the constructivist perspective that human cognition depends on the operation of internal representations, symbolic cognitive structures that model events and objects in the world. Unlike the constructivist perspective, however, internally represented information is not acquired in qualitative stages but is continuously and quantitatively built up over time. In addition to the structures that represent objects and events, the information-processing approach places emphasis on the roles of strategies and *metacognition* (cognition about cognition) that control the use of cognitive structures when reasoning about particular problems. An example is a person using a particular set of rules to perform a GIS procedure on several data layers. The information-processing approach is inspired by traditional rule-based digital computing, and is represented by work in formal/computational modeling and symbolic AI (*e.g.*, Newell & Simon, 1976). *Fuzzy logic* and *qualitative reasoning* have been influential within formal/computational modeling (*e.g.*, Zadeh, 1975).

Another perspective that, like the information-processing approach, has been popular with computational modelers is that of *connectionism* or *neural networks*. Stemming from Hebb's (1949) idea of *cell assemblies*, the connectionist perspective suggests that cognition operates by the activation of complexly interconnected networks of simple neuron-like nodes. The output of a network is determined by the patterns of interconnecting links, and weights on these links, that affect output from one node to another, essentially by increasing or decreasing the chances that a particular node will become active or not (Rumelhart & McClelland, 1986). These patterns change over time as a result of feedback into the network from the results of the network's previous outputs or the outputs of other networks. The connectionist perspective is thus thought to offer a model of cognition that does away with the need for the symbolic cognitive structures of the constructivist and information-processing perspectives. It is claimed to be a model of cognition that explicitly ties mental activity to the operation of the brain and nervous system, or at least a neurologically plausible model of the nervous system. Cognitive neuroscientists directly investigate the emergence of cognition in the brain and nervous system (Gazzaniga, 2000).

Throughout much of the 20th century, the importance of language as a vehicle of cognition has been stressed by researchers in anthropology, linguistics, and philosophy. During the 1980s, this *linguistic* perspective has been popularized and extended in the work of Johnson and Lakoff (Lakoff & Johnson, 1980; Lakoff, 1987), and linguists such as Jackendoff and Landau (1991), Levelt (1984), Levinson (1996), and Talmy (1983). According to this perspective, linguistic structures are the critical vehicles for human cognition. This points to the culturally variable nature of cognition insofar as people from different cultures speak different languages; as is well known by anyone attempting to translate ideas across languages, concepts in one language are only approximately similar to concepts in other languages. The "Whorfian Hypothesis" (among other names for this idea) states that language determines or at least influences the nature of cognition as it is practiced by members of different linguistic groups. According to "image-schemata" theory, language expresses meaning via the metaphorical extension of some modestly sized set of *image schemata*, cognitive structures that capture essential concrete relations in the world in ways that allow their application to all meaning, including very abstract meaning. An example of this is the extension of the concept of a "path" connecting two places to any situation where entities are sequentially connected in time or space, such as the path through a computer menu system.

A sixth perspective that has recently become popular also stresses the role of culture, in particular the way that cognition takes place within a context of situations and artifacts partially determined by one's culture. This is the perspective of *situated cognition*. Recently popularized in the English-language scientific literature, but originating early in the 20th century, Vygotsky (1934/1962) suggested that cognitive development is socially mediated and depends critically on language. More recently, others have popularized the insight that cognition serves to solve culturally-specific problems, and operates within contexts provided by culturally-specific problem-solving situations and task settings. Researchers such as Norman (1990) and Hutchins (1995) have stressed that cognition is actually embedded in structure provided by culturally devised tools and technologies. Thus, it is incorrect, according to this perspective, to identify cognition as residing only in the brain or the mind. It also resides in the human body, the surrounds, and in what might be called "cognitive instruments." A simple example is using one's fingers to do arithmetic. A more complex example is the way a computer interface structures thinking and information processing.

Quite recently, a seventh perspective is gaining currency among some cognitive scientists. An *evolutionary* perspective takes issue with the information-processing and connectionist notions that the mind is a general purpose problem-solver. It also differs from the culturally-specific focus of the linguistic and situated-cognition perspectives. Instead, cognition is richly shaped by an innate cognitive architecture that has evolved over the hundreds of thousands of years of human biological evolution (Tooby & Cosmides, 1992). This architecture is posited to consist of several "domain-specific" modules that are specialized to solve certain classes of universally important cognitive problems. Good examples of such problems are finding a mate or finding one's way through the environment. Importantly, the evolutionary perspective suggests that humans from any cultural background will tend to reason in certain universal ways about particular problems. Advances in pedagogy or technology must be compatible with or must overcome these fundamental ways of knowing; compatible advances will work faster and more naturally for humans.

These seven major perspectives, and variations thereof, provide ample theoretical and conceptual raw material for interpreting past research on cognitive issues in geographic information science, and for providing directions for future research. Like theories in any developed science, empirical evidence provides support for some perspectives and argues against other perspectives. For example, the ecological notion that cognition is direct, without involving internally represented information in some form, is untenable if taken literally. Similarly, the mind as a general-purpose problem solver versus a collection of interconnected domain-specialized modules is hotly debated today. But it is no longer very reasonable to argue for the idea that the mind is a *tabula rasa*, whose structures and processes develop entirely from experience after conception, without some significant innate contributions. However, these multiple perspectives are not entirely contradictory by any means. To some degree, they simply focus on different aspects of cognition, perhaps on lower-level rather than higher-level components. A connectionist perspective, for instance, may be about the lower-level neural representations of symbolic structures favored by the information-processing approach. Similarly, whatever the nature of internally represented information, one can appreciate the fact that these representations derive in part from experiences in a particular culture and operate in particular situations where the environment provides information to solve problems. Although not the focus of most perspectives, few explicitly exclude the possibility of an innate architecture that guides and structures the operation of human cognition, as described by the evolutionary perspective.

3.3 LITERATURE REVIEW

3.3.1 Spatial and Environmental Cognition

Cognitive research about space and place has focused on several issues: the responses of sensory systems that pick up spatial information, the development of spatial knowledge from birth to adulthood (ontogenesis) and upon first exposure to a new place (microgenesis), the accuracy and precision of knowledge about distances and directions, spatial language, cognitive structures and processes used during navigation, and perceptual and cognitive issues in cartography, and very recently, GIS. With the advent of new technologies like GIS, new questions about spatial perception and cognition develop, and old questions (both basic and applied) become focused in new ways.

One of the most basic concepts in this area is that of the *cognitive map*. Introduced by Tolman (1948) in his work with rat spatial behavior, the cognitive map is a mental representation, or set of representations, of the spatial layout of the environment. According to Downs and Stea (1973), "cognitive mapping is a process composed of a series of psychological transformations by which an individual acquires, stores, recalls, and decodes information about the relative locations and attributes of phenomena in his [or her] everyday spatial environment" (p. 9). The cartographic map thus serves as a metaphor for spatial and environmental knowledge. Other metaphors have been offered as well, from topological schemata to cognitive collage (see Montello & Freundschuh, 1995). GIS and virtual reality provide our latest metaphors for environmental knowledge.

Cognitive researchers are interested in comparing various sources of geographical knowledge. Montello and Freundschuh (1995) review the characteristics of acquiring knowledge from direct environmental experience, static pictorial representations such as maps (see Thorndyke & Hayes-Roth, 1982), dynamic pictorial representations (movies, animations), and language (see Taylor and Tversky, 1992). Montello and Freundschuh listed eight factors that may play roles in differentiating these sources of geographic information: sensorimotor systems involved, static vs. dynamic information, sequential vs. simultaneous acquisition, the arbitrariness of symbols, the need for scale translations and their flexibility, viewing perspective, precision of presented information, and the inclusion of detail varying in relevance.

It is commonly thought that spatial knowledge of the environment consists of three types of features: knowledge of discrete landmarks, knowledge of routes that connect landmarks into travel sequences, and configurational or survey knowledge that coordinates and metrically scales routes and landmarks. In fact, inspired by Piagetian theory, it has often been suggested that these features represent a necessary learning sequence (Siegel & White, 1975; for an opposing view, see Montello, 1998). Landmarks in particular are thought to play an important role as anchorpoints or reference points for the organization of environmental knowledge (Sadalla, Burroughs, & Staplin, 1980; Couclelis, Golledge, Gale, & Tobler, 1987).

Spatial cognition researchers have studied human navigation and orientation (Golledge, 1999). *Navigation* is coordinated and goal directed movement through space. It may be understood to consist of both locomotion and wayfinding processes. *Locomotion* refers to perceptual-motor coordination to the local surrounds, and includes activities such as moving towards visible targets and avoiding obstacles. *Wayfinding* refers to cognitive coordination to the distant environment, beyond direct sensorimotor access, and includes activities such as trip planning and route choice. Humans navigate and stay oriented both by recognizing landmarks (*piloting*) and by updating their sense of location via *dead reckoning* processes (Gallistel, 1990; Loomis, Klatzky, Golledge, & Philbeck, 1999). Some of these processes are relatively automatic (Rieser, Pick, Ashmead, & Garing, 1995), while others are more like conscious strategies (Cornell, Heth, & Rowat, 1992). A fundamental issue about human orientation concerns the systems of reference that people use to organize their spatial knowledge. Various possible systems have been discussed, including those that encode spatial relations with respect to the body, with respect to an external feature with or without differentiated appearance, or with respect to an abstract frame like latitude-longitude (Hart & Moore, 1973; Levinson, 1996). Several researchers have investigated reference systems within the context of verbal route directions (Allen, 1997).

A central effort in cognitive research on any task or skill domain, whether playing chess or solving calculus problems, is a characterization of the knowledge structures and processes involved in that domain. The same is true of research on spatial/environmental cognition. What is the nature of knowledge that results from exposure to environments or representations such as maps? How should we characterize the form or structure of that knowledge? What cognitive processes, such as encoding or image manipulation, are brought to bear on this knowledge during its use to navigate or give verbal directions?

Cognitive researchers have applied a variety of techniques to answering questions about the content of knowledge and how it may change with training and experience. Since the early 1970s, eye-movement studies have been conducted that record the direction and duration of the map reader's gaze while viewing maps (summarized by Steinke, 1987). Perhaps a more direct research strategy for uncovering the content of knowledge is the use of memory tasks or protocol analysis (*e.g.*, Pick, Heinrichs, Montello, Smith, Sullivan, & Thompson, 1995). A common strategy for elucidating the form or structure of knowledge is to examine distortions or systematic biases in the performance of tasks involving the knowledge. One of the most striking findings in this area is the repeated demonstration that spatial knowledge is not stored simply as a "map in the head" which is read. The map metaphor is quite misleading in some ways (Kuipers, 1982; Tversky, 1992). Researchers interested in spatial knowledge structures and processes have noted the occurrence of systematic distortions in spatial knowledge. The cognitive map has holes, is compressed or enlarged in different areas, may fail to preserve metric information, and shows regularization effects. Spatial knowledge is stored in multiple formats, including spatial, mathematical, and linguistic structures. Nonpictorial cognitive structures (*i.e.*, rules or heuristics) are used to organize one's knowledge of the environment, presumably because they decrease memory load and typically (but not always) support adaptive problem-solving.

Cognitive regionalization is an important example. The more or less continuous landscape is stored as discrete regions, and organized hierarchically, or at least partially so (Hirtle & Jonides, 1985; McNamara, 1992). Stevens and Coupe (1978) first suggested this with their finding that most people distorted the direction between San Diego, California and Reno, Nevada, indicating that Reno was east of San Diego (it is actually west). The authors attributed this to the notion that knowledge of city locations will be stored hierarchically within knowledge of state locations (California is mostly west of Nevada). Maki (1981) reached a similar conclusion from her response-time data showing that people were faster to identify the east-west relations of pairs of cities if they were in different states (see also McNamara, Hardy, & Hirtle, 1989).

Evidence for the operation of other simplifying heuristics for remembering spatial information has been gleaned from patterns of distortion. Tversky (1981) offered the heuristics of "rotation" and "alignment" to explain patterns of distortions she demonstrated. Both heuristics refer to phenomena wherein the remembered orientation or location of a feature learned from a map is distorted in order to more closely align the feature with another feature, or a feature and the global system provided by the cardinal directions. For instance, people typically underestimate how far north Europe is of the United States, instead remembering the two as being aligned with one another along the east-west dimension, and thus incorrectly answering questions about the relative north-south locations of cities in Europe and the United States (see also Mark, 1992). Recent work by Friedman and Brown (2000) suggests that these types of distortions in estimates of latitudes and longitudes ("psychological plate tectonics") are more conceptual than perceptual in origin. Their *plausible-reasoning approach* states that estimates will be based on a combination of multiple types of relevant knowledge, including prior beliefs, new information, and the context of the task. They demonstrated this in an interesting way by showing how estimates of the locations of world cities could be changed in systematic ways by providing subjects with "seed" locations for particular cities.

3.3.2 Cognition of Maps and Geographic Visualizations

One of the oldest areas of research in the cognition of geographic information is the study of cognitive and perceptual aspects of cartographic communication. Maps function to store and communicate information, and to support analysis and problem-solving with this information. Communication and problem-solving are, in part, mental and behavioral activities of individuals. Because maps are composed of sometimes complex systems of signs and symbols whose interpretation depends in profound ways on the prior knowledge and learning experiences of individuals, there are many interesting and subtle questions for researchers interested in the cognition of maps and map use (theoretical overviews may be found in Olson, 1979; Eastman, 1985; Blades & Spencer, 1986; MacEachren, 1992; Lloyd, 1993).

As a research topic, the cognition of maps has roots in the early 20th century. It began with a concern for map education (Gulliver, 1908; Ridgley, 1922), a concern that continues to this day (Blades & Spencer, 1986; Freundschuh, 1997). A second research focus on empirically evaluating and improving map design developed during the 1950s and 1960s. This body of work heralded the beginnings of what become known as *cognitive cartography*. Most of this research has dealt with questions about the perception of map symbols, such as graduated circles, legend symbols, and topographic relief symbols (for reviews, see Potash, 1977; Board, 1978; Castner, 1983). Petchenik (1983) provided an interesting and trenchant critique of this research enterprise. Among other points, she contrasted the analytic goals of research with the synthetic goals of mapmakers, and questioned the ability of research to accommodate the idiosyncratic nature of map users, map tasks, and map designs. Although Petchenik's critique probably moderated enthusiasm for map design research, the motivation to improve maps and map communication continues to inspire researchers (*e.g.*, Eley, 1987; Gilmartin & Shelton, 1989; MacEachren & Mistrick, 1992; Slocum & Egbert, 1993). But in the last couple decades, map-design research has been augmented with work that looks at reasoning and decision-making with maps. Here, we review two such areas—the effects of map orientation during use, and the cognitive development of map skills in children.

3.3.2.1 Map Orientation

Clear scientific evidence now confirms the intuitive understanding of many people that maps are easier or harder to use for tasks such as navigation if you orient them to face in particular directions. Maps are thus said to demonstrate *orientation specificity*: They are most accurately and quickly used when viewed in one specific orientation. If the map is turned to any other orientation, the increased errors and time involved in their use are known as *alignment effects*. When used during navigation, the most commonly preferred orientation for a map is with the top of the map being the direction one is facing in the world. This is variously called "track-up" or "forward-up" alignment. Levine and his colleagues (*e.g.*, Levine, Marchon, & Hanley, 1984) have convincingly demonstrated our preference for this orientation in the case of "you-are-here" (YAH) maps. Robust confusion results when using a YAH map whose top is not the direction one is looking when viewing the map. These researchers also documented the great frequency with which YAH maps in New York City are in fact designed (or placed) in such a misaligned way; it is likely that readers will find it easy to document this for themselves in their own hometowns.

Why does this alignment effect occur? It is clear that left and right on a properly oriented YAH map will directly correspond to left and right in the world, obviating the need for cognitively expensive mental rotation or manipulation. Furthermore, it may be relatively easy to metaphorically treat "forward" in the visual field as "up" on a map because the landscape does in fact "rise" in our visual fields as it stretches out in front of us (Shepard & Hurwitz, 1984). For most people, therefore, navigation maps will be easiest to use when they are oriented to the world in a track-up alignment. A more detailed discussion of map displays in In-Vehicle Navigation Systems is presented below.

However, maps are used for many other tasks than navigation. Thematic and statistical maps are used for scientific analysis, for example. Small-scale maps that depict large areas, such as world maps, are almost always used for purposes other than navigation. In these cases, the cognitive need for alignment with an immediate surrounds is no longer present. Instead, the preferred map orientation depends on learned conventions about how maps are designed and displayed, "north-up" in many cultures (*e.g.*, Evans & Pezdek, 1980). Some research with airplane pilots even indicates that a fixed alignment such as north-up is preferred by trained experts performing specialized and highly practiced navigation tasks (Aretz, 1991). But it bears emphasizing that while there are certainly instances in which track-up alignment is not preferred, research has consistently shown that maps are most easily used in a single preferred orientation for a given task. This fact is likely an instance of the importance of figural orientation in pictorial perception and cognition (Rock, 1974).

3.3.2.2 Education and Development of Map Cognition

The applied interest in map education mentioned above has been accompanied by a focus on basic-science questions about the development of children's map skills (Presson, 1982; Uttal 2000). One of the major cognitive abilities this research has highlighted is the ability to understand representational correspondence in maps, including the confusion sometimes surrounding iconic similarity (as when children believe a red line on the map is a red road in the world). These researchers have also considered the abilities required to understand the shift or rotation involved in interpreting oblique and vertical perspectives, and to use maps to perform planning and determine routes in the environment.

An intriguing debate has emerged about the development of map skills and the degree to which children are inherently equipped to understand maps. In brief, one side of the issue takes the position that young children's (ages 3–5) success at understanding aerial photographs and simple map-like representations indicates an inherent and "natural" ability to comprehend maps as semiotic systems (related claims are made by Landau, 1986; Blaut, 1991). The other side of the debate points to the empirical difficulties and confusions demonstrated by children attempting to understand maps, and takes Piagetian theories about the protracted development of spatial concepts as support for the notion that only rudimentary components of map skills are "natural" (Liben & Downs, 1989, 1993). In fact, this side argues, the full development of map skills is the result of specialized practice and training with maps over many years.

Although there is now agreement that young children can deal with map-like representations to an extent greater than was traditionally believed, and that early education with maps is desirable, the debate continues (Blaut, 1997; Liben & Downs, 1997). It appears that children must be exposed to a somewhat extended developmental and educational process to fully appreciate the more sophisticated significations of maps (such as contour lines). This point becomes most obvious when the complete diversity of map types and uses is recognized.

Liben (1997) presents a six-level, progressive typology for mastering external spatial representations such as maps "which begins with the straight-forward ability to respond to referential content depicted in presentations, and ends with the sophisticated ability to reflect upon the creation and utility of various kinds of representations" (p. 2). According to her model, children first identify the referential meaning of the representation, then the denotative meaning of the representation. Following that, children can distinguish between representation and referent, and intentionally attribute meaning to the representation. Children then come to appreciate that some, but not all attributes of the representation are motivated by attributes of the referent, and that some, but not all attributes of the referent motivate graphic attributes of the representation. After that, children extend their prior understanding of attribute differentiation to develop understanding of the formal representation and geometric correspondences between representation and referent. Finally children are able to reflect upon the mechanisms by which, and the purposes for which, graphic representations are created.

Studying the early emergence of map skills helps clarify how adults use and understand cartographic displays. A developmental perspective seeks to shed light on the basic, core processes that are involved in map comprehension. A systematic comparison of adults and children of various ages should inform our understanding about what aspects of maps and spatial representations are relatively difficult to comprehend and which are relatively easy. A developmental perspective gives us a fuller appreciation of the difficulties adults have in understanding some of the more advanced map concepts, and what experiences promote such understanding.

3.3.2.3 From Maps to Geographic Visualizations

The traditional map is being supplemented by newer forms of geographic information displays, or geographic visualizations (MacEachren, 1995). These include various types of remote imagery, multivariate data displays, movies and animations, sound displays (sonifications), and virtual displays. In their review of psychological factors in remote sensing, for instance, Hoffman and Conway (1989) discuss the issue of the best way to utilize color in graphic displays of imagery. A good example here is the custom of using red instead of green to represent lush vegetation, a practice that violates the natural expectations of novice viewers but is probably easily understood by experienced viewers. Other research questions involving imagery include feature search, the effects of clutter, and the interpretation of scale relations. Research on the effectiveness of geographic visualizations other than remotely sensed imagery is ongoing as well. An example is Evans' work (1997) examining the effectiveness of dynamic displays of data uncertainty. Nelson and Gilmartin (1996) performed an evaluation of multivariate point symbols such as glyphs, Chernoff faces, and multivariate histograms. Monmonier (1992) has considered cognitive questions about the design of graphic scripts, which consist of dynamic sequences of maps, graphs, text, and other displays. These examples and other recent work like them only scratch the surface, however. Cognitive studies on geographic visualizations will clearly be a major focus of research for some time to come.

3.3.3 Geographic Ontologies: Entities, Features, and Concepts

Barring an extreme rejection of realism, it is safe to say that entities on the earth have an objective existence. However, identifying and labeling these entities is a construction of human mind and culture; the objective reality of earth features alone does not determine what people notice, remember, talk about, and theorize about. Both experts and lay people dissect the world into discrete entities, separating reality into classes, verbally labeling instances of these classes, and theorize about the formation and properties of these classes. The construction of *ontologies*, systems of concepts or classes of what exists in the world, is a cognitive act as well as a reflection of objective reality.

As a traditional branch of philosophy, ontology and epistemology make up metaphysics. Ontology deals with the question of the nature of that which exists; epistemology deals with the question of *how we know about* the nature of that which exists. There is recent work on geographical ontology in the traditional philosophical sense, including a nontraditional tendency to model the nature of what exists in formal or computational terms. A particularly interesting example is the attempt to model features or regions that have fuzzy or indeterminate boundaries (Burrough & Frank, 1996; Smith & Varzi, 1997).

To a cognitive scientist, however, ontology concerns the study of what exists according to the cognitive systems of intelligent beings. Thus, the cognitive approach combines traditional ontology and epistemology. There is a growing body of work on geographic ontologies in this sense. Perhaps the most straightforward is work that attempts to characterize the classes of features in the world that some community of people conceptualize as existing on the earth. If this community consists of lay people, their conceptualization of the earth and its features has been called naïve or commonsense geography (Egenhofer & Mark, 1995). An example might be the belief that the world is flat. Vosniadou and Brewer (1992) studied the development of commonsense understanding of the earth by children; Samarapungavan, Vosniadou, and Brewer (1996) extended this to the sun and moon ("common sense cosmology"). At a more human scale, Tversky and Hemenway (1983) investigated the conceptual structure of environmental scenes.

The study of geographic ontologies is also concerned with the concept-ualizations of experts or experienced geographic information scientists of various types. Hoffman and Pike (1995) claim that understanding how expert terrain analysts conceptualize topographic features will help us develop expert systems to perform automated terrain analysis. They developed the Terrain Analysis Data-base, a compendium of perceived and labeled terrain features, based on standard reference works on terrain analysis and an extensive interview with a leading aerial photo interpreter. Montello, Sullivan, and Pick (1994) analyzed the terrain features identified in environmental-scene and topographic-map recall tasks by experienced topographic map readers.

In the geographic information sciences, cognitive ontology might be quite important to GIS and remote sensing. Images are analyzed, areas of the earth's surface are grouped into regions, and discrete features are identified. Hoffman and Conway (1989) recognized that studying the way expert image interpreters identify land use categories is needed in order to more effectively automate image analysis. They discuss earlier work by Hoffman in 1984 in which think-aloud protocols of image interpreters were collected while they attempted to identify features on a radar image. Similarly, Hodgson (1998) did an experiment on the optimal window size for image classification. He provided a simple cognitive model for how humans classify land use/land cover categories (p. 798). Lloyd and his colleagues (Lloyd & Carbone, 1995; Lloyd, 1997) have investigated neural network models of categorization of geographic features, such as climate or land use categories. In the words of Hoffman and Conway: "Whenever an interpreter sits down in front of a computer graphic display or a set of satellite photos and maps, then perception, learning, and reasoning processes will all play a critical role" (p. 3).

Much of the work on the cognitive ontologies of geographic entities has been inspired by cognitive and linguistic category theory, in particular the notions of prototypes and basic-level categories (Rosch & Mervis, 1975; Peuquet, 1988). According to Usery (1993): "A geographical feature is an intellectual concept, and is established by selecting attributes and relationships relevant to a particular problem and disregarding characteristics considered to be irrelevant ... selection based on a conceptual framework of basic objects in natural categories will maximize analytical utility and data transfer in feature-based GIS" (p. 8). Mark (1993) discussed the problem of cross-linguistic translation of geographic feature names such as lake and lagoon. The task of translating feature names is difficult because the categorical structure of apparently synonymous terms from different languages are not exactly the same. Gray (1997) also discussed the application of cognitive category theory to geographic information. An interesting application of Lakoff and Johnson's *image-schemata* to the problem of wayfinding in public spaces may be found in the work of Raubal, Egenhofer, Pfoser, and Tryfona (1997).

Work that applies *fuzzy logic* (Zadeh, 1975) is an important area related to cognitive category theory. Humans commonly use fuzzy concepts in order to communicate about the world. Unlike formal languages, natural languages used in everyday speaking and writing frequently refer to *ill-defined* categories and concepts that do not have precise referents and are not delimited by sharp semantic boundaries. Furthermore, and unlike formal concepts such as those of Euclidean geometry, exemplars of fuzzy natural language concepts vary in their degree of category membership—that is, they are probabilistic rather than deterministic (Smith & Medin, 1981; Lakoff, 1987). Researchers such as Wang (1994) and Wang & Hall (1996) believe fuzzy logic will allow the formal modeling of imprecise spatial language terms such as near and large, and fuzzy regions such as downtown; this modeling is necessary to develop automated systems that will allow GIS to communicate with people in natural languages such as English.

3.3.4 Formal and Computational Modeling of Geographic Cognition

Recently, researchers from several cognitive science disciplines have concentrated on developing and evaluating formal and computational models, both deterministic and stochastic, of geographic cognition. The neural network modeling of classification and category development discussed above is an example. Two additional approaches to formal/computational modeling have been especially active: (1) qualitative reasoning about spatial and temporal relations, and (2) formal models of cognitive mapping and navigation.

3.3.4.1 Qualitative Reasoning

One of the most active approaches in AI has been the development of *qualitative* models of cognition. Qualitative models represent spatial and temporal information using nonmetric or imprecise metric geometries. Generally, they also try to incorporate simple reasoning procedures rather than complex rules. For example, Egenhofer and Al-Taha (1992) present a model of topological relations between geographic features. The inspiration for qualitative modeling is the belief that it captures human cognition more faithfully than traditional quantitative models, and thus holds a key to modeling human spatial and temporal cognition. Qualitative modelers have noted several difficulties with information processing in the real world, including perceptual imprecision, temporal and memory limitations, the availability of only approximate or incomplete knowledge, and the need for rapid decision-making (Dutta, 1988). One of the attractive properties of such approaches is that they may provide a way to incorporate both the metric skills and metric limitations of human spatial behavior without positing separate metric and topological knowledge structures.

Models based on fuzzy logic (discussed in the Ontology section) provide an example of this approach. For instance, Dutta (1988) provides a fuzzy model of spatial knowledge in which a statement about distance and direction is modeled as two fuzzy categories, each category consisting of a center value, and left and right intervals of spread. The statement "object A is about 5 miles away", for example, is modeled as having a center of 5 miles and 1 mile ranges around 5 miles. The statement essentially says that the distance is between 4 and 6 miles. The statement "object A is in a north-easterly direction" is modeled as having a center at 45° and 10° ranges around 45°. The statement essentially says that the direction is between 35° and 55°. In both cases, the correct value is modeled as having some nonzero probability of falling within the category range.

Probably most of the work on qualitative metrics has focused on knowledge of directions in the environment necessary for navigation and spatial communication. Although the details of these proposals vary, they agree in positing a model of directions which consists of a small number of coarse angular categories, commonly four 90° categories (front, back, left, right) or eight 45° categories (front, back, left, right, and the four intermediate). Frank (1991) provides good examples of such approaches. His models consist of either 4 or 8 "cones" or "half-planes" of direction. Values along the category boundaries are considered "too close to call" and result in no decision about direction. He also provides a set of operators for manipulating these values. Other writers provide similar models of directional knowledge (Freksa, 1992; Ligozat, 1993). Some models of qualitative distance exist as well (Fisher & Orf, 1991; Zimmerman, 1993) Allen and Hayes (1985) provide a very influential model of qualitative temporal reasoning.

3.3.4.2 Models of Cognitive Mapping and Navigation

Several disciplines have been involved in developing formal/computational models of cognitive mapping and navigation. Most attempts to model cognitive mapping and navigation have been carried out in the field of robotics. Some of the earliest and most influential work of this type is by Kuipers (1978, 2000). An extension and clarification of his TOUR model is described in his Spatial Semantic Hierarchy (SSH). It posits four distinct and somewhat separate representations or levels for knowledge of large-scale space; the four are simultaneously active in the cognitive map, according to Kuipers. The four are: (1) the Control level—this is grounded in sensorimotor interaction with the environment, and is best modeled in terms of partial differential equations that describe control laws specifying continuous relations between sensory inputs and motor outputs; (2) the Causal level—this is egocentric like the control level, but discrete, consisting of "views" defined by sensory experience and "actions" for moving from one view to the next. The views and actions are associated as schemas and are best modeled using 1st order logic; (3) the Topological level—this includes a representation of the external world, but only qualitatively, including places, paths, regions and their connectivity, order, containment. First order logic is appropriate here too; and (4) the Metrical level— this representation of the external world includes distance, direction, and shape to the topological level, as well as frames of reference. This is best modeled by statistical estimation theory, such as Bayesian.

Additional work in robotic modeling is found in Brooks (1991); Chown, Kaplan, and Kortenkamp (1995); Gopal, Klatzky, and Smith (1989); McDermott and Davis (1984); Yeap (1988); and Yoshino (1991). All of these models share certain concerns or ideas. First, they all posit multiple representations of space which vary in the degree to which they are dependent or independent of each other; as in Kuipers' SSH, some models suggest that different computational approaches or ontologies are most appropriate for different types of representations. All models include bottom-up processing from sensorimotor information, though the models vary in the degree to which they explicitly model perception-action processes derived from sensorimotor information rather than taking them as given. All posit the importance of landmarks that are noticed, remembered, and used to help organize spatial knowledge. In some way, all models concern themselves with the derivation of three-dimensional maps from two-dimensional views of the world. Further, they consider the derivation of allocentric (externally centered) world models from egocentric (self- or viewpoint-centered) apprehension of the space; related to this is the construction of both local and global maps of the space. The different approaches vary in the degree of metric knowledge of distances and directions they posit in addition to topological knowledge; the metric knowledge is frequently modeled as being qualitative or fuzzy. The models all recognize the problem of integrating spatial information encoded in multiple frames of reference, and they generally employ some type of hierarchical representation structure such as graph trees to encode hierarchical spatial and thematic relations in the world.

3.4 FUNDAMENTAL RESEARCH QUESTIONS

Research on the cognition of geographic information addresses a host of fundamental issues in geographic information science. How do humans learn geographic information, and how does this learning vary as a function of the medium through which it occurs (direct experience, maps, descriptions, virtual systems, *etc.*)? What are the most natural and effective ways of designing interfaces for GIS? How do people develop concepts and reason about geographical space, and how does this vary as a function of training and experience? Given the ways people understand geographic concepts, do some models for representing information in digital form support or hinder the effective use of that information? How do people use and understand language about space, and about objects and events in space? How can complex geographical information be depicted to promote comprehension and effective decision-making, whether through maps, models, graphs, or animations? What are the contents of people's beliefs and value systems about places and features in built and natural environments? How and why do individuals differ in their cognition of geographic information, perhaps because of their age, culture, sex, or specific backgrounds? Can geographic information technologies aid in the study of human cognition? How does exposure to new geographic information technologies alter human ways of perceiving and thinking about the world? Several specific research questions can be identified as being of high priority at this time:

- Are there limitations of current data models that result from their inconsistencies with human cognitive models of space, place, and environment? What benefits could be derived from reducing these inconsistencies? Are there alternative data models that would be more understandable to novices or experts? How well can people understand common GIS operations such as buffer and overlay? Research on categorization indicates that humans understand what is essentially a continuous physical world in terms of discrete objects and places. How can the nature of human categories be incorporated into GIS? How do limitations of human categorization impact our ability to reason with geographic information? Self-report inventories and memory tests will help answer these questions, including sorting and category identification tasks.

- How can vehicle navigation system interfaces for wayfinding be designed and implemented in order to improve their effectiveness and efficiency for tasks such as route choice and the production of navigation information? Examination of errors and response times during the use of alternative systems will provide information on the strengths and weaknesses of particular designs.

- How can natural language be incorporated into GIS? How should it be? Issues to investigate include the interpretation of natural language queries, automated input of natural language data, and automated output of natural language instructions. Methods from linguistic and psycholinguistic studies can be focused on issues of geographic and spatial language.

- Spatial metaphors are frequently used to express nonspatial information ("spatialization"). For example, there is much interest in representing the semantic space of documents as a place or landscape. How can such metaphors best be used to represent and manipulate information? Both the speed and correctness of interpretations of spatializations can be tested, as well as the nature of the information browsing and searches they engender.

- How can GIS be used to represent and communicate important information in novel ways? Examples include information about error and uncertainty, scale and scale changes, and temporal information and process (as in animation). Performance measures can be collected on geographic tasks that require subjects to interpret the meanings of particular depictions of error, scale relationships, or temporal change.

- What are the possible applications of desktop, augmented, and immersive virtual-environment (VE) technologies to the exploration of information with GIS? What is the relationship of a VE format to traditional cartographic representations? Understanding the impact of such new media requires both systematic comparison to existing media and strategies for understanding novel experiential situations. Again, knowledge tests can be administered after exposure to VE representations, and compared to exposure to traditional map or verbal representations.

- How can geographic information technology be used to improve education in geography, and other earth and space-related disciplines? Conversely, how does research on child and adult learning and development inform us about the nature of human cognitive models, which in turn may have implications for the design of information technologies? What are ways of educating adults and children so that they have a better understanding of geographic information concepts and better access to its technologies? A variety of education research methodologies would contribute to answering these questions.

3.5 A CASE STUDY EXAMPLE: COGNITIVE ASPECTS OF VEHICLE NAVIGATION SYSTEMS

An example of the relevance of cognitive research to geographic information science involves the design of In-Vehicle Navigation Systems (IVNS), part of the broader topic of Intelligent Transportation Systems (ITS). Recently, systems have been developed to present navigational information to automobile drivers via digital displays. As of the writing of this chapter, these systems have moved out of the "experimental" phase and may be ordered as options in some new cars. Global Positioning System (GPS) technology, inertial navigation technologies, and digital GIS (including digital cartography) are being applied to the age-old problem of finding one's way. But how should all of this information be supplied to the navigator, whether walking, driving, or piloting an airplane (Mark, Gould, & McGranaghan, 1987)? There is a real need to select information that is useful and relevant, and avoid presenting excess information that causes cognitive overload to the navigator. What is the best way to depict navigational information? All of these considerations must also take account of individual differences among navigators. Not everyone has the same abilities, preferences, or navigational styles. Cognitive research will improve our ability to properly tailor systems to individual users.

For example, Whitaker and CuQlock-Knopp (1995) examined these questions in the context of off-road navigation. They used naturalistic observation, interviews, and lab studies to attempt to identify the skills involved in off-road navigation, the features that are attended to, and the reasoning strategies used. They are attempting to apply this knowledge to the design of a useful electronic navigational aid (a prototype was called NAVAID).

Research has shown that the effectiveness of IVNS placed in automobiles depends on the modality and format in which information is depicted to the driver. Streeter, Vitello, and Wonsiewicz (1985) performed a study in which automobile drivers attempted to follow routes in an unfamiliar environment using either customized route maps, vocal directions (on a tape recorder), or both. The tape recorded verbal instructions presented about one instruction per turn, and did not include any information that was not shown on the route maps. On average, drivers using the verbal instructions drove for shorter distances, took less time, and made fewer errors than drivers receiving only route map depictions. Further research is needed to determine which types of features are most useful to be included in computer-generated verbal instructions and how these features should be described. Should the verbal instructions focus exclusively on landmarks and turn instructions? Or should information about distances be included? Is it beneficial to provide information about error correction or overshoots? Which features should be selected as landmarks (Allen, 1997)?

Providing map information to the driver in the visual modality is clearly a poor idea, if the driver attempts to read the map while steering the car. Maps are useful in certain circumstances, however, and preferred by some drivers. Research will help determine the best way to design these maps to optimize communication of geographic information to the automobile traveler. One important characteristic of in-vehicle maps is their orientation relative to the driver's direction of travel. As described above, most map users find it easiest to use maps during navigation when the map is oriented with its top being the forward direction of travel. Aretz and Wickens (1992) examined this preference, and the need to mentally rotate map displays that are not oriented in this manner. In addition to this rotation in the vertical plane, drivers mentally rotate map displays horizontally to bring them into correspondence with the forward view. These mental rotations have a cost, and produce slower and less accurate interpretations of electronic map displays. However, Aretz (1991) documents that a fixed map orientation, such as "north-up", while it requires mental rotation, better supports the development over time of a cognitive map of the surrounds. Software and hardware must be implemented to support a driver's choice of either a fixed map orientation or real-time realignment of digital maps during travel.

Aside from the questions of what information to supply to drivers, and how best to display it, there are other important questions about vehicle navigation systems that may be addressed by cognitive research in GIS. "Do we need them, in what situations do we need them, and what will be their ultimate effects on the experience of the driver?" Having navigational information available in rental cars to new visitors is likely to be of great value. Survey or observational research might find, however, that residents of a place very rarely need such a system. A driver familiar with the area may not use a vehicle navigation system enough to make such a system worth its cost. Assuming such systems become common, we might further conjecture about the effects they will have on the driver's experience and phenomenology of the world (Petchenik, 1990). Will the widespread use of such technologies impair our traditional abilities to navigate and learn space unaided by the technologies (Jackson, 1997)?

3.6 REFERENCES

Allen, G.L., 1997, From knowledge to words to wayfinding: Issues in the production and comprehension of route directions. In Hirtle, S.C. and Frank, A.U. (eds.), *Spatial information theory: A theoretical basis for GIS,* (pp. 363–372), Berlin: Springer-Verlag.

Allen, J.F. and Hayes, P., 1985, A common-sense theory of time. *Proceedings of the 9ᵗʰ International Joint Conference on Artificial Intelligence*, pp. 528–531.

Appleyard, D., 1969, Why buildings are known. *Environment and Behavior,* 1, pp. 131–156.

Aretz, A.J., 1991, The design of electronic map displays. *Human Factors*, 33, pp. 85–101.

Aretz, A.J. and Wickens, C.D., 1992, The mental rotation of map displays. *Human Performance*, 5, pp. 303–328.

Bartlett, F.C., 1932, *Remembering*, Cambridge: Cambridge University Press.

Blades, M. and Spencer, C., 1986, The implications of psychological theory and methodology for cognitive cartography. *Cartographica,* 23, pp. 1–13.

Blaut, J.M., 1991, Natural mapping. *Transactions of the Institute of British Geographers*, 16, pp. 55–74.

Blaut, J.M., 1997, Children can. *Annals of the Association of American Geographers*, 87, pp. 152–158.

Board, C., 1978, Map reading tasks appropriate in experimental studies in cartographic communication. *The Canadian Cartographer*, 15, pp. 1–12.

Brooks, R.A., 1991, Intelligence without representation. *Artificial Intelligence*, 47, pp. 139–159.

Burrough, P.A. and Frank, A.U. (eds.), 1996, *Geographic objects with indeterminate boundaries*, London: Taylor & Francis.

Castner, H.W., 1983, Research questions and cartographic design. In Taylor, D.R.F. (ed.), *Graphic communication and design in contemporary cartography,* (Vol. 2, pp. 87–113), Chichester: John Wiley & Sons.

Chown, E., Kaplan, S. and Kortenkamp, D., 1995, Prototypes, location, and associative networks (PLAN): Toward a unified theory of cognitive mapping. *Cognitive Science*, 19, pp. 1–51.

Cornell, E.H., Heth, C.D. and Rowat, W.L., 1992, Wayfinding by children and adults: Response to instructions to use look-back and retrace strategies. *Developmental Psychology*, 28, pp. 328–336.

Couclelis, H., Golledge, R.G., Gale, N. and Tobler, W., 1987, Exploring the anchor-point hypothesis of spatial cognition. *Journal of Environmental Psychology*, 7, pp. 99–122.

Cox, K.R. and Golledge, R.G., 1969, *Behavioral problems in geography: A symposium*, Evanston: Northwestern University.

Davies, C. and Medyckyj-Scott, D., 1994, GIS usability: Recommendations based on the user's view. *International Journal of Geographical Information Systems*, 8, pp. 175–189.

Davies, C. and Medyckyj-Scott, D., 1996, GIS users observed. *International Journal of Geographical Information Systems*, 10, pp. 363–384.

Downs, R.M. and Stea, D., 1973, Cognitive maps and spatial behavior: Process and products. In Downs, R.M. and Stea, D. (eds.), *Image and environment,* (pp. 8–26), Chicago: Aldine.

Dutta, S., 1988, Approximate spatial reasoning. *The First International Conference on Industrial & Engineering Applications of Artificial Intelligence & Expert Systems,* (Vol. 1, pp. 126–140), Tullahoma, TN: ACM Press.

Eastman, J.R., 1985, Cognitive models and cartographic design research. *The Cartographic Journal*, 22, pp. 95–101.

Egenhofer, M.J. and Al-Taha, K.K., 1992, Reasoning about gradual changes of topological relationships. In Frank, A.U., Campari, I. and Formentini, U. (eds.), *Theories and methods of spatio-temporal reasoning in geographic space,* (pp. 196–219), Berlin: Springer-Verlag.

Egenhofer, M.J. and Golledge, R.G., 1998, *Spatial and temporal reasoning in geographic information systems*, New York: Oxford University Press.

Egenhofer, M.J. and Mark, D.M., 1995, Naive geography. In Frank, A.U. and Kuhn, W. (eds.), *Spatial information theory: A theoretical basis for GIS,* (pp. 1–15), Berlin: Springer-Verlag.

Eley, M.G., 1987, Colour-layering and the performance of the topographic map user. *Ergonomics*, 30, pp. 655–663.

Evans, B.J., 1997, Dynamic display of spatial data reliability: Does it benefit the map user? *Computers & Geosciences*, 23(4), pp. 409–422.

Evans, G.W. and Pezdek, K., 1980, Cognitive mapping: Knowledge of real world distance and location information. *Journal of Experimental Psychology: Human Learning and Memory*, 6, pp. 13–24.

Fisher, P.F. and Orf, T.M., 1991, An investigation of the meaning of near and close on a university campus. *Computers, Environment, and Urban Systems*, 15, pp. 23–35.

Frank, A.U., 1991, Qualitative spatial reasoning about cardinal directions. In Mark, D. and White, D., (eds.), *Proceedings of Tenth International Symposium on Computer-Assisted Cartography*, Autocarto 10 (pp. 148–167), Bethesda, MD: American Congress on Surveying and Mapping, American Society for Photogrammetry and Remote Sensing.

Freksa, C., 1992, Using orientation information for qualitative spatial reasoning. In Frank, A.U., Campari, I. and Formentini, U. (eds.), *Theories and methods of spatio-temporal reasoning in geographic space,* (pp. 162–178), Berlin: Springer-Verlag.

Freundschuh, S., 1997, Research in geography, space and development: How can it inform geographic education? In Boehm, R. and J. Peterson (eds.), *The First Assessment: Research in Geographic Education*, Glbert M. Grosvenor Center for Geographic Education: San Marcos, Texas.

Friedman, A. and Brown, N.R., 2000, Reasoning about geography. *Journal of Experimental Psychology: General,* 129, pp. 1–27.

Gallistel, C.R., 1990, *The organization of learning*, Cambridge, MA: The MIT Press.

Gazzaniga, M.S. (ed.), 2000, *Cognitive neuroscience: A reader*, Boston, MA: Blackwell Publishers.

Gibson, J.J., 1950, *The perception of the visual world*, Boston: Houghton Mifflin.

Gibson, J.J., 1979, *The ecological approach to visual perception*, Boston: Houghton Mifflin.

Gilmartin, P. and Shelton, E., 1989, Choropleth maps on high resolution CRTs: The effects of number of classes and hue on communication. *Cartographica*, 26, pp. 40–52.

Golledge, R.G., 1999, *Wayfinding behavior: Cognitive mapping and other spatial processes*, Baltimore, MD: Johns Hopkins Press.

Golledge, R.G. and Stimson, R.J., 1997, *Spatial behavior: A geographic perspective*, New York: The Guilford Press.

Gopal, S., Klatzky, R. and Smith, T.R., 1989, NAVIGATOR: A psychologically based model of environmental learning through navigation. *Journal of Environmental Psychology*, 9, pp. 309–331.

Gray, M.V., 1997, Classification as an impediment to the reliable and valid use of spatial information: A disaggregate approach. In Hirtle, S.C. and Frank, A.U. (eds.), *Spatial information theory: A theoretical basis for GIS,* (pp. 137–149), Berlin: Springer-Verlag.

Gulliver, F.P., 1908, Orientation of maps. *Journal of Geography*, 7, pp. 55–58.

Hart, R.A. and Moore, G.T., 1973, The development of spatial cognition: A review. In Downs, R.M. and Stea, D. (eds.), *Image and environment,* (pp. 246-288), Chicago: Aldine.

Hebb, D.O., 1949, *Organization of behavior*, New York: John Wiley & Sons.

Hirtle, S.C. and Jonides, J., 1985, Evidence of hierarchies in cognitive maps. *Memory & Cognition*, 13, pp. 208–217.

Hodgson, M.E., 1998, What size window for image classification? A cognitive perspective. *Photogrammetric Engineering & Remote Sensing*, 64, pp. 797–807.

Hoffman, R.R. and Conway, J., 1989, Psychological factors in remote sensing: A review of some recent research. *Geocarto International*, 4, pp. 3–21.

Hoffman, R.R. and Pike, R.J., 1995, On the specification of the information available for the perception and description of the natural terrain. In Hancock, P.A., Flach, J., Caird, J.K. and Vicente, K. (eds.), *Local applications of the ecological approach to human-machine systems,* (Vol. 2, pp. 285–323), Hillsdale, NJ: Lawrence Erlbaum Associates.

Hutchins, E., 1995, *Cognition in the wild*, Cambridge, MA: The MIT Press.

Jackendoff, R. and Landau, B., 1991, Spatial language and spatial cognition. In Napoli, D.J. and Kegl, J.A. (eds.), *Bridges between psychology and linguistics: A Swarthmore Festschrift for Lila Gleitman,* (pp. 145–169), Hillsdale, NJ: Lawrence Erlbaum.

Jackson, P.G., 1997, *What Effect Will Listening to Route Guidance Information Have upon Drivers' Cognitive Maps?* Paper presented at the 93rd Annual Meeting of the Association of American Geographers in Fort Worth, Texas.

Kuipers, B., 1978, Modeling spatial knowledge. *Cognitive Science*, 2, pp. 129–153.

Kuipers, B., 1982, The "map in the head" metaphor. *Environment and Behavior*, 14, pp. 202–220.

Kuipers, B., 2000, The spatial semantic hierarchy. *Artificial Intelligence*, 119, pp. 191–233.

Lakoff, G., 1987, *Women, fire, and dangerous things*, Chicago: University of Chicago Press.

Lakoff, G. and Johnson, M., 1980, *Metaphors we live by*, Chicago: University of Chicago Press.

Landau, B., 1986, Early map use as an unlearned ability. *Cognition*, 22, pp. 201–223.

Levelt, W.J.M., 1984, Some perceptual limitations in talking about space. In van Doorn, A.J., van der Grind, W.A. and Koenderink, J.J. (eds.), *Limits in perception,* (pp. 323–358), Utrecht: VNU Science Press.

Levine, M., Marchon, I. and Hanley, G.L., 1984, The placement and misplacement of you-are-here maps. *Environment and Behavior*, 16, pp. 139–157.

Levinson, S.C., 1996, Frames of reference and Molyneux's question: Cross-linguistic evidence. In Bloom, P., Peterson, M.A., Nadel, L. and Garrett, M.F. (eds.), *Language and space,* (pp. 109–169), Cambridge, MA: The MIT Press.

Liben, L.S., 1997, Children's understandings of spatial representations of place: Mapping the methodological landscape. In Foreman, N. and Gillett, R. (eds.), *Handbook of spatial research paradigms and methodologies. Volume 1: Spatial cognition in the child and adult,* (pp. 41–83), East Sussex, U.K.: The Psychology Press (Tayor & Francis Group).

Liben, L.S. and Downs, R.M., 1989, Understanding maps as symbols: The development of map concepts in children. In Reese, H.W. (ed.), *Advances in child development and behavior,* (Vol. 22, pp. 145–201), San Diego, CA: Academic Press.

Liben, L.S. and Downs, R.M., 1993, Understanding person-space-map relations: Cartographic and developmental perspectives. *Developmental Psychology*, 29, pp. 739–752.

Liben, L.S. and Downs, R.M., 1997, Can-ism and Can'tianism: A straw child. *Annals of the Association of American Geographers*, 87, pp. 159–167.

Ligozat, G F., 1993, Qualitative triangulation for spatial reasoning. In Frank, A.U. and Campari, I. (eds.), *Spatial information theory: A theoretical basis for GIS,* (pp. 54–68), Berlin: Springer-Verlag.

Lloyd, R., 1993, Cognitive processes and cartographic maps. In Gärling, T. and Golledge, R.G. (eds.), *Behavior and environment: Psychological and geographical approaches,* (pp. 141–169), Amsterdam: North-Holland.

Lloyd, R., 1997, *Spatial cognition: Geographic environments*, (Vol. 39), Dordrecht, The Netherlands: Kluwer.

Lloyd, R. and Carbone, G., 1995, Comparing human and neural network learning of climate categories. *The Professional Geographer*, 47, pp. 237–250.

Loomis, J.M., Klatzky, R.L., Golledge, R.G. and Philbeck, J.W., 1999, Human navigation by path integration. In Golledge, R.G. (ed.), *Wayfinding behavior: Cognitive mapping and other spatial processes,* (pp. 125–151), Baltimore, MD: Johns Hopkins Press.

Lynch, K., 1960, *The image of the city*, Cambridge, MA: MIT Press.

MacEachren, A.M., 1992, Application of environmental learning theory to spatial knowledge acquisition from maps. *Annals of the Association of American Geographers*, 82, pp. 245–274.

MacEachren, A.M., 1995, *How maps work: Representation, visualization, and design*, New York: Guilford Press.

MacEachren, A.M. and Mistrick, T.A., 1992, The role of brightness differences in figure-ground: Is darker figure? *The Cartographic Journal*, 29, pp. 91–100.

Maki, R., 1981, Categorization and distance effects with spatial linear orders. *Journal of Experimental Psychology: Human Learning and Memory*, 7, pp. 15–32.

Mark, D.M., 1992, Counter-intuitive geographic "facts": Clues for spatial reasoning at geographic scales. In Frank, A.U., Campari, I. and Formentini, U. (eds.), *Theories and methods of spatio-temporal reasoning in geographic space*, (pp. 305–317), Berlin: Springer-Verlag.

Mark, D.M., 1993, Toward a theoretical framework for geographic entity types. In Frank, A.U. and Campari, I. (eds.), *Spatial information theory: A theoretical basis for GIS*, (pp. 270–283), Berlin: Springer-Verlag.

Mark, D.M., Freksa, C., Hirtle, S.C., Lloyd, R. and Tversky, B., 1999, Cognitive models of geographical space. *International Journal of Geographical Information Science*, 13, pp. 747–774.

Mark, D.M., Gould, M.D. and McGranaghan, M., 1987, Computerized navigation assistance for drivers. *The Professional Geographer*, 39, pp. 215–220.

McDermott, D. and Davis, E., 1984, Planning routes through uncertain territory. *Artificial Intelligence*, 22, pp. 107–156.

McNamara, T.P., 1992, Spatial representation. *Geoforum*, 23, pp. 139–150.

McNamara, T.P., Hardy, J.K. and Hirtle, S.C., 1989, Subjective hierarchies in spatial memory. *Journal of Experimental Psychology: Learning, Memory, and Cognition*, 15, pp. 211–227.

Medyckyj-Scott, D. and Hearnshaw, H., 1993, *Human factors in GIS*. London: Belhaven Press.

Monmonier, M., 1992, Authoring graphic scripts: Experiences and principles. *Cartography and Geographic Information Systems*, 19, pp. 247–260.

Montello, D.R., 1998, A new framework for understanding the acquisition of spatial knowledge in large-scale environments. In Egenhofer, M.J. and Golledge, R.G. (eds.), *Spatial and temporal reasoning in geographic information systems*, (pp. 143–154), New York: Oxford University Press.

Montello, D.R. and Freundschuh, S.M., 1995, Sources of spatial knowledge and their implications for GIS: An introduction. *Geographical Systems*, 2, pp. 169–176.

Montello, D.R., Sullivan, C.N. and Pick, H.L., 1994, Recall memory for topographic maps and natural terrain: Effects of experience and task performance. *Cartographica*, 31, pp. 18–36.

Nelson, E.S. and Gilmartin, P., 1996, An evaluation of multivariate, quantitative point symbols for maps. In Wood, C. and Keller, P. (eds.), *Cartographic design: Theoretical and practical perspectives,* (pp. 191–210), London: John Wiley & Sons.

Newell, A. and Simon, H.A., 1976, Computer science as empirical inquiry: Symbols and search. *Communications of the ACM*, 19, pp. 113–126.

Norman, D.A., 1990, *The design of everyday things*, New York: Doubleday.

Nyerges, T.L., Mark, D.M., Laurini, R. and Egenhofer, M.J., 1995, *Cognitive aspects of human-computer interaction for Geographic Information Systems*, Dordrecht, The Netherlands: Kluwer Academic.

Olson, J.M., 1979, Cognitive cartographic experimentation. *The Canadian Cartographer*, 16, pp. 34–44.

Petchenik, B.B., 1983, A map maker's perspective on map design research 1950-1980. In Taylor, D.R.F. (ed.), *Graphic communication and design in contemporary cartography,* (Vol. 2, pp. 37–68), Chichester: John Wiley & Sons.

Petchenik, B.B., 1990, The nature of navigation: An epistemological perspective on recent developments in vehicle navigation assistance technology. *International Yearbook of Cartography*, 30, pp. 13–24.

Peuquet, D.J., 1988, Representations of geographic space: Toward a conceptual synthesis. *Annals of the Association of American Geographers*, 78, pp. 375–394.

Piaget, J., 1926/1930, *The child's conception of the world*, New York: Harcourt, Brace & World.

Piaget, J. and Inhelder, B., 1948/1967, *The child's conception of space*, New York: Norton.

Pick, H.L., Heinrichs, M.R., Montello, D.R., Smith, K., Sullivan, C.N. and Thompson, W.B., 1995, Topographic map reading. In Hancock, P.A., Flach, J., Caird, J.K. and Vicente, K. (eds.), *Local applications of the ecological approach to human-machine systems,* (Vol. 2, pp. 255–284), Hillsdale, NJ: Lawrence Erlbaum Associates.

Potash, L.M., 1977, Design of maps and map related research. *Human Factors*, 19, pp. 139–150.

Presson, C.C., 1982, The development of map reading skills. *Child Development*, 53, pp. 196–199.

Raubal, M., Egenhofer, M.J., Pfoser, D. and Tryfona, N., 1997, Structuring space with image schemata: Wayfinding in airports as a case study. In Hirtle, S.C. and Frank, A.U. (eds.), *Spatial information theory: A theoretical basis for GIS*, (pp. 85–102), Berlin: Springer-Verlag.

Ridgley, D.C., 1922, The teaching of directions in space and on maps. *Journal of Geography*, 21, pp. 66–72.

Rieser, J.J., Pick, H.L., Ashmead, D.H. and Garing, A.E., 1995, Calibration of human locomotion and models of perceptual-motor organization. *Journal of Experimental Psychology: Human Perception and Performance*, 21, pp. 480–497.

Robinson, A.H., 1952, *The look of maps*, Madison, WI: University of Wisconsin Press.

Rock, I., 1974, *Orientation and form*, New York: Academic Press.

Rosch, E.H. and Mervis, C.B., 1975, Family resemblances studies in the internal structure of categories. *Cognitive Psychology*, 7, pp. 573–605.

Rumelhart, D.E. and McClelland, J.L. (eds.), 1986, *Parallel distributed processing: Explorations in the microstructure of cognition, Vol. 1: Foundations*, Cambridge, MA: The MIT Press.

Saarinen, T.F., 1966, *Perception of drought hazard on the Great Plains*, (Department of Geography Research Paper No. 106). Chicago: University of Chicago.

Sadalla, E.K., Burroughs, W.J. and Staplin, L.J., 1980, Reference points in spatial cognition. *Journal of Experimental Psychology: Human Learning and Memory*, 6, pp. 516–528.

Samarapungavan, A., Vosniadou, S. and Brewer, W.F., 1996, Mental models of the earth, sun, and moon: Indian children's cosmologies. *Cognitive Development*, 11, pp. 491–521.

Shepard, R.N. and Hurwitz, S., 1984, Upward direction, mental rotation, and discrimination of left and right turns in maps. *Cognition*, 18, pp. 161–193.

Siegel, A.W. and White, S.H., 1975, The development of spatial representations of large-scale environments. In Reese, H.W. (ed.), *Advances in child development and behavior*, (Vol. 10, pp. 9–55), New York: Academic.

Slocum, T.A. and Egbert, S.L., 1993, Knowledge acquisition from choropleth maps. *Cartography and Geographic Information Systems*, 20, pp. 83–95.

Smith, B. and Varzi, A.C., 1997, Fiat and bona fide boundaries: Towards an ontology of spatially extended objects. In Hirtle, S.C. and Frank, A.U. (eds.), *Spatial information theory: A theoretical basis for GIS*, (pp. 103–119), Berlin: Springer-Verlag.

Smith, E.E. and Medin, D.L., 1981, *Categories and concepts*, Cambridge, MA: Harvard University Press.

Steinke, T.R., 1987, Eye movement studies in cartography and related fields. *Cartographica*, 24, pp. 40–73.

Stevens, A. and Coupe, P., 1978, Distortions in judged spatial relations. *Cognitive Psychology*, 10, pp. 422–437.

Streeter, L.A., Vitello, D. and Wonsiewicz, S.A., 1985, How to tell people where to go: Comparing navigational aids. *International Journal of Man/Machine Studies*, 22, pp. 549–562.

Talmy, L., 1983, How language structures space. In Pick, H.L and Acredolo, L.P. (eds.), *Spatial orientation: Theory, research, and application,* (pp. 225–282), New York: Plenum Press.

Taylor, H.A. and Tversky, B., 1992, Descriptions and depictions of environments. *Memory & Cognition*, 20, pp. 483–496.

Thelen, E. and Smith, L.B., 1994, *A dynamic systems approach to the development of cognition and action*, Cambridge, MA: The MIT Press.

Thorndyke, P.W. and Hayes-Roth, B., 1982, Differences in spatial knowledge acquired from maps and navigation. *Cognitive Psychology*, 14, pp. 560–589.

Tolman, E.C., 1948, Cognitive maps in rats and men. *Psychological Review*, 55, pp. 189–208.

Tooby, J. and Cosmides, L., 1992, The psychological foundations of culture. In Barkow, J.H., Cosmides, L. and Tooby, J. (eds.), *The adapted mind: Evolutionary psychology and the generation of culture,* (pp. 19–136), New York: Oxford University Press.

Tversky, B., 1981, Distortions in memory for maps. *Cognitive Psychology*, 13, pp. 407–433.

Tversky, B., 1992, Distortions in cognitive maps. *Geoforum*, 23, pp. 131–138.

Tversky, B. and Hemenway, K., 1983, Categories of environmental scenes. *Cognitive Psychology*, 15, pp. 121–149.

Usery, E.L., 1993, Category theory and the structure of features in geographic information systems. *Cartography and Geographic Information Systems*, 20, pp. 5–12.

Uttal, D.H., 2000, Seeing the big picture: Map use and the development of spatial cognition. *Developmental Science*, 3, pp. 247–264.

Vosniadou, S. and Brewer, W.F., 1992, Mental models of the earth: A study of conceptual change in childhood. *Cognitive Psychology*, 24, pp. 535–585.

Vygotsky, L.S.,1934/1962, *Thought and language*, Cambridge, MA: MIT Press.

Wang, F., 1994, Towards a natural language user interface. An approach of fuzzy query. *International Journal of Geographical Information Systems*, 8, pp. 143–162.

Wang, F. and Hall, G.B., 1996, Fuzzy representation of geographical boundaries in GIS. *International Journal of Geographical Information Systems*, 10, pp. 573–590.

Whitaker, L.A. and CuQlock-Knopp, V.G., 1995, Human exploration and perception in off-road navigation. In Hancock, P.A., Flach, J., Caird, J.K. and Vicente, K. (eds.), *Local applications of the ecological approach to human-machine systems,* (Vol. 2, pp. 234–254), Hillsdale, NJ: Lawrence Erlbaum Associates.

White, G.F, 1945, *Human adjustment to floods,* (Department of Geography Research Paper No. 93), University of Chicago.

Yeap, W.K., 1988, Towards a computational theory of cognitive maps. *Artificial Intelligence*, 34, pp. 297–360.

Yoshino, R., 1991, A note on cognitive maps: An optimal spatial knowledge representation. *Journal of Mathematical Psychology*, 35, pp. 371–393.

Zadeh, L.A., 1975, Fuzzy logic and approximate reasoning. *Synthese*, 30, pp. 407–428.

Zimmermann, K., 1993, Enhancing qualitative spatial reasoning—combining orientation and distance. In Frank, A.U. and Campari, I. (eds.), *Spatial information theory: A theoretical basis for GIS,* (pp. 69–76), Berlin: Springer.

CHAPTER FOUR

Scale

Nina Lam, Louisiana State University
David Catts, U.S. Geological Survey
Dale Quattrochi, NASA–Global Hydrology and Climate Center
Daniel Brown, University of Michigan
Robert McMaster, University of Minnesota

4.1 INTRODUCTION

This research priority calls attention to the multidisciplinary issues related to scale in the spatial as well as spatiotemporal domains. The objective of this chapter is to clarify and assess what scale and scale effects are, and through recognizing the fundamental existence of scale effects, determine how one can realistically approach and mitigate them. Primary issues therefore center on gaining a better understanding of how to effectively and efficiently measure and characterize scale; how to use scale information in judging the fitness of data for a particular use; how to automate scale change and simultaneously represent data at multiple scales; how scale and change in scale affect information content, analysis, and conclusions about patterns and processes.

The issue of scale is not new, nor is it a concern restricted to geographic information scientists. Scale variations have long been known to constrain the detail with which information can be observed, represented, analyzed, and communicated. Changing the scale of data without first understanding the effects of such action can result in the representation of processes or patterns that are different from those intended. For example, research has shown that reducing the resolution of a raster land cover map (going to larger cells) can increase the dominance of the contiguous classes, but decrease the amount of small and scattered classes (like wetlands in some locations) in the representation (Turner *et al.*, 1989). The spatial scaling problem presents one of the major impediments, both conceptually and methodologically, to advancing all sciences that use geographic information. Likewise, temporal scaling, a separate but related issue, is not well understood and thus difficult to formalize. In an information era, massive amount of geographic data are collected from various sources, often at different scales. Before these data can be integrated for problem solving, fundamental issues must be addressed.

Recent work on the scaling behavior of various phenomena and processes (*e.g.*, research in global change, ecological modeling, and environmental health) has shown that many processes do not scale linearly. The implication is that in order to characterize a pattern or process at a scale other than the scale of observation, some knowledge of how that pattern or process changes with scale is needed so that the scaling process can be adjusted accordingly. Attempts to describe scaling behavior by fractals or self-affine models, which mathematically relate complexity and scale, have been made. However, because the fractal model and its realizations have been interpreted and used differently by different researchers (*e.g.*, strict versus statistical self-similarity), the results from applying the fractal model are inconsistent and the approach receives mixed review (Lam and De Cola, 1993). Multifractals have shown some promise for characterizing the scaling behavior of some phenomena, but it is more likely that fractals will offer only a partial model (Pecknold, *et al.*, 1997). Alternative models such as the multi-resolution based wavelet approach are needed to understand the impacts that changes in scale have on the information content of databases (*e.g.*, Mallat, 1989). Examining the sensitivity of processes and analytical models to scale will help scientists validate hypotheses and reduce model uncertainties, which in turn will improve geographic theory-building.

Despite a longstanding recognition of the implications of scale on geographic inference and decision making, many questions remain unanswered. The transition from analog (*e.g.*, maps) to digital representations of geographic information forces users of those data to formally deal with these conceptual, technical, and analytical questions in new ways. It is easy to demonstrate by isolated examples that scale poses constraints and limitations on geographic information, spatial analysis, and models of the real world. The challenge is to articulate the conditions under which scale-imposed constraints are systematic and to develop geographic models that compensate or standardize scale-based variation. Mishandling or misunderstanding scale can bias inference and reasoning and ultimately affect decision-making processes. New types of analyses, for example the Geographical Analysis Machine (GAM) proposed by Openshaw *et al.* (1987) and the spatial Scan statistic by Kulldorff (1997), offer methods that explore scale effects more easily and may be less sensitive to scale than traditional quantitative techniques.

The widespread adoption of geographic information science has made the scale problem more acute and the need to develop solutions to cope with the scale effects more pressing. Geographic information systems (GISs) facilitate data integration regardless of scale differences. Common problems are when we try to use coarse aggregate data (*e.g.*, statewide or countywide data) and to compare those data with less coarse or disaggregate data (*e.g.*, data defined by census tract or by individual survey). The capability to process and present geographic information "up" and "down" from local, regional, to global scales has been advocated as a solution to understanding the global systems of both natural (*e.g.*, global climate change) and societal (*e.g.*, global economy) processes and the relationships between the two. Fundamental scale questions will benefit from coordinated research efforts among geographic information scientists with various interests and domain experts. Information systems of the future can sensitize users to the implications of scale dependence and provide scale management tools once we develop useful models of scale behavior, an improved understanding of the effects of scale, novel methods for describing the scale of data, and intelligent automation methods for changing scale.

In the following, we first clarify the scope of scale by describing its various definitions and meanings used in related fields. A widespread adoption of standardized terminology is necessary for better communications between researchers within and among various fields and disciplines. We will then illustrate some concrete examples of how scale might affect our understanding of geographical phenomena and hence our decision making in a number of application areas, including cartography, environmental policy, global land-cover change, and environmental health. Emphasis is placed on what researchers/decision makers have done to cope with the scale problem. This is followed by a brief synopsis of what is left to be done and the research challenge lies ahead.

4.2 MEANING AND SCOPE OF THE SCALE ISSUE

The term "scale" has been used in different manners under different contexts, making the comparison and communication among researchers and research results across subfields and disciplines more difficult. Typically, when an issue of scale is involved in a scientific paper or study, an explanation of the term is often provided at the beginning of the paper to avoid confusion in subsequent reading. While it may be counterproductive to uncover exhaustively the various meanings of scale, it is important to clarify what we mean and what the scope is when we are referring to the issue of scale in geographic information science.

We follow, with a minor revision, the definitions of scale outlined in Lam and Quattrochi (1992) and Cao and Lam (1997). For the sake of simplicity, our discussion and examples below focus on the spatial domain. However, the same meanings and scope will apply to the temporal as well as spatial-temporal domain. Four meanings of scale that are commonly used in geographic information science can be identified (Figure 4.1).

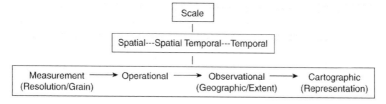

Figure 4.1 The meanings of scale (modified from Cao and Lam, 1997)

(1) The *observational* or geographic scale refers to the size or spatial extent of the study. Under this usage, a large-scale study covers a larger study area, as opposed to a small-scale study that encompasses a smaller study area. For example, a study of a disease pattern by county for the entire United States is a large-scale study, compared with a study of local health effects of a hazardous waste site in a neighborhood, which focuses only on a specific urban area and is a small-scale study.

(2) The *operational* scale refers to the spatial (or temporal) extent at which certain processes operate in the environment, and it may take from just several pixels to hundreds of pixels (*e.g.*, from meters to hundreds of meters). Some researchers refer operational scale as the "scale of action" (scale at which the pattern manifests the maximum variability), and methods have been suggested to find the "scale of action". This concept can be easily extended to both natural and social-political processes. Determining the operational scale of a phenomenon is an important step, because it can help suggest both the spatial extent and the resolution (defined below) needed to observe the patterns resulting from the process.

(3) The *measurement* scale, or commonly called *resolution*, refers to the smallest distinguishable parts of an object (Tobler, 1988), such as pixels in a remote sensing imagery or sampling intervals in an ecological study. In the ecological literature, resolution is often referred as "grain". In the past, because of limitation of data storage capacity, studies of large spatial extent were often associated with coarse resolution, and fine resolution was characteristic of small-scale studies. In this information age, however, when data storage has become a lesser problem, fine-resolution data in large-scale studies are increasingly common.

(4) The *cartographic* or map scale refers to the ratio between the measurements on a map and the actual measurements on the ground. A large-scale map covers a smaller area and the map generally has more detailed information. On the other hand, a small-scale map covers a larger area and the map often has less detailed information about the area. Unlike the last three definitions of scale, which refer to data characteristics, cartographic scale refers to data representation. Therefore, it is worth mentioning that once spatial data are encoded digitally, their resolutions are fixed, and that zooming in or out during display can only increase or decrease its cartographic scale but not its measurement scale (*i.e.*, resolution).

These four meanings of scale are closely related. Ordered from small to large, measurement scale is the smallest, followed by operational scale and observational scale, respectively. For example, a measurement scale of 10 meters results in a pixel resolution of 10 meters, but it often takes a number of pixels (operational scale) to recognize a feature (water body, stadium, residential area). It will take even more number of pixels (observational scale/extent) to understand the spatial pattern of and the process leading to the feature. The cartographic scale is then a representation of the data and results that most people will ultimately rely on for deriving interpretation, conclusion, and policy decision. As shown in the next section, decisions on cartographic scale and generalization are very much based on the data characteristics themselves, such as spatial and thematic resolution.

In defining the meanings and scope of scale, it is important to note not only that consistent terminology must be developed, but also researchers must adopt them widely. The "science of scales" will not progress until some fundamental building blocks, in this case lexical meanings of scale, have been laid out and agreed upon.

4.3 SCALE IN SELECTED APPLICATION AREAS

4.3.1 Cartography and Cartographic Generalization

There are five scale-related ingredients that affect the decisions made in cartographic design and presentation: cartographic scale, measurement scale (resolution), data model (representation), phenomena, and temporal scale. Each of these elements has unique characteristics and graphic resolutions, and all of these elements interact to challenge the graphic abstraction in cartography.

4.3.1.1 Cartographic Scale

Map scale is often neglected in many day-to-day maps. We encounter maps each day and unfortunately many of these do not have any reference of map scale. In some cases, the scale is purposefully distorted to promote a particular point of view (Monmonier, 1991), and the reader more often presumes the consistency of scale. The effective use of perspective views in cartography is an example that complicates the conventional design notion of adhering to uniform map scales.

The overall purpose of selecting a map scale is to provide map integrity, either by specifying the "exact proportions" for measurement or by maintaining relative geospatial relationships of geophysical features. Other cartographic elements of integrity are the use of an appropriate projection and datum, and providing reference information and some notations about the date and source of the reference materials used in the map design. By providing a "coverage diagram" or "source statement", the map designer can exactly specify the source material used in various areas of the map compilation.

The determining factors in selecting an appropriate map scale are geographical extent and final map size (Robinson *et al.*, 1995). The output media available may also predetermine the scale of a map. By selecting one of the various digital output options available to cartography, we are no longer restricted by the artistic possibilities of the paper and pen but by the size and resolution of the digital media. Paper printers vary in resolution generally from 72 dots-per-inch (DPI) to 1,200 DPI. Film recorders approach 1,200–2,400 DPI in resolution. The effective design of Internet maps is a more difficult challenge as screen resolutions and color qualities vary. In each design case, the resolutions and colors of the targeted media place technological restrictions on the size, colors, and overall quality of the map. These graphic restrictions place limits on what scale and content can be presented effectively.

There are various examples that illustrate how paper size or ease of measurement conventions predetermines the map scale. For example, the USDA Forest Service uses a scale of 1:126,720 for general reference maps of National Forests, and this proportion results in one inch of map distance equaling exactly 2 miles. The U.S. Geological Survey (USGS) produces over 25,000 1:24,000-scale maps of the United States, and all but a selected number are composed and revised at this standard series scale of 1 inch equaling 2,000 feet. In each case, the conventions of using a standardized map scale alter the symbol density and our perceptions of a feature's importance (Figure 4.2).

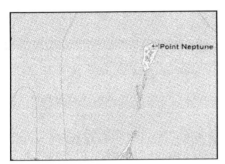 

Figure 4.2 Standardized scale series complicate the design for sparse and congested symbolization. Shown here are comparative graphic areas of feature content for a fixed scale map series (USGS 1:24,000 Freemason Island, LA (left) and Central Park, NY)

4.3.1.2 Measurement Scale (Resolution)

It is generally accepted that data compilation should occur at a finer resolution than the final map. The detail of compiled features is then selected, prioritized, and generalized as the map is compiled and assembled. Feature selection and content generalization are integrated into the cartographic process, and the term "cartographic license" is used to denote the composition and balancing of various cartographic components such as content, color, and symbols of the map.

The integration of geographic information systems throughout our society has provided the means for all of us to be portrayers of geospatial data. This ability to select and integrate data of disparate resolution and source expands the meaning of "cartographic license" beyond one of strictly map composition. The data sources used in the design may vary in map scale, source, content density, and resolution. It is the map design process that strives to balance the variety of these data characteristics in a meaningful and truthful way. GIS and the associated technologies allow all of us to "slam" data sets together to produce a result that may appear truthful or promotes integrity. The technological abilities of GIS underscore the warning of mismatched scales, as data of varying scales can be merged, warped, and conflated to present an intended message.

The determining factor in cartographic design is, "what is the purpose of the map?" From a content perspective, unimproved dirt roads may seem unimportant for a regional transportation map, but these access routes are extremely important at any scale when the map purpose is to show the potential impact of these access routes on undeveloped or roadless areas. In a hydrologic example, removing the lowest Strahler stream order class might reduce the graphic density of the drainage pattern. In a realistic situation, the subsurface geology of the area might produce perennial streams on one side of a ridge that flow year round, and these isolated streams become an important water source to be represented throughout a wider range of map scale. The geophysical relationships between data themes and within the entire context of the map's purpose need to be established for better map design.

4.3.1.3 Scale of the Data Model (Representation)

The integrity of the data resolution, and its scaleable characteristics, can be redefined by the data model used in storing, processing, and representing features. For example, the exact ground coordinate values can be truncated further by GIS systems that use disparate data precision to store the data internally; therefore the GIS data model and not the data itself further redefine the resolution of the data.

Changing scales of the data representation requires cartographic generalization. The challenge of the map design is to balance the map content to what is the threshold of sufficient information. If there is not enough space for the most important information, we find ways to redesign and repackage the map symbology, or we change the size of the design parameters. In representing distinct occurrences of phenomena across various scales, we use "representative patterns" of symbols to define the collective occurrence.

In the digital overposting method of map symbols or text, we take the "first come, first served" approach. A bitmap or index of the graphic area is active in the background, and as features are placed, the index recounts the placement of previous symbols. When a conflict occurs, the new symbol is replaced in various alternative positions, or it is eliminated (Figure 4.3). As conflicts in spatial relationships force a resolution in feature symbology, the relationships between symbol placement must be exaggerated. The rules governing symbol and text offsetting are based on a placement priority and placement rules (Figure 4.4).

In order to add intelligence to these decisions, we can integrate some rules of hierarchy into the "bitmap" approach. A rule base of feature importance is generated as the basis of this intelligent selection. To do this effectively, we must understand the physical nature and extent of the phenomena and we must understand

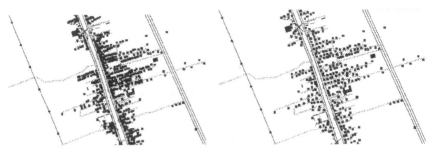

Figure 4.3 The culture pattern of housing is represented by using an overposting symbol priority

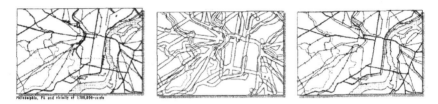

Figure 4.4 Line symbols are offset from a hierarchical schema. In this case, roads are offset from railroads, then streams are offset from all other features to resolve symbol conflicts

the interrelationships and variability of the phenomena. By encoding these attributes into the determining processes, we can more intelligently place, and replace, map text and symbols in a more effective way.

In all cases of map design, we must understand the scale thresholds of the various data themes and features that we use. In each situation, the geometric properties of individual features can be simplified and stored at various scale thresholds (Douglas & Peucker, 1973). Hierarchical classification of feature elements can control feature representations at various scales (Cromley, 1991). Approaches using convex hulls can address feature simplification and topological changes in addition to topological conflicts of the results (Muller, 1990). These computer-based solutions are generally reliable and consistent in their results. In selected examples, relationships between features and dimensional thresholds are addressed as part of the simplification process (Nickerson, 1988; Catts, 1990) (Figures 4.5 and 4.6). Dimensional thresholds are best resolved prior to, rather than after, the feature generalization process. In Figure 4.6, polygons whose areas are beneath an area threshold determined by the map scale are opened or compressed based on linear attributes (Catts, 1990). Overall, these evaluation processes identify specific characteristics of feature representations, but do not compensate for variety and interrelationships for determining dimensional thresholds of features that govern our asystematic and artistic decisions in map composition.

Figure 4.5 The dimensional characteristics and transitions of area hydrography
are resolved by using a minimum threshold (from Nickerson, 1988)

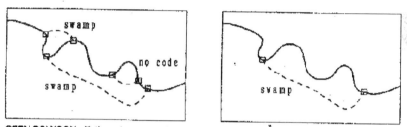

OPEN POLYGON - If the polygon contains an arc with apparent limit or closure line attribute, the specific arc and the label point are deleted. This opens the polygon up to an adjacent area.

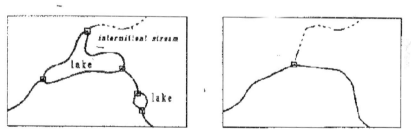

COMPRESS POLYGON - If the previous situations do not apply, the polygon is resolved by compressing the elements at the label point position.

Figure 4.6 Small area hydrographic features are eliminated, and the linear network is resolved
to maintain topological feature relationships before feature simplification (from Catts, 1990)

In order to effectively portray map content across scales, the two schools of thought are that map features can be represented as distinct and static representations of scale-specific content (example: scale series of 1:24,000; 1:100,000, 1:250,000), or that spatial features can be intelligently analyzed to generate the selection interactively in the generalization process. In selected examples, these approaches are integrated into a unique process of scaleable representations and functions within the data structure itself (Buttenfield, 1984), such as the Geographic Data Files (GDF) developed in Europe (Figure 4.7).

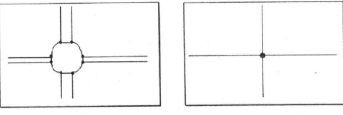

Level 1 Representation of a roundabout ------------> Level 2 Representation

Figure 4.7 Geographic data files contain multiple representations
of features for use at various scale thresholds (ERTICO, 1997)

As part of Project 615, the Forest Service is developing feature generalization techniques that allow for multiscale representations of its geographic data (Rodriguez, 1998). This activity addresses a comprehensive feature and symbol generalization for both the Forest Service 1:24,000 and 1:126,720 scaled series from a single geographic database. The complexity of addressing both feature simplification and symbol resolution will provide an integrated solution and knowledge base for developing multiscale representations from a single data source.

4.3.1.4 Scale and Dimensional Thresholds of Phenomena

What the feature scaling process presents to cartographic design are dimensional changes of the symbolized features. A simple example is that a cased-road might become a single line as the scale decreases, but clusters of polygons may interact in various ways. A collection of swampy areas may be distinctly mapped at a large scale, and as scale decreases these individual polygon areas might combine to become a "swampy area" in a larger sense (Figure 4.8). Eventually, these features may shift to another dimensional representation or disappear from the map.

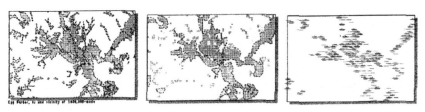

Figure 4.8 As the scale decreases, the symbolization of swampy areas shift in dimension
from unique areas, to selectively aggregate areas, to point symbols in a pattern

The correlation of geophysical phenomena and the techniques used for symbolizing features and phenomena across scales can be combined to provide intelligence for selectively altering or reducing feature content as the scale changes. Techniques for determining the dimensional shifts of geographic phenomena can be founded on data characteristics such as resolution and fractal dimension of the mapped phenomena (De Cola and Buttenfield, 1994).

The GIS technologies are in need of data and metadata interpreters. In general, image data mapped at too large a scale appears pixelated, whereas mapped at too small of a scale obscures hidden detail. Vector data does not have these inherent "visual clue" thresholds, so we must provide accompanying metadata documentation containing appropriate information on scale. As we get better about maintaining "data about data", we should more accurately use this information to optimize the use of data within appropriate scales and transitional thresholds. For example, what is the meaning of the term "Abscissa of Resolution," which is an item included in the Federal Geographic Data Committee metadata standard? Is this a reflection of the data resolution or the scale of data representation? How does this parameter relate to other spatial systems, such as AutoCAD or Arc/Info? How do we translate the various parameters and tolerances used in processing spatial information? How should the singular "Abscissa of Resolution" be integrated into a GIS software system to set values for parameters and tolerances, such as those used by Arc/Info to merge and reduce features (*e.g.,* WEED, GRAIN, SNAP, FUZZY, *etc.)*? As data representations become more sophisticated and the mapping solutions become more integrated, information about the individual geographic phenomena and its digital portrayal become more important to the decision processes. Metadata, or the provided information about data characteristics, is an integrated need in the cartographic process.

4.3.1.5 Temporal Scale

The use of GIS technologies has provided the means to integrate a variety of data sources regardless of resolution, accuracy, and temporal context. Tools used to integrate and map geographic data have allowed us to integrate various disparate data and provide what appears to be a meaningful and truthful representation (Figure 4.9). Temporal notations in the margin of this printed map explain to the reader that features may not agree in relative position. In the absence of such explicit notations, readers are left to interpret implicit clues of the temporal scale of the map features (MacEachren, 1995).

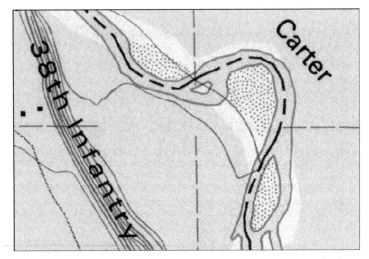

Figure 4.9 A disparity in temporal characteristics of related data may lead
to graphic conflict. Shown here is a map published in 1994. In this example,
hydrography was updated using 1990 photography and printed with
topography dated 1947 (Nisqually, WA)

4.3.1.6 Research Priorities

The relationships between real phenomena and their graphic representation generate
complex interrelationships between elements of map scale, data and model characteristics, and geophysical phenomena across time. The GIScience technologies have
provided us with the means to integrate disparate data resolutions and data types,
but without a good understanding of the unique characteristics and interrelationships
that may exist, we are faced with sets of incongruous data that could jeopardize the
traditional cartographic license of balancing technology and integrity.

In order to improve the efficiency and effectiveness of map communication
in this information age, spatial sciences must concentrate on developing research
priorities that address the overall model of integrating data in consideration of
natural and physical interrelationships. These sciences must integrate the effective technologies of data visualization into data analysis, and develop tools to aid
in development of effective computer graphics and maps. We should develop guidelines for presenting data effectively and truthfully, by selecting appropriate data
and referencing and presenting sufficient data source information. We must continue to promote the development of metadata; but we have to progress to using
metadata to determine the scaleable factors in map design and data presentations.
As we become more diligent and efficient about collecting data about the data,
we need to integrate the interpretation of natural phenomena and data characteristics into the map design and decision process. As the presentation of geographic

data in a virtual setting become more sophisticated, we need to develop visual clues to the varieties of scale and temporal characteristics of the data. We should develop knowledge of the dimensional thresholds of natural phenomena and the graphic abstractions with which they are represented. Multiple representations of features across scale and symbology thresholds should be promoted. Overall, we must promote the integrity of geographic data visualization in the variety of new cartographic technologies and products that are available to society.

4.3.2 Environmental Policy and Decision Making

The Earth's surface environmental properties are characterized in a manner that, ultimately, gives the appearance of spatial pattern. Because of the structure of these environmental properties, however, the scale (geographical and operational) at which these properties can be identified, observed, or measured is limited by the resolution of the measurement technology. One primary aim in geographical analysis is to identify and describe the pattern or scale of variation in spatial phenomena at the level of resolution of interest. This level of resolution can vary immensely depending upon what the objectives are for observation or measurement, as well as what the overall characteristics are of the spatial object or objects under study (*i.e.*, the operational scale). For example, in analyzing an environmental property such as soils, the data resolution (*i.e.*, measurement scale) can range from a soil core sample, a soil series, a field plot, an agricultural field, a county, state or other municipal boundary, including a country, or even the entire globe. At any of these levels, there is likely to be one or more spatial resolutions at which most of the variation occurs (Oliver, 2001). To detect this spatial variability, the level of observation must relate to the spatial (or even temporal) scale at which most of this variation becomes evident (*i.e.*, operational scale). It is important to note as a caveat, however, that what is observed as pattern or spatially correlated variation at one resolution can appear as "noise" or uncorrelated variation at another (Oliver, 2001).

As most environmental properties vary continuously over the Earth's surface, the areas involved in their studies or analyses are large. Information about such properties is either directly derived, or extrapolated from, smaller areas or sampling locations that are separated by much larger areas. Consequently, specific or detailed information can be sparsely distributed, where there may be relatively large intervening spaces about which nothing is known. This is problematic to users of environmental information because they usually want to know what the properties are like everywhere; *i.e.*, from a generalized perspective. Thus, there is a need to be able to predict either "real" or "inferred" values of environmental properties at unsampled places, but this is difficult in most cases because of the complex nature of spatial variation (Oliver, 2001).

Despite the seemingly inherent complexity of variation in environmental or landscape properties, such as soil moisture distribution or urban sprawl, one

theoretical tenet of geographical analysis is that the values of a spatial phenomenon that are in close proximity to one another are usually more similar than
those that are further apart; *i.e.*, the Tobler's Law of geography (Tobler, 1970).
This relationship between distance (or proximity) and the characteristics of
spatial phenomena provides the association that is required to discern the pattern,
arrangement, or orientation of these phenomena. Such information can be used to
guide sampling intensity or intervals, and to provide clues about the factors and
causal processes of spatial variability. The patterns observed through such spatial
analysis or interpretation may also form the basis for suggesting the need for different
forms of management of the environment (*e.g.*, process-response attributes,
cause-effect interactions, land-cover/land-use impacts). Thus, the nature of the
environment poses questions such as:

- At what spatial scale should the properties be studied or investigated?
- What kind of sampling scheme for observation should be used?
- How many samples should be taken?
- What sampling interval should be used?
- How should values be predicted or inferred at intervening places?

As a result of the wide range of spatial scales (extent and resolution) over
which environmental properties can vary, there is an obvious need to consider
what the level of scale of interest is at the start of an investigation. Moreover, the
detection of spatial variation in environmental phenomena requires adequate sampling that is representative, as well as the identification of an appropriate method
of estimation or prediction through spatial analysis methodology given the objectives of the investigation.

Consideration of scale (geographic extent and operational scale), spatial variability, and resolution can be a confounding and even overbearing problem, however,
when attempting to collect, manipulate, analyze, or interpret data for environmental
decision-making. This has a number of causes, but perhaps two of the most important issues are a misunderstanding of the meaning of scale and the explosion of data
that are available at different spatial scales (both resolution and extent). As discussed
in the beginning of this chapter, there has been much recent discussion of what the
definitions of scale and scale concepts are from a mostly quasi-theoretical or academic perspective. However, there has been perhaps a less than compelling focus on
the necessity for considering scale implications and their overall importance to environmental public policy or environmental decision-making. This is not to say that individuals involved with making sound decisions on a host of environmental issues are
not aware of the scale issue and its relation to spatial variability. They have a broad
understanding of scale, and in many cases, have a good tacit spatial knowledge.
However, as noted by Quattrochi (1993), this tacit perception or understanding of
scale causes confusion when inferred or implied spatial concepts are applied without due thought to what actually is meant by "scale". In short, with such things as
the very high-resolution remote-sensing data available today, in conjunction with

widespread adoption of geographic information science and technologies, the proverbial "trees" can literally be separated from the "forest" with pinpoint accuracy, without due thought to why this is necessary within the purview of environmental decision-making. The abundance of available spatial data from a plethora of sources with different resolutions, in concert with the increased sophistication of analytical tools (*e.g.*, GISs, image processing software), contributes to the scale problem.

Scale and the interpretation of scale (*i.e.*, what "scale" means and how perceived notions of "scale" are applied by policy and decision-makers or the general public) are central to the measurement of urban functions. Usually, the "functionality" of urban areas is observed at broad geographical scales (as well as coarse resolutions) with a focus on deriving a general rule and functional attributes for the entire urban area. As a result of this coarseness, these definitions or measurements of functionality are limited in usefulness for describing or initiating detailed physical or environmental planning policy (Mesev and Longley, 2001). On the contrary, at a very fine scale of analysis (higher resolution and smaller geographical extent), the physical structure of urban attributes allows for the discrimination of discrete characteristics, such as streets and buildings, that provides a spatial perspective of the activity space requirements necessary for urban living. Such information may be of utility for detailed site analysis and planning, but it does not contribute to a broader indication of the overall pattern, arrangement, or orientation of the variability of individual structures or parcels as related to overall urban spatial composition. Consequently, environmental policy or decision-makers could operate at either inappropriate scales or make decisions based on assumptions drawn from using either too coarse or too fine scales. These relationships can be illustrated as follows.

Figure 4.10 is a false color infrared satellite image of Washington, D.C. obtained from the Advanced Spaceborne Thermal Emission and Reflection Radiometer (ASTER). The ASTER instrument is flown onboard the National Aeronautics and Space Administration (NASA) Terra spacecraft (see the Terra/ASTER web site at *http://terra.nasa.gov/About/ASTER/about_aster.html* for more information). The area of this image covers 14 x 13.7 kilometers with a pixel spatial resolution of approximately 20 meters. At this spatial resolution, it is possible to discern both coarse and fine patterns or features inherent to the urban structure and fabric of the Washington, D.C. area, such as forest extent, built up areas, residential areas, parks, roadways, streets and clusters of buildings. Figure 4.11 is an image of the Washington Monument acquired by the QuickBird commercial remote sensing satellite (*http://www.digitalglobe.com*). This is a natural color image obtained at 0.61 meter spatial resolution. It is obvious that the very high spatial resolution of Figure 4.11 is useful, but for a much different set of urban planning or environmental assessment needs than the moderate resolution provided in Figure 4.10. That is, the "trees" (identification or measurement of discrete surfaces) are entirely evident in Figure 4.11, as opposed to the "forest" (identification of general urban patterns of land covers) that is seen from the perspective of Figure 4.10. The urban planner or environmental policy or decision-maker is thus faced with the quintessential spatial question of which scale to use for analysis of environmental characteristics or processes?

Figure 4.10 ASTER false color infrared satellite imagery of Washington, D.C. acquired at a pixel spatial resolution of approximately 20 meters, shown here in black and white (Credit: United States/Japan ASTER Science Team)

Figure 4.11 QuickBird panchromatic color enhanced satellite imagery of the Washington Monument acquired at a spatial resolution of 0.61 meters (Credit: DigitalGlobe, *http://www.digitalglobe.com*)

Current research in the GIS and remote sensing communities has helped to reduce the vexing problem of scale that is faced by environmental decision-makers and planners. One active research focus that has emerged is in developing both the theory and attributes of integrated GISs (IGIS) that can be used to solve complex space and time scale issues related to environmental assessment (see the National Center for Geographic Information and Analysis (NCGIA) web site at *http://www.ncgia.ucsb.edu/* for a host of references to research related to IGIS development). Here, data of different types and of different structures can be integrated together to permit querying or modeling of spatiotemporal features and characteristics across a range of scales.

Other research has led to developing a more concise definition of scale and scale characteristics in reference to remotely sensed and non-remotely sensed data. Quattrochi (1993) introduces the concepts of absolute and relative scales, the examination of the differences between the two scales may be of particular consequence to policy and decision-makers in attempting to identify which scales to use for assessment of processes or functionality, versus the pattern, arrangement, or orientation of environmental phenomena (Quattrochi, 1993; Quattrochi and Goel, 1995; Montello and Golledge, 1998). Absolute scale describes actual distance, direction, shape, and geometry as defined by a grid-type system, as well as the size of the area under investigation (*e.g.*, local, regional, global) (Meentemeyer and Box, 1987; Meentemeyer, 1989; Quattrochi, 1993). Relative scale refers to a scale transformed from absolute scale that describes the relative distance, direction, and geometry of phenomena, predicated on a functional relationship (*i.e.*, a process-oriented spatial relationship). In relation to Figures 4.10 and 4.11, the features present in these images can be defined from both absolute and relative scale perspectives. Depending upon what the need or desire is for a specific decision, Figure 4.10 may be more appropriate for defining a process (*i.e.*, it gives a broader or coarser view of the urban landscape) more so than the identification of the geometry or location of discrete objects as given in Figure 4.11.

Additionally, the recent emphasis on the applications of geostatistical techniques for analysis of spatial data has produced new insights on how scale can be interpreted or measured. For example, work by Lam and Quattrochi and their colleagues on the application of fractal analysis to remote sensing data has provided fruitful results that can be used for assessment of spatial data (Quattrochi *et al.*, 1997, 2001; Emerson *et al.*, 1999; Qiu *et al.*, 1999). Fractals and other geostatistical methods offer a potentially robust method for measuring both the variability present in remote sensing data obtained at different space and time scales, as well as for quantifying the homogeneity or heterogeneity of the land surface or environmental attributes (Quattrochi and Goodchild, 1997; Lam *et al.*, 1998). As a consequence, the applications of fractals and similar geostatistical measures are poised to become important tools in the applications arena, whereby they can be used to produce new or improved information on scale and the scaling of data, that, in turn, can be employed to make better and more sound environmental decisions by policy or decision-makers.

4.3.3 Land-Use and Land-Cover Change

Major national and international efforts, as well as many regional and local activities, have focused on understanding, measuring, and modeling the areal extents, magnitudes, rates, and impacts of land-use and land cover change. The international joint program of the International Geosphere-Biosphere Program (IGBP) and the International Human Dimensions Program (IHDP) on Land-Use and Cover Change (LUCC) and NASA's research program on Land-Cover and Land-Use Change (LCLUC) both seek to understand the role of land-use change in, and its interactions with, other global environmental changes. The LUCC research plan outlines the major research questions and issues for this endeavor, and presents a plan for how a broad community of researchers might approach these issues (Turner *et al.*, 1995). These efforts make significant use of geographic information science and technologies in both the measurement, often using remote sensing, and the modeling of change. Spatial and temporal scale issues in GIScience are critical for how these observations are made and how the models are built. Indeed, the science plan by Turner *et al.* (1995) highlights scalar dynamics as a key integrating activity for the LUCC program.

Why is scale so important for LUCC investigations? Scale effects are driven by the standard concepts of resolution and extent of observations, but also by the interaction of spatial scale with thematic detail and with the processes depicted (*i.e.*, operational scale), and by the temporal resolution of observations and frequencies of update within dynamic models. The next two subsections outline, respectively, how scale can affect the ability to effectively monitor LUCC and the nature and appropriateness of models of various types.

4.3.3.1 Spatial Scale and Resolution in Observations

Much progress has been made in recent years in the application of satellite remote sensing to the regional and global monitoring of land-use and cover change patterns. The interaction between the grain or resolution of imagery used to monitor LUCC, the thematic detail observable, and the size of land-use and cover patches detected can produce some unintended consequences in the analysis of landscape change. The first effect to consider is the link between spatial resolution and thematic detail. As the image resolution decreases (*i.e.*, the grid cell size gets larger), the ability of the analyst to distinguish thematic classes is reduced and, depending on the classes and the imagery, a point (*i.e.*, scale threshold) is reached at which the number of classes must be reduced. Second, as the resolution of the imagery decreases there is a tendency to detect fewer and fewer of the small patches of land covers—essentially these get merged with surrounding cover types (Moody and Woodcock, 1995). At coarse resolutions, the area of cover types that tend to occur in smaller patches will, therefore, be underestimated, whereas the area of cover types that tend to occur in large patches will be overestimated. The interactions are complicated by the influence that differences in the spectral characteristics and definitions of the various cover types have on their detectability, but the general pattern of resolution effects tends to hold.

These scale impacts have affected some very important scientific and policy debates that center on LUCC. A famous example of the effects image resolution can have on LUCC analysis concerns the rate of deforestation in the Brazilian Amazon, which holds about 30 percent of the world's tropical forests and has critical importance to estimates of both the global carbon budget and rates of biodiversity loss. Early estimates of deforestation between 1978 and 1988, made using the Advanced Very High Resolution Radiometer (AVHRR) meteorological satellite sensor with a resolution of 1 kilometer, were compared with an updated analysis, made using Landsat Multispectral Scanner (MSS) imagery with a resolution of 60 meters (Skole and Tucker, 1993). The results indicated that the coarser resolution imagery overestimated the area deforested by about 50 percent. Because models of carbon cycling and estimates of biodiversity loss both require estimates of LUCC rates, attempts to reconcile the global carbon budget and estimate biodiversity loss were affected by these scale and resolution effects.

Studies of urbanization and urban sprawl must contend with complex interactions between the scale of the phenomena, the resolution of the data used to study them, and the definitions of thematic classes represented. The fine-grained nature of the urban fabric reduces the effectiveness of Landsat-class (*e.g.*, resolution of 20–50 meters) satellite imagery in studies of urbanization, especially at the fringes of urban areas where most of the change is taking place (though, see the discussion in section 4.3.2 about the relative merits of coarse and fine resolution imagery for urban environments). The thematic classes used in LUCC studies involve land use, land cover, or both. The class "urban" refers to a land use, but it is composed of a fine-grained pattern of land covers. A coarse-resolution image (*i.e.*, coarser than the grain of the urban landscape) sees only a mixture of land covers in urban areas and that mixture must be interpreted as a land-use class. The ability to interpret detailed land-use classes is limited by the ability to map land cover mixtures to land use (Cihlar and Jansen, 2001). That ability is affected, in non-linear ways, by resolution.

Although urbanization studies have been supported by satellite remote sensing, other sources of information have been at least as useful. At the most detailed, LUCC studies in urban settings use aerial photography, and now high-resolution satellite imagery, because of the ability to detect individual structures and land-parcel level changes. Aerial photographs are often used in conjunction with detailed cadastral (tax parcel) records in a GIS. Another very important source of spatial information for urbanization studies is derived from censuses of population and housing. These data are usually provided for areas on the grounds that correspond to enumeration districts. In the U.S., enumeration districts take the form of census blocks, which are aggregated to form block groups and census tracts. Some planning-oriented analyses further combine census units into planning districts or traffic analysis zones. The resolution of the data in the case of areally aggregated data is determined by the level at which the data are summarized. Significant work on the modifiable areal unit problem tells us that attempts to characterize urban land use or urban land-use change using census-type data are affected by the size, shape, and position of the spatial units by which those data are aggregated. Different results (*i.e.*, patterns, rates, and policies) can be obtained using data based on different zoning systems.

4.3.3.2 Scale of Process and Dynamics

Central to the land-use and cover change research program is the identification of processes by which the observable patterns change. These processes are, themselves, distributed across a number of spatial, temporal, and administrative scales. Attempts are underway to place the processes within some multi-scale conceptual framework, like hierarchy theory that was developed in ecology (Allen and Starr, 1982). The relevance to GIScience is in the methods and models of analysis. For example, investigations are underway to examine relationships and processes across scales (*e.g.*, Walsh *et al.*, 2001) and to build spatial models that incorporate processes at multiple scales (deKoning *et al.*, 1999; Brown *et al.*, 2000).

Temporal resolution of the data (*i.e.*, how frequently data are updated or the temporal sampling interval) affects the ability to monitor land use and cover dynamics and, therefore, to parameterize models of these processes. Some processes are only observable on a seasonal basis, requiring seasonal updates, others annual, and yet others decadal. Just as the spatial scale of the data needs to match the operational scale of the target, the temporal resolution needs to match the frequency of the dynamics.

Models of LUCC processes might be roughly categorized in two types that have relevance to the types and scales of data required: *top-down models* based on observations of the patterns of change; and *bottom-up models* based on detailed representation of individual-level decision making. These model types function at different operational scales, which have direct effects on the spatial scales at which they are relevant.

Top-down refers to models that establish relationships between variables that describe the driving factors of LUCC and land changes. These models can be thought of as most appropriate to capturing coarse resolution, large extent dynamics, and are generally built in association with Focus II of the LUCC research program on land cover dynamics. Building, parameterizing, and validating top-down models require spatial data on both the drivers of change and on the changes. These data can be derived using remote sensing or areally aggregated data, each of which has scale implications (as described above). Example approaches that are top-down include spatially explicit Markov chains, logistic regression, and input-output models. Although these models can describe observed dynamics reasonably well, their utility is limited by the scale (resolution and extent) at which the data are collected and by their inability to capture the details of the various processes that make up those dynamics (Agarwal *et al.*, 2000).

Bottom-up refers to models that attempt to describe the important elemental components of the land-use change process, and build land-use and cover change descriptions up from these finer scale depictions. They are most commonly associated with work on Focus I of the LUCC research program on land-use dynamics. Although these models can be scaled to describe large areas, they require significant data and computational power to do so. Most bottom-up models to date cover relatively small extents. They rely on disaggregate data on individual-level decision making, where the individuals can be persons, households, parcels, and units of government. These data might include landscape perception of people, household wealth and means of livelihood, and farmers' evaluations of risk, *etc.* Cellular automata and agent-based models are examples of modeling approaches that address bottom-up dynamics. These models may offer a stronger platform for cross-scale integration, because the models can include processes at a variety of scales that impinge on agent behavior and can include agents that operate at a variety of scales.

Important scaling challenges involve the selection of appropriate models (*i.e.*, top-down vs. bottom-up) for particular questions, settings, and scales. Furthermore, because processes and patterns operate at a variety of scales, integration of the top-down and bottom-up dynamics with data on fine-scale behavior (*i.e.*, individual and collective human perceptions, behaviors, and decision making processes) and coarse-scale outcomes (*i.e.*, changes in land-use and cover patterns) is important to the LUCC effort.

4.3.4 Environmental and Public Health

The study of the environment in relation to health is inherently a geographical problem. As with other geographical problems, scale is a main factor contributing to the uncertainties of the analysis, hence our understanding of and decision-making about environmental and public health phenomena. This is especially true for those health risk assessment studies that involve the use of small areas. Small-area ecological studies in environmental health refer to studies that require data defined in fine spatial scale, such as census tracts, census blocks, and zip code districts. They typically involve data from disparate sources when intensive data processing and analysis is a characteristic. Therefore, they are amenable to GIS technology.

Small-area ecological studies explore the statistical connections between the frequency of a disease and the level of exposure to a particular agent in population groups rather than individuals (English, 1992). Disparate data sources, such as vital records, hospital discharges, disease registries, census, and estimates of exposure, are typically utilized in order to uncover relationships that most believe to occur and be detectable only at a finer scale (smaller spatial extent and finer spatial resolution). Historically, these types of geographical studies have provided important clues to the etiology of disease, as marked by the infamous example of the discovery of the cause of cholera in London by Dr. John Snow in 1890 (Elliott *et al.*, 1992). Unlike case-control studies, small-area ecological studies seldom lead to definite conclusions about the causes of disease. However, eco-logical studies are especially suited to forming and refining hypotheses about etiology, which in turn could provide useful guidelines for case-control studies.

Although using small areas is almost a norm in environmental and public health studies, it is worthwhile to note that geographical studies at an inter-national scale have been successful in identifying broad relationships because they exploit large differences in both the frequency of the disease and the preva-lence of exposure. The relationship between lack of sunlight and rickets is a good example (English, 1992). Using small areas to study rickets will unlikely reveal the relationship, as rates of disease and environmental exposures are likely to be homogeneous in a small geographical area. The large spatial extent (*i.e.*, large geographical scale) thus plays an important role in uncovering the cause of this disease. As discussed below, this example also points to the need for study using a multi-scale approach, so that patterns and relationships can be determined with higher degree of certainty.

In general, small-area studies focus on four areas of inquiry: geographical correlation, mapping and visualization, the analysis of disease risk around a putative point source, and the assessment of spatial clustering. We provide below an example in each area of inquiry to illustrate how scale affects our under-standing of the phenomena and how policy making about the phenomena is subsequently impacted.

4.3.4.1 Geographical Correlation

In a geographical correlation study of leukemia in Wisconsin, Cleek (1979) pointed out that correlations are more easily affected than regression coefficients by changes in data resolution and aggregation. As the individuals are aggregated into county level, socioeconomic measures are generalized. The coefficient of variation (standard deviation/mean) for mortality rates from leukemia was 7.0 percent comparing states at the national level and 20.9 percent comparing counties within the state of Wisconsin. Similarly the coefficient of variation of mortality rates for cancer of the nasopharynx is 188.9 percent for Wisconsin at the county level and 24.4 percent for the United States at the state level. Colon cancer, however, has higher state-level variation (25.8 percent) than county-level variation (17.3 percent). Cleek suggested using regression coefficients in report-ing and treating correlation coefficients with care (Meade and Earickson, 2000).

This study demonstrates that patterns of association are different at different spatial resolutions and spatial extents, and it is an error to infer an association from one level of scale to another. Especially in environmental and public health studies where the public is sensitive to those issues, caution must be taken in reporting and communicating the results to the public, and policy making based on these results will need to consider the various results at multiple scales. Cleek's study further shows that not all the variables respond the same way as spatial resolution decreases, such as colon cancer in the study, where higher aggregation level (*i.e.*, coarser spatial resolution) does not necessary result in smoother distribution (or lower variation). The interaction between resolution, spatial extent, analysis method, and the phenomenon itself is complex, and strategies must be developed to systematically uncover such interaction.

4.3.4.2 Disease Mapping

Mapping disease patterns is an effective means of summarizing and communicating potential relationships. The production of cancer maps for many countries testifies to the appeal of this approach. A well-known example is the Chinese atlas of cancer mortality published in 1981, which showed distinct regional variation in cancer mortality for many cancer sites (*The Atlas of Cancer Mortality in the People's Republic of China*, 1981; Lam, 1986). In the atlas, the sex-specific mortality rates of the nine most common cancers by county, based on data collected in 1973–75, were generalized and shown in the form of choropleth maps. There are many issues related to the mapping method itself; here we focus on the issues related to scale. In the case of esophagus cancer, the second most common cancer in China, the county-level map for male (Figure 4.12) shows distinct clustering in the central part of China. Clusters of high rates were also observed in remote counties located in the northern and northwestern part of China. However, when the mortality rates were aggregated and mapped at the provincial level (Figure 4.13), the spatial pattern becomes more generalized, and the clustering in the remote counties disappears in the provincial map. These counties could be important clues for studying the etiology of the disease.

In a related study that compares cancer mortality patterns at the provincial level for the entire country and a small region at the commune level (of finer resolution), Lam (1990) showed that the spatial autocorrelation statistics computed for the two map patterns were quite different. The spatial autocorrelation statistic computed for the esophagus cancer for male at the provincial level was found to be not significant at the 0.05 significance level. But when mapped at the commune level (on average, a county has about 30 communes), the spatial autocorrelation statistic indicates that the pattern is positively autocorrelated at the 0.05 significance level, implying the existence of a clustered pattern.

In an attempt to detect at which range of scales the relationship are alike and at what scale the patterns change, the fractal and variogram techniques were introduced to analyze the cancer mortality patterns in China (Lam *et al.*, 1993). Variogram plots of three cancer patterns (esophageal, stomach, and liver) for the Taihu Region in China show that esophageal and liver cancers have the same self-similar scale ranges (roughly between 6 and 150 kilometers), and they are larger than that of stomach cancer (roughly up to 100 kilometers). These self-similar scale ranges were determined based on the portion of the curves that remain linear (Figure 4.14). These self-similar scale ranges could suggest that, if the

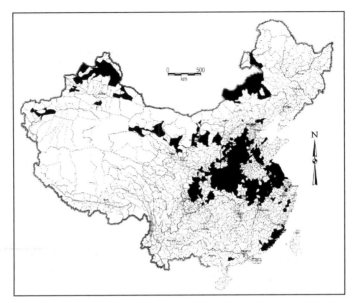

Figure 4.12 Esophagus cancer mortality (1973–75) for male by county.
Shaded counties are those that have high and significant mortality rates
(data source: The Atlas of Cancer Mortality in China, 1981)

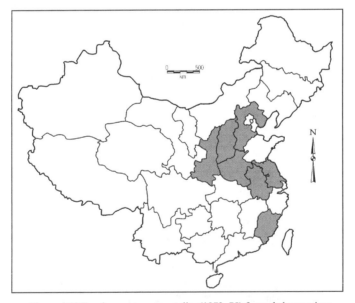

Figure 4.13 Esophagus cancer mortality (1973–75) for male by province.
Shaded provinces are those that have high and significant mortality rates
(data source: The Atlas of Cancer Mortality in China, 1981)

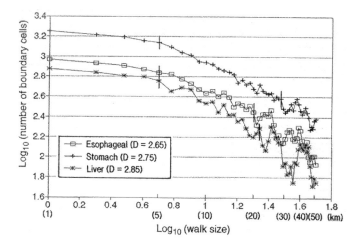

Figure 4.14 Variogram plots of esophageal, stomach, and liver cancer mortality patterns for the Taihu Region, China. The bars on each curve indicate the portion of the curves that remain linear, or in fractal terminology, the range of scales that are self-similar (from Lam *et al.*, 1993)

communes are aggregated to a distance greater than the scale ranges, the map patterns could look very different, which could lead to a different interpretation of the underlying processes. It was also speculated that whatever the underlying controlling factors (*e.g.*, climate, topography, water source), they are likely to operate in ranges of scales similar to that of the cancer patterns.

4.3.4.3 Risk Assessment around a Source

A common question in this area of inquiry is: do certain facilities or industrial sites (*e.g.*, hazardous waste treatment plants) pose an adverse effect on the health of people living close by? This question has been asked over and over again. However, uncertainties involved in this type of health risk assessment studies are high, of which scale contributes an important part. Conflicting results can be generated from various factors, including definitions of data in different spatial and time scales. Consequently, conclusive statements on the impacts of hazardous industrial sites on human health are often lacking.

For example, the famous controversy on whether significant increase in leukemia incidence is related to nearby nuclear installation in England during the late 1980s has generated numerous studies and a disparity of results (Forman *et al.*, 1987; Gardner, 1989). The techniques used in these studies varied: some used incidence while others used mortality; some used rates and compared observed with expected numbers while others used Poisson probabilities; some used existing administrative boundaries while others used circles of arbitrary sizes drawn around a presumed source. Disease clusters can be easily made to appear or disappear, because of these various ways of data manipulations (Glass *et al.*, 1968; Openshaw *et al.*, 1988).

In a study of the health effects of a hazardous waste site on its nearby population in Louisiana, arbitrary circles of 1-mile and 2-mile radii were drawn around the waste site (Lam, 2001). If the 1-mile proximity zone is used, then the region is found to have significantly elevated rates in certain cancers. But if the 2-mile proximity zone is used, then the results are reversed, and the cancer rates defined within the 2-mile wide circle are significantly lower. The dilemma is clear: which results should one adopt for reporting to the public and for deriving policies?

4.3.4.4 General Cluster Detection

Detecting clusters of disease when there is no suspected source is a central, but probably most controversial, research problem in geographical epidemiology. There are many methods and models to detect spatial, temporal, and spatio-temporal clusters. The goal is to identify clusters of areas that are significant and merit further investigation, and at the same time, avoid unnecessary costly epidemiologic (*e.g.,* case-control) studies if the clusters are found to be spurious (Marshall, 1991).

Figure 4.15 is a hypothetical example illustrating how spatial scale and boundary delineation can impact cluster detection. Consider a study region with nine small areas (*e.g.,* census blocks), where each small area is assumed to have roughly the same population and same number of disease occurrence. Given this configuration with only aggregate data provided, no cluster will be detected, as the incidence rates are the same for all small areas. The obvious spatial cluster located at the lower left will therefore never be detected if aggregate data are used.

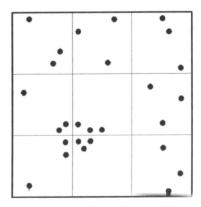

Figure 4.15 A hypothetical study region with nine small areas, each having the same population and same number of disease occurrence. An obvious cluster is observed in the lower-left portion of the map, but the cluster will not be detected if incidence rates by small area are used, as the rates are equal among all small areas

4.3.4.5 Strategies for Mitigating the Scale Effects

As mentioned earlier, scale effects exist and can never be eliminated. Strategies must be developed to mitigate or reduce the scale effects. Two approaches to mitigating the scale effects in health applications can be identified. First, emphasis is placed on developing techniques to detect at which range of scales the relationships are alike and at which other range of scales do things appear and act differently. Variogram, correlogram, and fractal analysis are some of the spatial techniques that have been proposed to detect the range of scales that yield the most information (*i.e.*, scale of actions) (Lam *et al.*, 1996; Cao and Lam, 1997; Tate and Atkinson, 2001).

The other approach, which receives increasing attention from researchers, is to employ multi-scale analysis so that scale effects can be examined and uncertainties due to scale can be reduced. The Centers for Disease Control and Prevention (CDC, 1990) called for a systematic approach to investigating health outcomes. In response to the CDC call, Schneider *et al.* (1993) studied cancer clusters in New Jersey using data at multiple geographical scales. They concluded that analysis of cancer incidence data and cancer clusters at multiple geographic scales provides confidence of the results, alleviates fears of the general public, and prevents costly and unwarranted epidemiologic studies. Ozdenerol (2000) studied infant low birth weight in a typical American city (Baton Rouge, Louisiana) using data aggregated at three spatial scales–census tracts, block groups, and blocks. The results showed that spatial clustering existed in the same general location (inner city neighborhood) for all three scales. This type of multi-scale analysis provides more confidence on the findings. Lam (2001) proposed the use of an integrated spatial analysis framework to reduce uncertainties associated with health risk assessment studies, and scale analysis is a key component of the proposed framework.

4.4 RESEARCH PRIORITIES

Issues of scale affect nearly every GIS application and involve questions of scale cognition, the scale or range of scales at which phenomena can be easily recognized, optimal digital representations, technology and methodology of data observation, generalization, and information communication. These are very different types of questions, and have been addressed quite abundantly in different subfields and disciplines, including geography (Hudson, 1992), remote sensing (Quattrochi and Goodchild, 1997; Tate and Atkinson, 2001), cartography (Buttenfield and McMaster, 1991), spatial statistics (Wong and Amrhein, 1996), hydrology (Sivapalan and Kalma, 1995), and ecology (Ehleringer and Field, 1993).

Effective research in the area of scale will require interdisciplinary efforts of geographers, geostatisticians, cartographers, remote sensing specialists, domain experts, cognitive scientists, and computer scientists. Scale research in many institutes, agencies, and in the private sector began in an *ad hoc* fashion. Motivated both by practical needs as well as theoretical development, recent attention is focused on formalizing the study of scale, on developing theory, and on exploring robust methods for information representation, analysis, and communication across multiple scales.

It has become clear that global and regional processes have implications for local places and that individual and local decisions collectively have global and regional implications. Therefore, scientific information about global and regional patterns and processes must be understood on a local level and vice versa. As the policy-making and scientific communities come to terms with these relationships, systematic understanding about spatial and temporal variations in scale gain importance. Geographical information plays an ever larger role as we move to an increasingly automated information economy. Our understanding of scale and the management of data at various scales must keep pace. Ultimately data and information must inform and must produce better decisions.

Based on the discussion above, a number of research priorities on scale can be outlined as follows.

- *Definitions of scale concepts.* There has been much confusion and misuse of the term "scale." A summary of the meanings and scope of scale is provided in Section 4.2. However, more thorough work is needed to help clarify all connotations of scale in a multidisciplinary context. A good understanding of the conceptual relationship between these connotations should benefit the geographic information users in conjunction with other standardization efforts in geographic information use.

- *Systematized bases for scale-related decision making.* Basic research on the effects of scale on information content will yield practical information for the many users of geographical information. We need to ask, for example, if we lose a significant portion of our explanatory power when we represent global population trends by country as compared with representations at the level of primary divisions within countries. In an increasing array of management and policy settings, decisions about the appropriate scales of analysis are made every day. Identification of critical scales and of scale-invariant data sets or modeling procedures can make those decisions explicit and better informed.

- *Practical guidance on data integration and use.* Often, data at suboptimal or disparate scales are the best available. Using imperfect data that are available is in many cases preferable to using no data at all, but there are implications for the validity of results. For the user community, creation of knowledge about scale provides principles to improve a data set's fitness for use and guidelines by which to discount model results when necessary. For example, we already know that analysis using aggregate data cannot be used to impute the behavior of individuals (called the ecological fallacy). A better understanding of the problem of conflation (the procedure of merging the positions of corresponding features in different data layers) will enable the fusing of data sets, produced at different scales or produced at the same scale but from different sources. Conflation currently poses a serious impediment in the map overlay process, a critical component of GIS. A better understanding of conflation is also necessary for the integration of data sets produced by different agencies.

- *New methods for quantifying and compensating for the effects of scale in statistical and process models.* Scientists and land managers apply a variety of analytical tools to answer geographic questions. Many existing methods do not allow the user to adjust or compensate for the effects of the data scale on the analysis results. Methods that are sensitive to scale will allow the inclusion of scale correctives, much like the correctives that compensate for inflation of the significance of statistical relationship in the presence of spatial autocorrelation.

- *Intelligent automated generalization methods.* GIS tools can be expanded to provide users with methods for intelligently changing the scale of their data. Basic research is still needed to understand how scale changes are perceived, and this in turn can inform interface design. Intelligent generalization will permit the encoding of raw observations in digital form to derive more responsive, application-specific representations.

- *Improved understanding of cognitive issues of scale.* Many scale questions involve human cognition (*i.e.*, how humans perceive the world and information representing it). This issue is explicit especially during human-computer interaction and must be dealt with technically during interface development. It ultimately affects a chain of decisions. Basic research lays a foundation for answering how humans perceive scale changes and its relationship with the conceptual and technical questions about the proper use of spatial and temporal scales in geographic information processing.

4.4.1 Potential Projects

From the above broader research priorities, a number of potential projects are suggested:

- Develop more consistent definitions of scale and its associated attributes that can be uniformly understood or perceived across multiple disciplines. The challenge here is how to make the "standardized" lexicon be adopted and utilized multidisciplinarily.

- Encourage the development of a scale analysis module to be included in major environmental analytical and measurement methods, so that sensitivity analysis, robustness testing, and the effects of scale on the analytical findings can be assessed.

- It is important to point out that many existing scale studies rely heavily on the resampling methods to generate multiscale data for analysis, and as such, the findings on the scale effects may not really be due to the scale effects, but rather they may be an artifact attributable to the use of different resampling methods (Weigel, 1996). The effects of resampling methods on the scale studies are an aspect requiring further research.

- More work needs to be done to identify, develop, and compare efficient spatial/geostatistical techniques for assessing and characterizing the scale effects. This includes refinement of existing techniques, such as fractals, spatial autocorrelation, Shannon index, geographical variance, and local variance, as well as exploration of new techniques such as wavelet analysis, local fractals, and multifractals (Mallat, 1989; Pecknold *et al.*, 1997; Hou, 1998; Myint, 2001; Zhao, 2001). The indices derived from these techniques, if sufficiently discriminatory and information-rich, could be included as part of the metadata.

- Identify the ranges of scale over which an encoded attribute classification scheme is valid. A project of this type will help build a better understanding of the linkages between the scale of a spatial representation and the appropriate corresponding attribute detail. For example, land cover data represented as 30-meter pixels should include more thematic detail (*i.e.*, more number and types of classes) in the land cover classification than similar data represented as 1-kilometer pixels. Studies of these relationships in specific settings and for specific data sets will need to be performed with the aim of building theoretical bases for understanding these relationships more generally.

- Identify the optimal scales of analysis for common data sets, applications, and needs, and critical scales at which the content or structure of phenomena change suddenly (*e.g.*, in vegetation or population). This work will involve the application of sensitivity analyses and spatial statistical tools for describing those sensitivities. Ultimately the goal is to allow users of geographic data to determine the appropriate scales prior to data collection or analysis.

- Develop methods for cost-benefit analysis comparing the use of pregeneralized data (*e.g.*, soil polygons) versus automated generalization of raw data (*e.g.*, soil data from collection points). "Cost" could refer to data collection costs and/or to computational cycles; "benefits" can similarly serve as a metaphor for data validity; each will depend on the needs of the particular application. As automated generalization methods become readily available (many are now in commercial GIS packages), users may wish to access data in the rawest form possible as opposed to using data generalized for a specific purpose. Research of this type would have implications for data collection and archival agencies as well as users.

- Develop improved methods for data generalization. A well researched, yet still unsolved, problem associated with spatial generalization is the creation of multiple versions of databases. Research in cartographic generalization has taken several directions including algorithmic design and testing, the design of models and conceptual frameworks, the application of expert systems, and the modeling of cartographic features. Thus far, most of the work in generalization has focused on what is termed "cartographic" generalization, which involves the graphical considerations associated with scale change. A second, less researched, area is in "model" generalization, by which generalization operators (simplification, smoothing, aggregation, agglomeration, and others) are applied to an original digital landscape model (DLM) in order to create secondary representations of the database, called digital cartographic models (DCM). These terms, DLM and DCM, are taken from the European cartographic literature, where a significant amount of work has been completed in this area.

- Develop improved methods for incorporating knowledge about scale into metadata. We need to model descriptive data about data sets (*i.e.*, metadata) to assess the consequences of downloading data at a finer resolution than is needed for a particular GIS application. The question is to determine if the choice of appropriate scale can be made on the basis of metadata alone. With the expansion of global computer networks and the use of those networks for geographic data transmission, efficient modes of communicating data content are needed. We need to determine which modes are most appropriate for representing scale and to evaluate their relative effectiveness. The provision of information about scale aids in the education of users on how to assess appropriate scales for a given application. Scale is clearly a fundamental component of any metadata report.

- Design and develop a multi-scale database. We still lack the database management tools and associated functions needed to effectively and efficiently store multi-scale data, perform multi-scale analysis, and intelligently change scale. This project should be aimed at the development of a software product that would be useful in GIS applications.

4.5 REFERENCES

Agarwal, C., G.L. Green, Grove, M., Evans T. and Schweik, C., 2000, *A Review and Assessment of Land-Use Change Models: Dynamics of Space, Time, and Human Choice*, Bloomington, IN: CIPEC, Indiana University.

Allen, T.F.H. and Starr, T.B., 1982, *Hierarchy: Perspectives for Ecological Complexity*, Chicago: University of Chicago Press.

The Atlas of Cancer Mortality in the People's Republic of China, 1979. Beijing: China Map Press.

Brown, D.G., Pijanowski, B.C. and Duh, J.D., 2000, Modeling the relationships between land-use and land-cover on private lands in the upper Midwest, USA. *Journal of Environmental Management,* 59: 247–163.

Buttenfield, B., 1984, *Line Structure in Graphic and Geographic Space*, Unpublished Ph. D. Dissertation, University of Washington.

Buttenfield, B.P. and McMaster, R.B. (eds.), 1991, *Map Generalization: Making Rules for Knowledge Representation*, New York: Longmont Scientific and Technical.

Cao, C.Y. and Lam, N.S.-N., 1997, Understanding the scale and resolution effects in remote sensing and GIS. In Quattrochi, D.A. and Goodchild, M.F. (eds.), *Scale in Remote Sensing and GIS,* Boca Raton, FL: CRC/Lewis Publishers, pp. 57–72.

Catts, D., 1990, Generalization of GIS digital maps. *Papers and Proceedings of Applied Geography Conference*, Binghamton, NY: Applied Geography Conference, Inc.

Centers for Disease Control and Prevention, 1990, Guidelines for investigating cluster of health events. *Morbidity and Mortality Weekly Report*, 39: 1–23.

Cihlar, J. and Jansen, L.J.M., 2001, From land cover to land use: A methodology for efficient land use mapping over large areas. *The Professional Geographer,* 53(2): 275–289.

Cleek, R.K., 1979, Cancer and the Environment. The Effect of Scale. *Social Science and Medicine*, 13D: 241–247.

Cromley, R.G., 1991, Hierarchical methods of line simplification. *Cartography and Geographic Information System,* 18: 125–131.

De Cola, L. and Buttenfield, B., 1994, Multiscale mapping for the NSDI: Data modeling and representation. *Proceedings of GIS/LIS annual conference.*

De Koning, G.H.J., Veldkamp, A. and Fresco, L.O., 1999, Exploring changes in Ecuadorian land use for food production and their effects on natural resources. *Journal of Environmental Management,* 57. 221–237.

Douglas, D. and Peucker, T., 1973, Algorithms for the reduction of the number of points required to represent a digitized line or its caricature. *Canadian Geographer*, 10(3): 112–122.

Ehleringer, J.R. and Field, C.B. (eds.), 1993, *Scaling Physiological Processes, Leaf to Globe.* New York: Academic Press.

Elliot, P., Cuzick, J., English, D. and Stern, R. (eds.), 1992, *Geographical and Environmental Epidemiology: Methods for Small-Area Studies,* Oxford: Oxford University Press.

Emerson, C.W., Lam, N.S.-N. and Quattrochi, D.A., 1999, Multi-scale fractal analysis of image texture and pattern. *Photogrammetric Engineering and Remote Sensing,* 65(1): 51–61.

English, D., 1992, Geographical epidemiology and ecological studies. In Elliot, P., Cuzick, J., English, D. and Stern, R. (eds.), *Geographical and Environmental Epidemiology: Methods for Small-Area Studies,* Oxford, UK: Oxford University Press, pp. 3–13.

ERITO, 1997 (last update), *http://www.ertico.com/links/gdf/gdfintro/gdfincon.htm.*

Forman, D., Cook-Mozaffari, P., Darby, S. and others, 1987, Cancer near nuclear installations. *Nature,* 329: 499–505.

Gardner, M.J., 1989, Review of reported increases of childhood cancer rates in the vicinity of nuclear installations in the UK. *Journal of Royal Statistical Society A,* 152: 307–325.

Glass, A.G., Hill, J.A. and. Miller, R.W., 1968, Significance of leukemia clusters. *Journal of Pediatrics,* 73: 101–107.

Hou, R.-R., 1998, *A Local-Level Approach in Detecting Scale Effects on Landscape Indices,* Baton Rouge, Louisiana: Louisiana State University. M.S. Thesis.

Hudson, J., 1992, Scale in space and time. In Abler, R.F., Markus, M.G. and Olson, J.M. (eds.), *Geography's Inner Worlds: Pervasive Themes in Contemporary American Geography,* New Brunswick, NJ: Rutgers University Press, pp. 280–300.

Kulldorff, M., 1997, A spatial scan statistic. *Communications in Statistics: Theory and Methods,* 26: 1,481–1,496.

Lam, N.S.-N., 1986, Geographical patterns of cancer mortality in China. *Social Science and Medicine,* 23(3): 241–247.

Lam, N.S.-N., 1990, *The role of geographical scale in analyzing cancer mortality patterns in China.* Paper presented at the 1990 IGU Regional Conference on Asian Pacific Countries, Beijing, China.

Lam, N.S.-N., 2001, Spatial analysis for reducing uncertainties in human health risk assessment. *Acta Geographica Sinica,* 56(2): 239–247. (In Chinese with English abstract).

Lam, N.S.-N., Qiu, H.L., Zhao, R. and Jiang, N., 1993, A fractal analysis of cancer mortality patterns in China. In. Lam, N.S.-N and De Cola, L. (eds.), *Fractals in Geography,* Englewood Cliffs, NJ: Prentice Hall, pp. 247–261.

Lam, N.S.-N. and Quattrochi, D.A., 1992, On the issues of scale, resolution, and fractal analysis in the mapping sciences. *The Professional Geographer,* 44(1): 88–98.

Lam, N.S.-N. and De Cola, L. (eds.), 1993, *Fractals in Geography,* Englewood Cliffs, NJ: Prentice Hall.

Lam, N.S.-N., Fan, M. and Liu, K.B., 1996, Use of space-filling curves in generating a national rural sampling frame for HIV/AIDS research. *The Professional Geographer,* 48(3): 321–332.

Lam, N.S-N., Quattrochi, D.A., Qiu, H.-L. and Zhao, W., 1998, Environmental assessment and monitoring with image characterization and modeling system using multiscale remote sensing data. *Applied Geographic Studies,* 2(2): 77–93.

MacEachren, A., 1995, *How Maps Work,* New York: The Guilford Press.

Mallat, S.G., 1989, A theory for multi-resolution signal decomposition: the wavelet representation. *IEEE Transactions on Pattern Analysis and Machine Intelligence,* 11: 674–93.

Marshall, R.J., 1991, A review of methods for the statistical analysis of spatial patterns of disease. *Journal of Royal Statistical Society A,* 154, Part 3: 421–441.

Meade, M.S. and Earickson, R.J., 2000, *Medical Geography,* New York: The Guilford Press.

Meentemeyer, V., 1989, Geographic perspectives of space, time, and scale. *Landscape Ecology,* 3(3/4): 88–98.

Meentemeyer, V. and. Box, E.O., 1987, Scale effects in landscape studies. In Turner, M.G. (ed.), *Landscape Heterogeneity and Disturbance,* New York: Springer-Verlag, pp. 15–34.

Muller, J., 1990, The removal of spatial conflicts in line generalization. *Cartography and Geographic Information Systems,* 17: 141–149.

Monmonier, M., 1991, *How to Lie With Maps.* Chicago: University of Chicago Press.

Montello, D.R. and Golledge, R.G., 1998, *Summary Report: Scale and Detail in the Cognition of Geographic Information,* NCGIA Report of Specialist Meeting held under the auspices of Project Varenius, May 14–16, 1998. University of California at Santa Barbara, 65 pp.

Moody, A. and Woodcock, C.E., 1995, The influence of scale and the spatial characteristics of landscapes on land-cover mapping using remote sensing. *Landscape Ecology,* 10(6): 363–379.

Myint, S.W., 2001, *Wavelet Analysis and Classification of Urban Environment Using High-Resolution Multispectral Image Data,* Ph.D. Dissertation, Louisiana State University.

Nickerson, B., 1988, Automated cartographic generalization for linear features. *Cartographica,* 25(3): 15–66.

Oliver, M.A., 2001, Determining the spatial scale of variation in environmental properties using the variogram. In Tate, N.J. and Atkinson, P.M. (eds.), *Modelling Scale in Geographical Information Science,* Chichester, UK: John Wiley & Sons, pp. 193–220.

Openshaw, S., Charlatan, M., Wymer, C. and Craft, A., 1987, A Mark 1 geographic analysis machine for the automated analysis of point data sets. *International Journal of Geographic Information Systems,* 1(4): 335–358.

Openshaw, S., Craft, A.W., Charlton, M.G. and Birch, J.M., 1988, Investigation of leukemia clusters by use of a geographical analysis machine. *Lancet* Feb. 6: 272–273.

Ozdenerol, E., 2000, *A Spatial Inquiry of Infant Low Birth Weight and Cancer Mortality in East Baton Rouge Parish, Louisiana.* Ph.D. Dissertation, Louisiana State University.

Pecknold, S., Lovejoy, S., Schertzer, D. and Hooge, C., 1997, Multifractals and resolution dependence of remotely sensed data: GSI to GIS. In Quattrochi, D.A. and Goodchild, M.F. (eds.), *Scale in Remote Sensing and GIS,* Boca Raton, FL: CRC/Lewis Publishers, pp. 361–394.

Quattrochi, D.A., 1993, The need for a lexicon of scale terms in integration of remote sensing data with geographic information systems. *Journal of Geography,* 92(5): 206–212.

Quattrochi, D.A. and Goodchild, M.F. (eds.), 1997, *Scaling in Remote Sensing and GIS,* Boca Raton, FL: CRC/Lewis Publishers, Inc.

Quattrochi, D.A. and Goel, N.S., 1995, Spatial and temporal scaling of thermal infrared remote sensing data. *Remote Sensing Reviews,* 12: 255–286.

Quattrochi, D.A., Emerson, C.W., Lam, N.S-N. and Qiu, H-L., 2001, Fractal characterization of multitemporal remote sensing data. In Tate, N.J. and Atkinson, P.M. (eds.), *Modelling Scale in Geographical Information Science,* John Wiley & Sons: Chichester, UK, pp 13–34.

Qiu, H-L., Lam, N.S-N., Quattrochi, D.A. and Gamon, J.A., 1999, Fractal characterization of hyperspectral imagery. *Photogrammetric Engineering and Remote Sensing,* 65(1): 63–71.

Robinson, A., Morrison, J., Muehrcke, P., Kimerling, A.J. and Guptill, S.C., 1995, *Elements of Cartography,* 6th edition, New York: Wiley.

Rodriguez, A., 1998–99, USFS GSTC Salt Lake, Utah, personal commun.

Schneider, D., Greenberg, M.R., Donaldson, M.H. and Choi, D., 1993, Cancer clusters: The importance of monitoring multiple geographical scales. *Social Science and Medicine*, 37(6): 753–759.

Sivapalan, M. and Kalma, J.D., 1995, Scale problems in hydrology: Contributions of the Robertson workshop. *Hydrological Processes,* 9(3/4): 243–250.

Skole, D.L. and Tucker, C.J., 1993, Tropical deforestation and habitat fragmentation in the Amazon: Satellite data from 1978 to 1988. *Science,* 260: 1,905–1,910.

Tate, N.J. and Atkinson, P.M. (eds.), 2001, *Modelling Scale in Geographical Information Science*, Chichester, UK: John Wiley & Sons.

Tobler, W. 1970, A computer movie simulating urban growth in the Detroit region. *Economic Geography,* 46 (Supplement): 234–240.

Tobler, W. 1988, Resolution, resampling and all that. In Mounsey, H. and Tomlinson, R. (eds.), *Databases for Global Science*, London: Taylor & Francis, pp. 129–137.

Turner, B.L., Skole, D., Sanderson, S., Fischer, G., Fresco, L. and Leemans, R., 1995, *Land-Use and Land-Cover Change Science/Research Plan,* Joint publication of the International Geosphere-Biosphere Programme (Report No. 35) and the Human Dimensions of Global Environmental Change Programme (Report No. 7), Stockholm: Royal Swedish Academy of Sciences.

Turner, M.G., O'Neill, R.V., Gardner, R.H. and Milne, B.T., 1989, Effects of changing spatial scale on the analysis of landscape pattern. *Landscape Ecology,* 3: 153–162.

Walsh, S.J., Crawford, T.W., Welsh, W.F. and Crews-Meyer, K.A., 2001. A multi-scale analysis of LULC and NDVI variation in Nang Rong District, Northeast Thailand. *Agriculture Ecosystems and Environment,* 85(1-3): 47–64.

Weigel, S.J., 1996, *Scale, Resolution and Resampling: Representation and Analysis of Remotely Sensed Landscapes Across Scale in Geographic Information Systems,* Ph. D. Dissertation, Louisiana State University.

Wong, D. and Amrhein, C. (eds.), 1996, The Modifiable Areal Unit Problem. Special issue of *Geographic Information Systems* 3: 2–3.

Zhao, W., 2001, Multiscale Analysis for Characterization of Remotely Sensed Images, Ph.D. Dissertation, Louisiana State University.

CHAPTER FIVE

Extensions to Geographic Representations

May Yuan, University of Oklahoma
David M. Mark, State University of New York at Buffalo
Max J. Egenhofer, University of Maine
Donna J. Peuquet, Pennsylvania State University

5.1 INTRODUCTION

Representations play a key role in computational systems. "Coarsely speaking, a *representation* is a set of conventions about how to describe a set of things. A *description* makes use of the conventions of a representation to describe some particular things" (Winston, 1984, p. 21). Thus the representation defines, and limits, the power of any computational system. "Finding the appropriate representation can be a major part of a problem-solving effort" (Winston, 1984, p. 22). A good representation "makes important things explicit," while exposing "the natural constraints inherent in the problem" (Winston, 1984, p. 24). The manner in which geographic information is represented is a central issue for any field that studies phenomena on, over, or under the surface of the Earth. Geographic phenomena often embrace high spatial and temporal variations over a large area with varying degrees of details, yet representations of geographic phenomena can only signify certain geographic characteristics at particular levels of abstraction. For example, in studies of routing problems, spatial information is typically represented as links between places, and places are reduced to points. In studies of environmental problems, the pollutants in air, water, or soil tend to be represented simply as values at points in regular grids or lattices, while in other studies, these entities may be represented as polygonal objects that are defined by explicit boundaries.

A representation is a means to communicate geographic information, and is also a binary structure in a computer or electronic storage medium that corresponds with an object, measurement, or phenomenon in the world. The representation chosen for a geographic phenomenon has a profound impact on interpretation and analysis. The selection of information to be represented, and the choice of representational scheme, are often driven by the purpose of the analyses, although they might also be based on available data, or on an abstraction of the actual phenomena being represented (Mark, 1979). In turn, the results of an analysis can be greatly influenced by the way in which the phenomena under study are viewed. Drivers may be able to follow a strip map or route map more easily than an overall areal map, but a route map is of limited use to show the overall distribution of geographic features we encounter within a given area.

Geographic information systems (GIS) and spatial analysis are influenced by representations at three levels: data models, formalization, and visualization. Data modeling involves developing constructs, and organizing data accordingly, in order to represent a selected set of entities and relationships in a database. A data model is the conceptual core of an information system; it defines data object types, object relationships, operations, and rules to maintain database integrity (Codd, 1980). Formalization provides computational models or formal languages to support querying, reasoning, and computation. Depending on the chosen data model, formalization draws fundamental mathematical principles from algebra, topology, or set theory (Worboys, 1997). Visualization, on the user's end, offers a graphical means for data exploration, analysis, and interpretation. In particular, geographic visualization integrates cognitive and semiotic approaches to facilitate scientific understanding of geographic worlds (MacEachren *et al.*, 1999).

The three levels of representations inextricably facilitate conceptualizing, formalizing, and visualizing geographic phenomena. Since the real world has infinite complexity, selection of a representation circumscribes what information is accountable, computable, and visible in an analysis. Among the three levels of representations, data modeling determines the information that can be represented and thus can be computed and visualized. The effectiveness of a data model in representing geographic phenomena is, therefore, unarguably critical. The objectives of the UCGIS research challenge in extensions to geographic representations are to expand into volumetric (three-dimensional) and dynamic phenomena and to develop analytical approaches that support these extensions, especially in very large, distributed databases.

Volumetric and dynamic phenomena are by definition multi-dimensional, which means that they are conceptually and computationally challenging. The challenge becomes greater when we consider large-scale geographic processes. In many cases, simply introducing an additional orthogonal axis (Z) is convenient but insufficient, because important spatial and temporal characteristics and relationships may be indiscernible in this approach. Although visualization techniques for three or more dimensions have become popular in recent years, data models and formal languages have not yet fully developed to support advanced spatial and temporal analysis in multiple dimensions. Hence, data visualization is mainly working on fields of raw data. It lacks automatic procedures to query geographic features, and it lacks functions to analyze topological relationships. Location-based representations, for example, embrace geographic information as properties at locations. Since there is no data object representing a geographic process that changes over space and time, information about the behaviors of the process and their interactions with other processes is unavailable for quantitative analysis in visualization.

An additional challenge for volumetric and dynamic representations results from the shape of the Earth. While remotely sensed data inherit a grid structure from their collection technology, it is geometrically impossible to represent the spheroidal Earth with a single mesh of uniform, rectangular cells (Dutton, 1983). Although other geometries, particularly the triangular mesh, do not exhibit this problem and have other well-known favorable properties (Peuquet, 1984), converting grid-based data to other shapes of geometric primitives is a complicated task and inevitably results in information loss. Along with the concern of the spheroidal Earth, spatial analysis at a continental and global scale needs to account for spherical distances and orientations on a terrain surface rather than rely on Cartesian planar geometries inherent in current commercial GISs (Willmott *et al.*, 1997).

Furthermore, requirements for volumetric representations and analytical methods vary with domain applications. The dimensionality of oceanographic data, for example, poses a different challenge than is typically found for terrestrial data. While terrestrial data are based on a datum in relation to a defined ellipsoid, bathymetric data are referenced to a water level datum (Li, 1999), which varies at locations and time. Compared with terrestrial data, oceanographic data are sparse in the horizontal dimension but are dense in the vertical dimension. Even in the horizontal plane, data are highly anisotropic, since sampling is dense along ship-tracks while trackline spacing is comparably large. Adequate spatial interpolation methods are important for the analysis of three-dimensional oceanographic data (Wright and Goodchild, 1997). While oceanographic data are samples of thematic attributes (salinity, sea floor depth, *etc.*), representations of oceanographic features are necessary to enable effective feature-based queries to search and retrieval of objects and continuous fields (Li *et al.*, 1995).

To cope with the dynamic nature of both the terrestrial and oceanographic environments, advanced remote sensing and survey technologies have flooded us with multi-terabytes of data arriving at an unprecedented rate. Representational support for volumetric and dynamic phenomena and for the development of related analytical methods is increasingly important as the size and complexity of geospatial data have been growing significantly. With the growth, comes an increased need to distill useful information from very large geospatial data sets. Medical data, for example, will be more valuable if information about movements of individuals in space and time, called geospatial lifelines (Mark and Egenhofer, 1998), can be intersected with environmental data to detect spatiotemporal correlation between illness and environmental events. Likewise, extracting events from weather or climatological data to reveal spatiotemporal behaviors of these events will leverage the usefulness of the data to scientific understanding. Such information support requires new geographic representations and analytical methods to formulate information from a large geospatial database, especially those with temporal information, more than two spatial dimensions, or both.

As the conceptual core of a geographic information system, geographic representations determine what information is available for communication, exploration, and analysis. Hence, research in extensions to geographic representations is critical to advancing geographic information science. In the remainder of this chapter, we review major research progresses in geographic representations in the next section, and then we outline the limitations of the current geographic representations, elaborate fundamental research questions in the area, and present a case study in geographic representations. After that, the chapter proposes a research agenda including research topics of short-term, intermediate, and long-term scopes. Lastly, we summarize key ideas and arguments and specify conditions for success in this ambitious agenda.

5.2 MAJOR RESEARCH PROGRESSES IN GEOGRAPHIC REPRESENTATIONS

Representing geography is one of the focal research areas in the development of geographic information systems. From the spaghetti to topological models, geographic representations inherit the traditional cartographic paradigm with paper (or parchment) as a representation medium. With some ingenious exceptions, maps historically have been limited predominantly to a flat two-dimensional static view of the Earth. This view is also at a single scale, with assumed exactitude and with no capability for dynamic interaction by the user. Peuquet (1984) gives an insightful analysis of these two-dimensional conceptual frameworks used in GIS to model geographic features as static and geometrically fixed objects. Although scalar properties of geographic features remain constant, geographic representations of these geographic features may vary with the scale of observation. For example, a city may be represented as a point at one scale but may be represented as a polygonal feature at a larger map scale. However, a map can only represent geographic features in the particular geometry significant to a particular scale at some time. In order to represent a geographic feature at multiple scales, more than one map is necessary to represent the same feature. Consequently, such multiple representations of a single geographic feature results in problems with data redundancy and consistency (Buttenfield, 1990).

In such a cartographic view of the world, two major schemes used to represent geographic space are regular and irregular tessellation models (Frank and Mark, 1991). The regular tessellation models subdivide space into cells of regular size and shape, while the irregular tessellation models represent space as irregular subdivisions of zero-, one-, or two-dimensional cells. Distinguished from information of other kinds, geographic information consists of spatial and thematic components, and both of them need to be fully represented. With the two spatial frameworks, there are three approaches to incorporate thematic information. A layer approach separates geographic information into independent layers of themes. This approach suits well for thematic mapping. A space-composite approach integrates all thematic information into the largest areas such that all attribute values are spatially homogeneous within each area. The third approach uses objects to describe individual geographic entities with spatial and non-spatial properties.

Both regular and irregular tessellation approaches are map-based representations, which greatly facilitate overlay operations to reveal spatial relationships among geographic variables but do not support three-dimensional analysis or dynamic modeling. Three-dimensional visualization techniques cannot fully solve the problem because a true three-dimensional application requires information that can only be derived from analyzing three-dimensional topological relationships beyond simple visualizing of the data volume. This is because topological integrity forms the basic operations to manipulate and analyze data in a two-, three-, or four-dimensional GIS (Egenhofer and Al-Taha, 1992; Egenhofer and Herring, 1994; Egenhofer and Mark, 1995; Hazelton, 1998). For example, a GIS must have capabilities to compute information about adjacency in the vertical space to answer a three-dimensional query for areas where sandstone lies upon shale layers to identify areas of landslide potential.

Incorporation of temporal components into a geographic representation presents an additional challenge because time has properties distinct from those of space. During the last decade, researchers in both GIS and database management have been examining ways to incorporate time into information systems (Langran, 1992; Tansel *et al.*, 1993; Egenhofer and Gollege, 1998). In relational databases, temporal information is often incorporated by time-stamping tables (Gadia and Vaishnav, 1985), tuples (Snodgrass and Ahn, 1985), or cells (Gadia and Yeung, 1988). Similarly, time-stamping techniques have been applied in accordance with the three approaches of incorporating spatial and thematic information; there are three common approaches to incorporate time into spatial data models. The layer approach incorporates time by a collection of snapshots. The snapshot approach shows the states of a geographic theme at different time instances, and there is no explicit relationship among objects on any two snapshots (Armstrong, 1988). The second approach represents the world as a set of space-exhausting, spatially homogenous and temporally uniform areas, named space-time composites (Langran and Chrisman, 1988). The space-time composite model is a result of spatial joins of snapshot layers, and it consists of the largest common units in attribute, space, and time dimensions. The third approach, the spatio-temporal object model, extends time to an independent dimension orthogonal to the space (Worboys, 1992). It represents the world as a set of discrete objects composed of spatiotemporal atoms. Spatiotemporal atoms are the largest spatio-temporal homogeneous units in which properties hold in both space and time. While a spatiotemporal object corresponds to a geographic entity, the atoms constituting the spatiotemporal object describes the history of the geographic entity.

Besides the three approaches to incorporate time into spatial data models, a number of representations have been proposed for domain specific applications. The Potential Path Area (PPA) layers (Miller, 1991) and Temporal Map Set (TMS, Beller *et al.*, 1991) combine ideas from the snapshot and object models to represent activities in space and time. A PPA layer is designed to reason about individuals' time budgets for travel and activity participation, in which each PPA describes a spatial extent to which an individual can reach by a certain transportation medium during a certain period of time. Hence, PPA layers extend the snapshot model by representing probable activities of individuals under spatial and temporal constraints on a layer. Similarly, the TMS model bundles binary snapshots of a geographic event to represent the development of the event over time. Both PPA and TMS models provide data frameworks applicable for spatiotemporal intersection (STIN) analysis of spatiotemporal relationships among objects in space and time (Lin and Calkins, 1991; Lin, 1992). In addition, there are numerous ways to incorporate time into spatial databases. Langran (1992) provides a comprehensive discussion on GIS representations and management of spatiotemporal data. Peuquet (1994) further elaborates geographic conceptualization about space and time and proposes a triad framework to integrate geographic themes, location, and time.

All time-stamping approaches have difficulty representing dynamic information, such as transition and motion. Geographic information cannot be extracted from a system in which the information cannot be represented. Hence, data models developed using the time-stamping approaches are incapable of supporting spatiotemporal queries about the dynamic characteristics of geographic processes, including movement, rate of movement, frequency, and interactions among processes. Geographic representations and analyses *"must deal with actual processes, not just the geometry of space-time"* (Chrisman, 1998, p. 91). Recent research in GIS representations has emphasized representing dynamic processes, including the Smith *et al.* (1993, 1994) modeling and database system (MDBS), Peuquet and Duan's (1995) event-based spatiotemporal data model (ESTDM), Raper and Livingstone's (1995) geomorphologic spatial model (OOgeomorph), and Yuan's (1994, 1997) three-domain model. MDBS takes a domain-oriented approach to support data processing in hydrological applications. It incorporates time through sequential snapshots. ESTDM represents information about changes at predefined cells as a result of the passage of an event. The change-based approach has shown its efficiency and capability to support spatial and temporal queries in raster systems. OOgeomorph, on the other hand, adopts an object-oriented design to form processes by integrating point data that are observations at particular space and time. Alternatively, the three-domain model provides a framework that incorporates both the space-time composite model and ESTDM model to enable representing histories at locations as well as occurrences of events in space and time (Yuan, 1999).

5.3 LIMITATIONS OF CURRENT GEOGRAPHIC REPRESENTATIONS

The cartographic paradigm poses four major limitations to geographic representations: (1) volumetric and temporal objects; (2) heterogeneous types of geospatial data from an integrated global perspective and at multiple scales; (3) dynamic geographic processes and their interactions; and (4) data quality and uncertainty. These representational limitations also constrain GIS support for information query, analysis, and visualization.

5.3.1 Limitations on Representing Volumetric and Temporal Objects

While current GIS techniques for representing data are capable of recording complex associations among multiple variables, these techniques still generally depict static situations on a two-dimensional plane surface at one specific scale. Many of these two-dimensional representations can be extended conceptually to accommodate volumetric applications (*e.g.*, representing pollutant concentrations in air and groundwater), but integration of operational capabilities for visualizing and analyzing three-dimensional data has been realized only recently in general-purpose, commercially available geographic information systems. While some volumetric geographic data handling systems already are in use for graphical and specialized analytical applications, such systems do not have the representational flexibility and power that are needed for addressing complex, global-scale analyses. Nor can they handle three-dimensional spatial and topological relationships that are critical to true three-dimensional applications.

Likewise, current GIS representations are able to support simple spatiotemporal query, analysis, and modeling (Langran, 1992; Peuquet, 1994), but support for complex cases is problematic (Yuan, 1999). Temporal information is by large incorporated into GIS databases by time-stamping layers (the snapshot model in Armstrong, 1988), attributes (the space-time composite model in Langran and Chrisman, 1988), or spatial objects (the spatiotemporal objects model in Worboys, 1992). Such time-stamping approaches inherit a static view of geographic worlds, in that time is represented as an attribute of spatial objects and in a discrete fashion (ordinal time points). However, there are many types of time (Frank, 1998), each of which signifies a distinct temporal perspective and conceptualization. A simple discrete time model lacks the abilities to represent cyclic intervals and accumulative behaviors that require models of cyclic time and continuous time. Dynamic events and processes, therefore, are not well represented in the time-stamp approaches.

5.3.2 Limitations on Representing Heterogeneous Types of Data from an Integrated Global Perspective and at Multiple Scales

Current spatial data storage and access techniques are not designed to handle the increased complexity and robustness needed for representing heterogeneous types of data for a wide range of analytical and application contexts, as is currently envisioned for handling these same earth-related problems. Earth-related data are being collected in digital form at a phenomenal rate, far beyond anything we have experienced before. For Spot Image satellite data to provide a single, complete coverage of the Earth's surface at 10-meter pixel resolution would require approximately 1.5×10^{13} pixels. If we assume that a single data value for a single pixel can be stored in one byte, then 1.5×10^{13} bytes (or 15 terabytes) of storage would be required for that single, complete coverage. Querying and analyzing such massive amounts of data challenges computer memory allocation and algorithm design for data access and retrieval.

To cope with the vast influx of data, various Federal agencies are cooperating in the development of a "global spatial data infrastructure." The infrastructure includes the agreements, materials, technology, and people necessary to acquire, process, store, maintain, and provide access to most of the Earth-related data collected and maintained by the Federal government. However, hurdles to the development of an integrated global database are significant. In addition to massive data volume, development of an integrated global database needs to overcome the need for a framework that can accommodate the spheroidal Earth and integrate marine and terrestrial data. Remote sensing technology has been the major means for acquiring large-scale geospatial data in regular gridded arrays. However, it is geometrically impossible to represent the spheroidal Earth with a single mesh of uniform, rectangular cells (Dutton, 1983). Other geometries, especially the triangular mesh, will provide better coverage for the spheroidal Earth, but converting grids to triangular meshes will inevitably introduce uncertainty. Besides a global framework that counts for the spheroidal Earth, GIS integration of marine and terrestrial data must address representation issues on multiple dimensionality, data dynamism, and spatial data references that vary their relative positions and values over time (Wright and Goodchild, 1997).

Without significant extensions to current representational techniques, however, the usefulness of geospatial data remains limited for data intensive research at a global scale. We therefore need to develop highly flexible, yet highly efficient data models (*i.e.*, concepts addressing structure and format) for handling Earth-related data of this range and magnitude. Severe tradeoffs in the capabilities of representational techniques exist, usually between representational power and efficiency. Hence, it is an urgent research challenge to develop representational techniques for GISs in order to handle complex, multi-scale volumetric data for interactive analytical and modeling applications.

5.3.3 Limitations on Representing Dynamic Geographic Processes

Although many efforts have been made to integrate GISs with dynamic modeling, most of the efforts are limited to the development of an interface between two separate types of software systems. Modeling software tends to operate within very narrowly defined domains and to use mathematical simulation, whereas GISs are employed primarily for preprocessing observational data and post-processing data for comparative display. The ability to represent and examine the dynamics of observed geographic phenomena within a GIS context, except in the most rudimentary fashion, is currently not yet available. Such capability is urgently needed to enhance the effectiveness of geographic analysis to an increasing variety of problems at local, regional, and global scales. The careful analysis of change through time and patterns of change is vital to understanding a range of problems, from urban growth and agricultural impacts to global change. Thus research to improve representational schemes is a high priority.

Building dynamic processing within a GIS is difficult because the current GIS data models are geared toward static situations. A number of characteristics of space-time data make the development of space-time representations more difficult than volumetric representations. First, unlike volume or geographic area, time cannot be measured in spatial units (*i.e.*, feet or meters). Second, the nature of time itself differs from that of space. At a given moment, everything everywhere is at the same point in time. Furthermore, unlike space, time is unidirectional, progressing only infinitely forward, yet time also embraces cyclic properties (Frank, 1998). For example, July 4, 1996 occurs only once, but every year has a summer season. Information contained within a geographic database may be augmented or modified over time, but the current GIS representation schemes cannot handle successive change or dynamics through time, except through some extremely simplistic methods (the most frequently used method is equivalent to a series of still snapshots).

The amount of information attainable from a GIS database is a function of the chosen representational scheme. Geographic representations must be able to support information needed in sciences and applications to ensure the usefulness of GIS data to modeling and decision making. Interactions between space-time processes are complex and must be represented in some way in GIS in order to support modeling of dynamic phenomena. Because spatial characteristics of geographic phenomena may vary with scale of observation and their interactions can occur across scales, geographic representations must incorporate processes to connect geographic phenomena and to support dynamic modeling at multiple scales in space and time.

5.3.4 Limitations on Representing Data Quality and Uncertainty

Given the rapidly increasing use of GISs for policy analysis and decision making, another urgent issue is how to represent data of varying exactness and varying degrees of reliability and then convey this additional information to the user. Data representation techniques need to embrace information about data quality and uncertainty, especially for examining multiple "what if" scenarios. The importance of this issue is underscored by the fact that the National Center for Geographic Information and Analysis designated accuracy in spatial databases as Initiative 1. Much work remains to be done on developing methods to handle fuzziness and imprecision—which are inherent in geographic observational data —within a digital database. Such methods are crucial for combining multiple layers of data from varying sources (Goodchild and Gopal, 1989). Also needed are methods to measure accuracy of geospatial data, such as digital elevation models, to facilitate decision-making (Kyriakidis *et al.*, 1999).

From a human standpoint, spatial relationships between geographical entities (cities, *etc.*) are often expressed in an imprecise manner that can be interpreted only within a specific context (*e.g.*, Is New York near Washington, D.C.?). Current methods of data representation and query, however, are limited to absolute and exact values and cannot handle inexact terms, such as "near" (Beard, 1994). Yet inexactness and context dependency is an integral component of human cognition and of the human decision-making process. In order for GISs to become truly useful and user-friendly tools, whether for addressing complex analytical issues such as global change or urban crime or for making day-to-day decisions, the data model used by GISs needs to accommodate such cognitive issues.

5.3.5 Representational Limitations and Information Query Support

In addition to analysis and modeling, query support is one of the most important capabilities for an information system. The faithfulness of a representation to which it attempts to represent is critical to assessing the effectiveness of the representation. Other factors include the amount of data need to be stored in order to comply with the representation, the degree to which data can be associated to resolve patterns and relationships, and the amount of information can be computed from the representation. As the aforementioned, current GIS representations have limited capabilities to depict volumetric and dynamic processes as well as their interactions at multiple scales. They also have limited support for queries that inquire information about spatial relationships in a three-dimensional environment, spatio-temporal behaviors of a process, and interactions among processes across scales.

Although many studies have directed to improve geographic representations (c.f. Langran, 1992; Peuquet, 1994), further research is needed to examine the essence of geographic processes to extend geographic representation. The extended representation needs to provide a framewok to hold data in ways that information can be inferred to characterize geographic processes that generate these data. Many research questions remain unanswered. How should a GIS handle processes of various kinds to ensure that salient characteristics and behaviors of these processes are represented? How does identification of entities and relationships relate to different problem domains? How can a representation accommodate different ways that geographic entities and relationships are identified to allow interoperability of data among applications? Research in extensions to geographic representation attempts to provide both conceptual and practical frameworks with which that geographic phenomena are represented in optimal ways to support GIS data analysis, dynamic modeling, and information query in multiple dimensions across various scales in space and time.

5.4 AN EXAMPLE OF GEOGRAPHIC REPRESENTATION RESEARCH: MOVEMENT OF POINT-LIKE OBJECTS IN GEOGRAPHIC SPACE

Movement of objects in geographic space represents an important class of dynamic geospatial phenomena. Data consisting of a series of discrete space-time samples over the domain of continuous movements, describing an individual's location in geographic space at regular or irregular temporal intervals, have been termed *geospatial lifelines* (Mark and Egenhofer, 1998). Reducing the moving object to a point simplifies many aspects of the conceptualization and geospatial computing, but of course also limits the application to a subset of geographic movement. Paths of objects moving in geographic space were referred to as life paths by Hägerstrand (1970), and in the spatial database literature, they are treated as moving points (Erwig *et al.*, 1997). Efficient methods for representing and analyzing geospatial lifelines is an important topic because, in the near the future, very large volumes of geospatial data will be collected through the deployment of global postioning system (GPS) equipped wireless telephones and other such devices. Much important geospatial lifelines information will refer to movements of people and the devices they carry. However, the same type of data is also recorded for animals, vehicles, and other devices. Geospatial lifeline data may be recorded at different resolutions, and the interaction between spatiotemporal sampling intervals, precision, and the nature of physical movement patterns at different scales will be critical to the design of analysis and reasoning procedures. For example, a list of residential addresses over an individual's life also defines a geospatial lifeline, with a much lower temporal resolution than the examples mentioned above. Lastly, although recording and analyzing lifeline data by service providers and other authorities might provide useful information to decision makers from the public and private sectors, such data also raise major implications for privacy and personal security.

Much of the basic theory for geospatial lifelines was provided some three decades ago by Torsten Hägerstrand in a conceptual model that became known as time geography (Hägerstrand, 1970). In that classic paper, Hägerstrand laid out very clearly the ways in which velocity and temporal schedules constrain the parts of space in which humans and animals can act. If two people must be at certain distinct places at particular times, this may severely constrain the times and places at which they could meet. Furthermore, the length of any meeting in such a situation depends on the constraints and the meeting location. Although Hägerstrand's model can readily be formalized, it was used mainly as a conceptual model in human geography, perhaps because of the limitations of programming environments available to geographers and planners in the 1970s. An exception is Miller's paper reporting on an implementation of some aspects of Hägerstrand's model using a commercial GIS (Miller, 1991). Recently some researchers in spatiotemporal databases have explicitly linked their work to Hägerstrand's model (Fauvet *et al.*, 1998, 1999; Dumas *et al.*, 1999).

Geospatial lifelines have a wide range of potential applications. Residential life histories and travel diaries can provide valuable insights for the social and behavioral sciences (Odland, 1998; Janelle *et al.*, 1998; Thériault *et al.*, 1999a, 1999b). The same data can be of great value to transportation planners and engineers (Mey and terHeide, 1997; Miller, 1999). Geospatial lifelines data at various scales can be used to assess or infer exposures in environmental health (Mark *et al.*, 1999) and in policing (Fyfe, 1992). They also have considerable potential in wildlife biology. Progress in this research area will be best achieved through multidisciplinary collaboration among computational reasoning experts, database researchers, and domain scientists.

5.4.1 Interactions between Scales of Movement and Sampling Intervals

Although there are many exceptions, a typical human being sleeps in the same place (home) almost every night and makes one particular place home for several years in a row. In developed countries, a typical adult has a job outside the home and spends 35–50 hours per week at some particular work place. Over the course of the week, the person might spend 60 percent of his or her time at home, 25 percent at the work place, and the other 15 percent either at other places, or moving between places. All of these sites, including places along a commuting path, provide possibilities for acquiring goods, for interacting with other people, and for exposure to environmental risks. The pattern of movement among two regular sites and a variable set of other places repeats through most of the year, but people in developed countries typically have 2–4 weeks of vacation each year when they do not go to work, and most people go away from home for some or all of that time. At a longer time scale, they may from time to time change their place of residence, or their place of work, either separately or simultaneously. At a shorter time scale, they move about with shorter spatial scales within their work place and in their home. Human movement has different characteristic spatial and temporal scales, depending on the purpose, nature, and mode of the movement. Data sampled at the wrong scale or interval may completely miss, or drastically distort, the effects of particular processes. For example, since most people move their home location only every few years, a record of home location or regular sleeping place would typically appear as long intervals of constant location (hundreds or thousands of days), interspersed with jumps that happen on a single day or over periods of just a few days. A 1-day sampling interval would capture this scale of spatiotemporal variation and would in fact be highly redundant. Research is needed to investigate the scales of various human and animal behavioral patterns and examine the interactions between these aspects of real-world spatiotemporal variability and the data acquisition characteristics for geospatial lifelines. Representations of geospatial lifelines must be able to capture these various concepts of resolution and scale in a form that can be a basis for inference.

5.4.2 Research Priorities for Geospatial Lifelines

Research on geospatial lifelines will be advanced through solution of the following sub-problems.

- Determining a comprehensive set of queries and reasoning procedures for lifelines that are application-independent but which satisfy all known application requirements;

- Designing and prototyping computational models that can deal efficiently with large sets of geospatial lifelines;

- Documenting the ways in which spatial and temporal sampling intervals and regularity interact with characteristics of various kinds of geospatial movement;

- Developing statistical methods for geospatial lifelines, including detection of clusters or hot spots in space-time, and appropriate null hypotheses for sets of lifelines; and

- Studying the ethical and legal implications of recording individuals' geospatial lifelines and establishing procedures for appropriate restrictions on data analysis and dissemination.

Research on this topic must begin with the development of an effective data model for geospatial lifelines, and continue with investigations of methods for querying and presenting lifeline information and for improved analysis. Methods must be implemented and tested using real data sets of realistic size and complexity.

5.4.3 Developing a Data Model for Geospatial Lifelines

Hägerstrand's concept of time geography provides a framework for modeling geospatial lifelines. The basic element of lifeline data is a space-time observation consisting of a triple < ID, location, time >, where ID is a unique identifier of the individual that is used for all observations of that individual's movements, location is a spatial descriptor (such as a coordinate pair, a polygon, a street address, a zip code, or some other locative expression), and time is the time stamp that indicates when the individual was at that particular location (such as a clock time in minutes or event time in years). Identities of entities may disappear and later reappear (Hornsby and Egenhofer, 1997) —for example, a toxic waste site may for a time be considered to be totally remediated and later be found to still be polluted. Identities may also disappear when objects are aggregated (Hornsby and Egenhofer, 1998). In most cases, the time stamp will refer to the real time at which an event occurred, but not when it was stored in a database (Snodgrass 1992); for some applications, however, such as in assessing liability or fault, it may be important to preserve both sorts of time within the database (Worboys, 1998). For example, legal responsibility might depend both on when the site was polluted and on when the pollution became known to company or government officials.

The data model will be influenced by a variety of factors. For example, geospatial lifelines typically will record discrete positions, whereas the phenomena they describe are typically continuous. For this reason, both temporal sampling intervals and interpolation methods should depend on the ontological characteristics of the movement in question. Tracking commuters as they pass through polluted areas might require data recorded at 5-minute or even 1-minute intervals, whereas data that recorded people's home addresses at 6-month intervals would require different interpolation methods.

5.4.4 Formalizing a Query Language for Geospatial Lifelines

The concepts of time geography also provide a framework for queries about geospatial lifelines. Possible queries about individual lifelines can be derived from the geospatial lifeline triple <ID, location, time>, distinguishing whether any combination of the three arguments is known or unknown. Examples include "What individuals were at location S at time T?", "Did X stay at location S at time T?", and "Which individuals have ever visited location S?" Through the use of aggregate operators as functions, additional information can be derived about durations of immobility or of trips. Such aggregate operations can then be used in combination with queries across several individuals, as in the case of, "What locations have been visited by more than n members of a given population within any single year?" Queries about geospatial lifelines involve more complex sets of constraints if the different travel speeds of individuals must be considered as well.

Some of these queries over space-time prisms resemble operations on OLAP ("On-Line Analytical Processing") data cubes (Gray *et al.*, 1996), such as slice, dice, roll-up, and drill-down; however, the semantics of the data cube operations are not (yet) well defined and lack the intersection operation, which is meaningful and important in the analysis of geospatial lifelines. Similar to the development of visual query languages for cubes (Frank, 1992a; Richards and Egenhofer, 1995), operations upon geospatial lifeline prisms lend themselves to the design of direct-manipulation user interfaces.

A formal language will provide an organizational framework for queries about lifelines and queries that relate lifelines to other objects. For certain application domains, a positive answer to some such queries would indicate a consistency violation (*e.g.*, a physical object cannot be located simultaneously at two different locations).

Research is needed to develop an algebra over geospatial lifelines based on the concept of lifeline prisms. This formalism would then lead to a spatio-temporal query language, enabling testing of space-time hypotheses. This would provide core concepts that are essential to all other aspects of computational models for geospatial lifelines.

5.4.5 Matching Geospatial Lifelines and Reasoning about Relations between Lifelines

If large volumes of geospatial lifeline data become available, methods for determining the similarity of two geospatial lifelines would be critical for making sense of such data. Exact matches of geospatial lifelines are easy to define, although efficient search for such matches in very large databases will require basic research. More challenging are cases of partial match, of matching with tolerance, since here degrees of match and mismatch would have to be weighted to come up with an over-all index of similarity. The development of similarity measures for geospatial lifelines will require examination of concerns of end users of information and the reasons that matches are being sought. For example, for cancer case-control sampling, if the latency of the particular form of cancer is known to be around 10 years, then similarity of geospatial lifelines around 10 years ago could be given maximum weight, while similarity the last 3–5 years might be ignored altogether.

An alternative method for matching lifelines would involve the comparison of qualitative descriptions of geospatial lifelines, particularly their qualitative spatiotemporal relations. Qualitative spatial relations and qualitative spatial reasoning methods have shown significant results in two-dimensional geographic space (Frank, 1992b; Hernández, 1994; Sharma *et al.*, 1994; Hornsby and Egenhofer, 1997), with a focus on two-dimensional, areal objects embedded in a two-dimensional plane. For example, models for topological relations between spatial regions (Egenhofer and Franzosa, 1991; Randell *et al.*, 1992) have been highly successful in GISs as a basis for querying and spatial reasoning. The data model for geospatial lifelines, however, suggests the embedding of a one-dimensional object (the lifeline) in a multi-dimensional space (location + time); therefore, extensions of the existing models will be needed. For example, the set of topologically-distinct relations between two undirected lines embedded in two- or three-dimensional space comprises 33 relations (Egenhofer and Herring, 1994). When applied to geospatial lifelines, the set of realizable relations is smaller, because geospatial lifelines must be monotonic in time. The significance of the ordering of the time axis, however, requires consideration of the direction as an integral part of lifeline relations so that "before" and "after" can be distinguished (Allen, 1983).

5.4.6 Processing of Incomplete Geospatial Lifeline Data

As noted above, data for geospatial lifelines often are recorded as sets of discrete space-time measurements. In theory, the temporal resolution of the recorded geospatial lifelines could be very dense, such that every time point queried would be available (Tansel *et al.*, 1993). There are, however, numerous reasons that speak against such an approach. From a system perspective, we would obtain extremely large data sets for even short timelines. On the other hand, even if the samples are very dense, there may be unforeseen queries that would require a higher resolution. A third argument against the assumption of the availability of dense time recordings is the expectation that occasionally the device for capturing space or time may be unavailable or out of order, producing gaps in the data. For example, geospatial lifelines relevant to environmental health studies usually will be a series of home addresses or work places, constructed from memory by patients or their relatives, perhaps containing errors of gaps. For example, relatives may have little or no idea of where loved ones traveled while serving overseas in the military during times of conflict.

In order to fill such gaps, interpolation methods may be useful to infer locations occupied, answering queries such as: "Given that we have no knowledge of the patient's residence or residences between two points in time, where might the person have been in between?" Or, "Given positions of a ship at two distinct times, where could it have traveled between those times?" The larger the time intervals and the faster the movement, the more imprecise we expect such interpolations to be. There are, however, a number of ways to add specificity into the interpolation methods if one has, for instance, some knowledge about the space, but not about the time; some knowledge about the time, but not about the locations; or some information about a possible travel mode. There may be other constraints that can improve interpolation, since topography or transportation infrastructure may constrain movements. For example, if a car travels through a city and its locations are tracked every minute, then the interpolation needs to take account of the fact that the car is probably moving along roads in a road network, or through parking places and garages. Such interpolations may be ambiguous if there are multiple possible paths between two locations; however, gaps may be filled by analyzing similar paths taken by the same individual at different times, from which a more likely intermediate location can be determined.

Sometimes, it may be appropriate to take a more conservative stance by treating gaps in the record simply as missing data. In order to support such a choice by a researcher, all of the query and analysis methods developed under this research priority must be able to deal properly with geospatial lifelines with gaps, despite the fact that, logically, the person or object in question must have been somewhere at every moment during the gap in the record.

Research must explore interpolation methods based on several properties: (1) the individual's immediate past, as described by the most recent part of the geospatial lifeline, (2) the constraints on movement present in the environment, and (3) the individual's typical behavior in the past in similar situations (*e.g.*, whether the individual followed the same route in the past and, if so, whether it consistently led to the same destination).

5.4.7 Intersecting Geospatial Lifelines with Environmental Data

For a variety of applications, including health problems induced by short-term exposures to environmental toxins, researchers might need to relate individual geospatial lifelines of people in a population potentially at risk with the spatio-temporal distribution of the hazard. This might even be done in nearly real time, in order to warn people about the possible consequences of their exposures. For example, many delivery trucks and emergency vehicles as well as some luxury cars currently are equipped with GPS receivers that continuously monitor their positions, and soon most cellular telephones may have the same. After an airborne toxic release near a highway, methods to be developed would identify all vehicles or cellular phones that passed through high concentrations of the toxin, allowing drivers to be contacted and advised to undergo testing. Progress on this topic would require the implementation of methods to represent diffusion, movement, and variable concentration of environmental toxins as three-dimensional fields in space time, where a three-dimensional field is a single-valued function of position in a three-dimensional space. Here, one of the three dimensions actually is time, and the field provides the concentration value of the toxin at each point in space-time. The second aspect of research in this area will involve algorithms for efficient comparison of individual geospatial lifelines, to quickly determine the maximum or cumulative field values that the particular lifeline encountered. Such spatiotemporal intersection procedures may need to be modified if they are to deal efficiently with comparisons of many geospatial lifelines to one spatiotemporal field.

5.4.8 Protecting Privacy

As noted above, potential for surveillance of locations and activities of people in space and time will increase dramatically over the next decade, as more and more people carry devices combining the functionality of pagers, cellular telephones, and GPS receivers. The example presented in the previous section shows how access to data about the movements of individuals may at times be useful to them. But such massive surveillance of individuals would likely be resisted by people concerned about personal privacy, even if there were potential health or safety benefits. How should individual privacies be protected when tracking data on people is made available to researchers, planners, or health and safety officials? Designers of technologies bear at least some of the social responsibility for the systems they design. In this case, if geospatial lifelines could be developed with capabilities or design characteristics such that certain intrusive applications could be avoided altogether, researchers and the software industry should so design them.

5.4.9 Summary

This section has reviewed issues regarding the representation and processing of data about movement of points in geographic space. Of course, real objects cannot be points, and the larger an object is compared with the precision with which its precision is recorded, the more dangerous it will be to ignore the shape of the object. Also, larger geographic objects such as storms or wildfire do not really move as solid bodies, but rather may change shape and size as they move. Thus the methods discussed in this section will need to be extended, modified, or replaced for dealing with motion of such extended geographic phenomena. Nevertheless, geospatial lifelines appear to be an important and core class of geo-spatiotemporal phenomena, and thus are a high priority area for extension of geographic representations.

5.5 A RESEARCH AGENDA TO EXTENDING GEOGRAPHIC REPRESENTATIONS

Extension to geographic representations is critical to the development of geographic information science, since a representation sets a framework to handle and analyze geographic phenomena. As discussed previously, the current representations inherit much limitation from a map paradigm that enforces a two-dimensional, geometry-centered framework. The map framework also limits ways to integrate global data, incorporate temporal information, and express data uncertainty. Extensions to geographic representations must overcome these hurdles.

A primary theoretical issue on extending geographic representations is to develop a new GIS representational approach that optimizes both the capabilities of the modern computing environment and representational techniques recently developed in a number of fields, including GIS, and that incorporates human cognition of geographic space. Addressing this issue involves decisions that range from the most philosophical (*e.g.*, determining how time differs from space and how those differences can be represented) to the most practical (*e.g.*, choosing high performance computing techniques for handling vastly increased data volumes). The complexity and broad range of research needs in geographic representations requires an interdisciplinary effort involving geographers, computer scientists (particularly those currently involved in database management, artificial intelligence, and high-performance computing), applied mathematicians, cognitive scientists, and experts from the application domains.

One promising approach to extending geographic representations is to combine geometries (rectangular, triangular, and hexagonal) for location-based representations, and to integrate location-based, feature-based, and time-based representations. Researchers within the GIS community as well as developers of commercial GISs generally recognize this multi-representational approach as the best method, although its deliberate use as a long-term solution for designing geographic databases is a recent development. Tools such as computational geometry and object-oriented design are becoming widely available to enhancing representational capabilities required for volumetric and dynamic geographic data. Recent attempts at extending current representational techniques to include time have served mostly to demonstrate the complexity of the problem (Langran, 1992; Peuquet, 1994). Several worldwide efforts are addressing the representation of geographic data, and separately, the representation of dynamics within database management systems (DBMS) (Tansel *et al.,* 1993). These efforts lay the groundwork for exploring how temporal DBMS techniques can be applied to combined space-time representations.

5.5.1 Priority Areas for Research

The research challenge of extending geographic representations calls for multi-disciplinary team efforts and multi-modal approaches. To meet the challenge, the following outlines of short-, medium-, and long-term objectives present steps to achieve the daunting goal of extending geographic representations to three-dimensional, temporal, and global geospatial data.

5.5.1.1 Short-Term Objectives

- Test new DBMS techniques, particularly temporal DBMS, to the geographic context; examine alternative ways of representing the temporal component, evaluate alternative temporal DBMS design, and identify aspects of time in geographic data that cannot be represented in existing DBMS.

- Apply high-performance computing techniques to the geographic context, examining methodologies for distributed databases and distributed processing that accommodate the spatial nature of both the data and potential retrieval queries.

5.5.1.2 Medium-Term Objectives

- Develop taxonomy of geographic primitives, analyze their thematic, spatial and temporal components, categorize spatial and temporal relations in the geographic context, and synthesize schemes for spatial and temporal reasoning.

- Develop new strategies and techniques that combine current approaches, such as the use of object-oriented programming techniques and the use of computational geometry for multiple modes of geographic representation.

- Develop a space-time data model that can represent dynamic processes and spatial interactions and support complex spatial and temporal queries in an effective manner.

- Develop new graphical interface techniques that utilize the increased capabilities needed for visualizing and analyzing large, multi-scale, heterogeneous data.

- Develop a new query language capable of handling the increased dimensionality of spatiotemporal data (*e.g.*, although standard query languages have been extended to handle spatial queries, research is still needed to make appropriate extensions to accommodate space-time phenomena).

5.5.1.3 Long-Term Objectives

- Develop a representational theory that closely reflects human cognition yet is also highly efficient and minimally complex from a computing standpoint.

- Develop characteristics based upon the new representational approach that allow geographic databases and associated analytical capabilities to be implemented with predictable characteristics.

- Enable full representational and computational support for spatial and temporal data mining and knowledge discovery in the geographic context.

In terms of research, the highest priority should be placed on the long-term efforts and the second highest priority on the medium-term efforts because many private GIS providers and government agencies are already funding or directly participating in the areas identified as short-term efforts. Areas identified as medium- and long-term efforts are areas in which the most work needs to be done and where the highest benefit will be derived. These areas are also least likely to gain support of private GIS providers or government agencies because of the length of time that sustained support is needed before concrete benefits to GIS would be realized.

Although much conceptual work is required to extend methods for geographic representations, the proof and practical refinement of any new data model lies in its implementation and empirical testing on real-world data. Such activities require significant investment in programming time and computing resources. Because methods of data representation are so fundamental to software design, data models can rarely be replaced within existing software. Instead, software for testing new data models needs to be custom-made. Each of these priority areas also requires the context of an example problem. The test problem should include multiple scales, multiple dimensions, and a diverse range of data types so that the research will focus on solutions that are directly useful and applicable. Possible test problems include the global water cycle, global carbon cycle, Central American forests, land-use change and social impacts, crime, dynamic changes in urban neighborhoods, geospatial lifelines, and emergency response.

5.5.2 Importance to National Research Needs/Benefits

The need is increasingly urgent to better understand the effects of human activities on the natural environment at all geographic scales. In natural resource management within the developed world, emphasis is shifting from inventory and exploitation to maintenance of the long-term productivity of the environment. Such maintenance requires interactive space-time analysis at multiple scales to clarify the complex interrelationships of environmental systems. As only one component of this analysis, Global Circulation Models (GCMs) are used to study climate dynamics, ocean dynamics, and global warming. We need to verify and refine these models. To do so requires sophisticated analysis of large volumes of multidimensional data, particularly the study of change through time and patterns of change through time around the entire Earth, including the oceans and the atmosphere.

In an urban context, an interactive and real-time means is necessary for solving problems in emergencies (*e.g.*, floods or wildfires) to preserve life and property. As populations and development have increased, the need for predicting human/environmental interactions through the use of multiple "what if" scenarios has become recognized. For all of these diverse uses of GISs, we must have the ability to perform interactive space-time analysis at multiple scales and to have data of known reliability.

Enormous amounts of data, already in digital form, are being collected for studies of a diverse range of urgent environmental, economic, and social problems. Nevertheless, current representational techniques for storing and accessing these data within GISs are not adequate. We need significant advancements in representational methods in order to access these data in forms that are useful for analysis and improve the science being done. The proposed research agenda reflects the needs to represent multiple dimensional and dynamic geographic phenomena in an ever growing, heterogeneous geospatial database. The short-, medium-, and long-term objectives outline the research efforts necessary to meet the challenge of extending geographic representations.

5.6 CONCLUSIONS

This chapter has outlined the importance of extending geographic representations to advancing geographic information science, milestones in the development of geographic representations, and limitations and challenges in representing geographic phenomena. Geographic representations define what can be represented in an information system and how the information is represented. Consequently, it determines the ways in which we can analyze and visualize information. Extensions to geographic representations, therefore, play a key role in enhancing data analysis, reasoning, and modeling in the geographic context. Many limitations have been recognized in representing multi-dimensional, multi-scale, heterogeneous, dynamic, and fuzzy geographic worlds. Much progress has been made to improve the capability of geographic representations. Conceptual and theoretical milestones are well synthesized by Langran (1992), Peuquet (1994), Egenhofer and Golledge (1998), and Peuquet (2002). However, research challenges remain, especially in representing three-dimensional and dynamic geographic phenomena and in integrating marine and terrestrial data to a seamless global database.

The research challenges are complicated by the fact that geographic representations relate to domain applications. This chapter presents an example of geographic representation research on geospatial lifelines that represent movement of point-like objects in geographic space. The case study applies Hägerstrand's theory of time geography (Hägerstrand, 1970) to model the point locations of an individual over time, through steps of developing a data model, formalizing a query language, reasoning about relations between lifelines, and projecting geospatial lifelines to environments. Numerous issues on extending point-based representations to the temporal dimension, on analytical and reasoning methods, and on societal ethics are elaborated to demonstrate the complexity of this research area.

A research agenda is proposed to meet the research challenge in extending geographic representations. The proposal advocates for multidisciplinary team efforts and multi-modal approaches. Objectives for short-, medium-, and long-term research range from incorporating DBMS techniques to developing a representation theory for geographic representations. As much attention from the industry and government already target to the short-term objectives, support is urgently needed to sustain medium- and long-term research. Powerful and effective geographic representations will enhance the usefulness of geospatial data to scientific inquiry and applications. Successful extensions to geographic representations for three-dimensional, dynamic, and heterogeneous geospatial data at a global scale will foster a better understanding of human-environment interactions on the living Earth.

5.7 REFERENCES

Allen, J. 1983, Maintaining knowledge about temporal intervals. *Communications of the ACM,* 26(11): 832–843.

Armstrong, M.P., 1988, Temporality in spatial databases. *Proceedings: GIS/LIS'88,* 2: 880–889.

Beard, K., 1994, Accommodating uncertainty in query response. *Proceedings, Sixth International Symposium on Spatial Data Handling,* Edinburgh, Scotland: International Geographical Union, Vol. 1, pp. 240–253.

Beller, A., Gilblin, T., Le, K., Litz, S., Kittel, T. and Schimel, D., 1991, A temporal GIS prototype for global change research. *Proceedings: GIS/LIS'91,* 2: 752–765, Atlanta, GA. ACSM-ASPRS-URISA-AM/FM.

Buttenfield, B.P., 1990, *NCGIA Research Initiative 3 Technical Report: Multiple Representation,* University of California at Santa Barbara.

Chrisman, N.R., 1998, Beyond the snapshot: Changing the approach to change, error, and process. In Egenhofer, M.J. and Golledge, R.G. (eds.), *Spatial and Temporal Reasoning in Geographic Information Systems,* Chapter 6, pp. 85–93.

Codd, E.F., 1980, Data models in database management. *Proceedings: Workshop on Data Abstraction, Databases, and Conceptual Modeling,* Brodie, M.L. and Zilles, S.N. (eds.), Pingree Park, CO, June 23–26 1980, pp. 112–114. ACM SIGPLAN Notices 16, No. 1 (January 1981); ACM SIGART News-letter No. 74 (January 1981); ACM SIGMOD Record 11, No. 2 (February 1981).

Dumas, M., Fauvet, M.C. and Scholl, P.C., 1999, *TEMPOS: A Temporal Database Model Seamlessly Extending ODMG,* Research-Report 1013-I-LSR-7, LSR-IMAG Laboratory, Grenoble, France.

Dutton, G., 1983, Geodesic modelling of planetary relief. *Proceedings, AutoCarto VI,* Ottawa, October 1983, Vol. 2, pp. 186–201.

Egenhofer, M. and Franzosa, R., 1991, Point-set topological spatial relations. *International Journal of Geographical Information Systems,* 5(2): 161–174.

Egenhofer, M.J. and Al-Taha, K.K., 1992, Reasoning about gradual changes of topological relationships. In Frank, A., Campari, I. and Formentini, U. (eds.), *Theories and Methods of Spatio-Temporal Reasoning in Geographic Space,* Lecture Notes in Computer Science, vol. 639, pp. 196–219. New York: Springer-Verlag.

Egenhofer, M. and Herring, J., 1994, Categorizing topological spatial relations between point, line, and area objects. In Egenhofer, M., Mark, D. and Herring, J. (eds.), *The 9-Intersection: Formalism and its Use for Natural-Language Spatial Predicates,* Santa Barbara, CA: National Center for Geographic Information and Analysis, Report 94-1.

Egenhofer, M.J. and Mark, D., 1995, Modeling conceptual neighborhoods of topological line-region relations. *International Journal of Geographical Information Systems*, 9(5): 555–566.

Egenhofer, M.J. and Golledge, R.G. (eds.), 1998, *Spatial and Temporal Reasoning in Geographic Information Systems*, New York: Oxford University Press.

Erwig, M., Güting, R.H., Schneider, M. and Vazirgiannis, M., 1997, *Spatio-temporal data types: An approach to modeling and querying moving objects in databases*, FernUniversität Hagen, Germany, Informatik-Report 224.

Fauvet, M., Chardonnel, S., Dumas, M., Scholl, P. and Dumolard, P., 1998, Analyse de données géographiques: application des bases de données temporelles. *Révue Internationale de Géomatique*, 8(1-2): 149–165.

Fauvet, M., Chardonnel, S., Dumas, M., Scholl, P.C. and Dumolard, P., 1999, Applying temporal databases to geographical data analysis. In Tjao, A.M., Cammelli, A. and Wagner, R. (eds.), *Tenth International Workshop on Database and Expert Systems Applications: Spatio-Temporal Data Models and Languages*, Florence, Italy, pp. 475–480, Los Alamitos, CA: IEEE Computer Society.

Frank, A.U., 1998, Different types of time in GIS. In Egenhofer, M. J. and Golledge, R.G. (eds.), *Spatial and Temporal Reasoning in Geographic Information Systems,* Chapter 3, pp. 40–62.

Frank, A.U., 1992a, Beyond query languages for geographic databases: data cubes and maps. In: Gambosi, G., Scholl, M. and Six, H.-W. (eds.), *Geographic Database Management Systems*, pp. 5–17, New York: Springer-Verlag.

Frank, A.U., 1992b, Qualitative spatial reasoning about distances and directions in geographic space. *Journal of Visual Languages and Computing,* 3(4): 343–371.

Frank, A.U. and Mark, D., 1991, Language issues for GIS. In Maguire, D.J., Goodchild, M.F. and Rhind, D.W. (eds.), *Geographical Information Systems: Principles and Applications*, vol. 1, pp. 147–163. Essex: Longman & Technical.

Fyfe, N.R., 1992, Space, time, and policing—towards a contextual understanding of police work. *Environment and Planning D,* 10(4): 469–481.

Gadia, S.K. and Vaishnav, J.H., 1985, A query language for a homogeneous temporal database. *Proceedings of the ACM Symposium on Principles of Database Systems*, pp. 51–56.

Gadia, S.K. and Yeung, C.S., 1988, A generalized model for a relational temporal database. *Proceedings of ACM SIGMOD International Conference on Management of Data*, pp. 251–259.

Goodchild, M.F. and Gopal, S. (eds.), 1989, *Accuracy of Spatial Databases*, London: Taylor and Francis.

Gray, J., Bosworth, A., Layman, A. and Pirahesh, H., 1996, Data cube: A relational aggregation operator generalizing group-by, cross-tab, and sub-total. In Su, S. (ed.), *Proceedings of the Twelfth International Conference on Data Engineering*, pp. 152–159, Los Alamitos, CA: IEEE Computer Society.

Hägerstrand, T., 1970, What about people in regional science? *Papers, Regional Science Association*, 24: 1–21.

Hazelton, N.W.J., 1998, Some operational requirements for a multi-temporal 4D GIS. In Egenhofer, M.J. and Golledge, R.G. (eds.), *Spatial and Temporal Reasoning in Geographic Information Systems*, Chapter 4, pp. 63–73. New York: Oxford University Press.

Hernández, D., 1994, *Qualitative Representation of Spatial Knowledge*, Lecture Notes in Artificial Intelligence 804, New York: Springer-Verlag.

Hornsby, K. and Egenhofer, M., 1997, Qualitative representation of change. In Hirtle, S. and Frank., A. (eds.), *Spatial Information Theory—A Theoretical Basis for GIS, International Conference COSIT '97*, Lecture Notes in Computer Science 1329, pp. 15–33, Berlin: Springer-Verlag.

Hornsby, K. and Egenhofer, M., 1998, Identity-based change operations for composite objects. In Poiker, T. and Chrisman, N. (eds.), *Proceedings: Eighth International Symposium on Spatial Data Handling*, Vancouver, Canada, pp. 202–213.

Janelle, D., Klinkenberg, B. and Goodchild, M., 1998, The temporal ordering of urban space and daily activity patterns for population role groups. *Geographical Systems*, 5(1 2): 117–138.

Kyriakidis, P.C., Shortridge, A.M. and Goodchild, M.F., 1999, Geostatistics for conflation and accuracy assessment of digital elevation models. *International Journal of Geographic Information Science*, 13(7): 677–708.

Langran, G. and Chrisman, N.R., 1988, A framework for temporal geographic information. *Cartographica*, 25(3): 1–14.

Langran, G., 1992, *Time in Geographic Information Systems*, Bristol, PA: Taylor & Francis.

Li, R., 1999, Data models for marine and coastal geographic information systems. In Wright, D. and Bartlett, D. (eds.), *Marine and Coastal GIS*, New York: Taylor & Francis, pp. 25–36.

Li, R., Qian, L. and Blais, J.A.R., 1995, A hypergraph-based conceptual model for bathymetric and related data management. *Marine Geodesy*, 18: 173–182.

Lin, H., 1992, *Reasoning and Modeling of Spatiotemporal Intersection*. Unpublished Ph.D. thesis, State University of New York at Buffalo.

Lin, H. and Calkins, H.W., 1991, A rationale for spatiotemporal intersection. *Technical Papers of 1991 ACSM-ASPRS Annual Convention*, vol. 2, pp. 204–213. Baltimore, Maryland.

MacEachren, A.M., Wachowicz, M., Edsall, R., Haug, D. and Masters, R., 1999, Constructing knowledge from multivariate spatiotemporal data: Integrating geographical visualization with knowledge discovery in database methods. *International Journal of Geographical Information Science*, 13(4): 311–334.

Mark, D.M., 1979, Phenomenon-based data-structuring and digital terrain modelling. *Geo-processing*, 1, pp. 27–36.

Mark, D. and Egenhofer, M., 1998, Geospatial lifelines, In Günther, O., Sellis, T. and Theodoulidis, B. (eds.), *Integrating Spatial and Temporal Databases*. Dagstuhl-Seminar-Report No. 228, *http://timelab.co.umist.ac.uk/events/dag98/report.html*, Schloss Dagstuhl, Germany.

Mark, D., Egenhofer, M., Bian, L., Rogerson, P. and Vena, J., 1999, *Spatio-Temporal GIS Analysis for Environmental Health*. National Institute of Environmental Health Sciences, National Institutes of Health, grant number 1 R 01 ES09816-01.

Mey, M. and terHeide, H., 1997, Towards spatiotemporal planning: Practicable analysis of day-to-day paths through space and time. *Environment and Planning B*, 24(5): 709–723.

Miller, H., 1991, Modeling accessibility using space-time prism concepts within a GIS, *International Journal of Geographical Information Systems*, 5(3): 287-301.

Miller, H., 1999, Measuring space-time accessibility benefits within transportation networks: Basic theory and computational procedures. *Geographical Analysis*, 31(2): 187–212.

Odland, J., 1998, Longitudinal analysis of migration and mobility: Spatial behavior in explicitly spatial contexts. In Egenhofer, M. and Golledge, R. (eds.), *Spatial and Temporal Reasoning in Geographic Information Systems*, pp. 238–259, New York: Oxford University Press.

Peuquet, D. J., 1984. A conceptual framework and comparison of spatial data models. *Cartographica*, 21(4): 66–113.

Peuquet, D.J., 1994, It's about time: A conceptual framework for the representation of temporal dynamics in geographic information systems. *Annals of the Association of American Geographers*, 84(3): 441–462.

Peuquet, D.J. and Duan, N., 1995, An event-based spatiotemporal data model (ESTDM) for temporal analysis of geographical data. *International Journal of Geographical Information Systems*, 9(1): 7–24.

Peuquet, D.J., 2002, *Representation of Space and Time*, New York: Guilford.

Randell, D., Cui, Z. and Cohn, A., 1992, A spatial logic based on regions and connection. In Nebel, B., Rich, C. and Swartout, W. (eds.), *Principles of Knowledge Representation and Reasoning, KR '92*, Cambridge, MA, pp. 165–176.

Raper, J. and Livingstone, D., 1995, Development of a geomorphological spatial model using object-oriented design. *International Journal of Geographical Information Systems*, 9(4): 359–384.

Richards, J. and Egenhofer, M., 1995, A comparison of two direct-manipulation GIS user interfaces for map overlay. *Geographical Systems*, 2(4): 267–290.

Sharma, J., Flewelling, D. and Egenhofer, M., 1994, A qualitative spatial reasoner. In Waugh, T. and Healey, R. (eds.), *Sixth International Symposium on Spatial Data Handling*, Edinburgh, Scotland, pp. 665–681.

Smith, T.R., Su, J., Agrawal, D. and El Abbadi, A., 1993, Database and modeling systems for the earth sciences. *IEEE Data Engineering Bulletin*, 16(1): 33–37, (Special Issue on Scientific Databases).

Smith, T.R., 1994, On the integration of database systems and computational support for high-level modeling of spatio-temporal phenomena. In Worboys, M.F. (ed.), *Innovations in GIS*, pp. 11–24. Bristol, PA: Taylor & Francis.

Snodgrass, R., 1992, Temporal databases. In Frank, A., Campari, I. and Formentini, U. (eds.), *Theories and Methods of Spatio-Temporal Reasoning in Geographic Space*, Lecture Notes in Computer Science 639, pp. 22–64, Berlin: Springer-Verlag.

Snodgrass, R. and Ahn, I., 1985, A taxonomy of time in databases. *Proceedings of ACM SIGMOD International Conference on Management of Data*, pp. 236–264.

Tansel, A., Clifford, J., Gadia, S., Jajodia, S., Segev, A. and Snodgrass, R., 1993, *Temporal Databases: Theory, Design, and Implementation*, Benjamin/ Cummings: Redwood City, CA.

Thériault, M., Claramunt, C. and Villeneuve, P., 1999, A spatio-temporal taxonomy for the representation of spatial set behaviour. In Böhlen, M., Jensen, C. and Scholl, M. (eds.), *Spatio-Temporal Database Management*, Lecture Notes in Computer Sciences 1678, Berlin: Springer-Verlag.

Thériault, M., Séguin, A.M., Aubé, Y. and Villeneuve, P., 1999, A spatio-temporal data model for analysing personal biographies. In Tjao, A.M., Cammelli, A. and Wagner, R. (eds.), *Tenth International Workshop on Database and Expert Systems Applications: Spatio-Temporal Data Models and Languages*, Florence, Italy, pp. 410–418, Los Alamitos, CA: IEEE Computer Society.

Willmott, C., Raskin, R., Funk, C., Webber, S. and Goodchild, M., 1997, *Spherekit: The Spatial Interpolation Toolkit*, National Center for Geographic Information and Analysis, (*http://whizbang.geog.ucsb.edu/spherekit/*).

Winston, P.H., 1984, *Artificial Intelligence* (Second Edition), Reading, Massachusetts: Addison-Wesley.

Worboys, M.F., 1992, A model for spatio-temporal information. *Proceedings: the 5th International Symposium on Spatial Data Handling*, 2:602–611.

Worboys, M.F., 1997, *GIS: A Computing Perspective*, London: Taylor & Francis.

Worboys, M., 1998, A generic model for spatio-bitemporal geographic information. In Egenhofer, M. and Golledge, R. (eds.), *Spatial and Temporal Reasoning in Geographic Information Systems*, pp. 25–39, New York: Oxford University Press.

Wright, D.J. and Goodchild, M.F., 1997, Data from the deep: Implications for the GIS community. *International Journal of Geographic Information Science*, 11: 523–528.

Yuan, M., 1994, Wildfire conceptual modeling for building GIS space-time models. *Proceedings: GIS/LIS'94*, pp. 860–869. Re-published in *Temporal Data in Geographic Information Systems*, compiled by Frank, A., Kuhn, W. and Haunold, P. (eds.), 1995. Department of Geoinformation, Technical University of Vienna. pp.47–56.

Yuan, M., 1997, Knowledge acquisition for building wildfire representation in Geographic Information Systems. *The International Journal of Geographic Information Systems,* 11(8):723–745.

Yuan, M., 1999, Representing Geographic Information to enhance GIS support for complex spatiotemporal queries. *Transactions in GIS*, 3(2):137–160.

Yuan, M., 2001, Representing complex geographic phenomena with both Object- and Field-like Properties. *Cartography and Geographic Information Science,* 28(2): 83–96.

CHAPTER SIX

Spatial Analysis and Modeling in a GIS Environment

Arthur Getis, San Diego State University
Luc Anselin, University of Illinois, Urbana–Champaign
Anthony Lea, Compusearch
Mark Ferguson, Compusearch
Harvey Miller, University of Utah

6.1 THE LINK BETWEEN SPATIAL ANALYSIS AND GEOGRAPHIC INFORMATION SYSTEMS (GIS)

As GIS technology advances in a number of directions, people who require geo-referenced data for answers to their research questions can use its many attributes to advantage. The main attributes of GIS include the ability to manipulate spatial data in many ways, store and retrieve large amounts of data, display research outcomes on maps and charts in a variety of ways, and do all of this quickly.

More and more GIS packages include the resources to model geographic relationships and processes and to provide tools for spatial and non-spatial analyses. At a meeting of the UCGIS in Columbus in 1996, a group of GIS and spatial modelers called for a concerted effort on the part of vendors and academic researchers to better wed the needs of spatial analysts with the ever-growing technology of GIS. Since that time, considerable progress has been made toward that objective, although some might argue that the gains have been modest. In this chapter, we provide some background to the issues that surround the link between analysis and GIS, including a discussion of barriers that stand in the way of spatial analytic research (Getis). Then we raise questions that ask how interested parties connect GIS with particular research domains (Anselin). This is followed by a discussion of a "representational" research agenda (Miller). Finally, we give an example of the current use of GIS for solving a real world problem that involves the optimum location of banks (Lea and Ferguson).

6.2 SOCIETAL NEEDS FOR NEW RESEARCH EFFORTS

Behind the need to use GIS in a research environment is the stimulus provided by the concerns of society for a deteriorating physical environment and a salubrious socio-economic environment. And, in some respects, it is the technology itself, which, by its existence and development, enables researchers to raise questions that heretofore could not have been answered easily. As a byproduct of this, GIS, as they become more popular and flexible, raise questions about their own impact on society.

At meetings in Columbus (1996), Bar Harbor (1997), and Park City (1998), UCGIS delegates identified a variety of research topics that not only are of considerable concern to societies around the world, but, on the surface, appear to be the kinds of issues to which GIS technology lends itself. These include such critical issues as the study of:

- The occurrence and transmission of disease,
- The location and abatement of crime,
- The development and testing of models of water, air and other types of environmental variables, with an eye toward reducing environmental pollution and ameliorating the likely effects of climate change,
- The management of automotive and other traffic,
- The changing physical landscape,
- The social, cultural, and economic trends that manifest themselves on the landscape,
- How to improve the accessibility to and equity of opportunities and services for society in general.

6.3 BARRIERS TO RESEARCH

With such strong motivation, there is no wonder that much money, effort, and time have gone into finding ways to explore these issues as deeply as possible. In the process, however, spatial analysts have identified a number of barriers that stand in the way of real progress. These barriers have as much to do with the shortcomings of current GIS packages as they have to do with currently available research methods. The response to the problems has resulted in a UCGIS agenda that sets high priorities on the need to:

- Develop methods for handling massive spatial data sets, including the development of computationally intensive procedures,
- Analyze the spatial, non-spatial, and temporal aspects of problems simultaneously,
- Develop statistics that are suited for testing hypotheses in both markedly spatially autocorrelated and spatially heterogeneous data set environments,

- Find the geographic scales for analysis that are appropriate for the phenomena being studied, and

- Accommodate operations research, spatial interaction, and geostatistical models for the development of optimal problem solutions.

In order to overcome these research barriers software developments must be integrated into the various GIS packages. As mentioned earlier, some progress has been made in wedding spatial analysis and modeling to GIS. Such research tools as spatial autocorrelation coefficients, kriging, and econometric modeling are now included in or easily interfaced with GIS packages, especially with the ARCVIEW package. Remote sensing applications have been improved by powerful image analysis functionality in such packages as IMAGINE. Nonetheless, the barriers to successful GIS-related research mentioned above remain. Briefly, these barriers are outlined below.

6.3.1 The Large Data Set Problem

Modern data collection methods, such as remote sensing, are capable of supplying data in amounts, detail, and combinations that literally boggle the mind. The increased availability of large, georeferenced data sets, and improved capabilities for visualization, rapid retrieval, and manipulation within a GIS all point to the need for new ways to approach data analysis. In recent years, exploratory spatial data analytic techniques have been augmented by data-mining routines. They are a good example of the possibilities and problems that a new technology generates. On the one hand, data mining is able to tease out of large datasets patterns of location and behavior that previously could not have been easily identified. On the other hand, however, without theoretical justification it is not always clear if the mined patterns are legitimate views of objective reality. New theories must be devised that provide understanding of relationships that allow for meaningful large-scale data analysis.

6.3.2 The Space-Time Problem

Existing analytic methods usually focus on a problem's spatial or temporal component. Much lip service is given to the need to study the environment within a time-space framework, but the number of temporal-spatial studies is unusually low. Part of the reason for this is that each of these dimensions has its own associated research techniques and conventions. What is the spatial analog of a temporal unit given as *one day*? Surely it is not 1 mile or 1 kilometer. Clearly, there is a need to adjust our models so that sample data given in time and space components can be properly balanced and emphasized.

6.3.3 The Spatial Autocorrelation Problem

Perhaps the most often expressed need of spatial analysts is to better understand the effect of distance on spatial association and interaction. Much progress has been made in finding appropriate statistics, parameters, and tests that identify spatial dependency and heterogeneity. What have not been found, however, are suitable justifications that defend modelers' representations of distance. Further research in visualization may be the key to making progress in this area.

6.3.4 The Scale Problem

This is yet another well-recognized problem that often receives lip-service in the research literature. When the problem of selecting the appropriate scale is recognized (which in and of itself is a step forward), most researchers find it too painstakingly tedious to search for either a scale-independent solution or a set of parameters dependent on scale that adjust research results accordingly. Again, the power of GIS to help come to grips with this problem is one of the more exciting possibilities in the next decade. In addition, the related problem of zoning arrangement and its effect on results can be better understood and controlled in a powerful computing environment.

6.3.5 Accommodating a Variety of Research Needs

Most research areas have their own sets of techniques and traditions. This feature of research stands as a barrier to an all-encompassing GIS and associated software packages. For example, an operations research specialist defines and measures clustering differently than does an ecologist or epidemiologist. The theory, model, and algorithms are different, although there may be common threads—most likely in the nature of the data that are used—that market-conscious GIS providers might discover. Otherwise, spatial researchers face an even more difficult communication problem in the future.

6.3.6 The Literature in This Area

A large literature has emerged on the techniques of spatial analysis where some sort of modern technology, such as supercomputers or GIS or networked interaction, is implied. For this area of research, at the end of this chapter we have supplied the names of articles, software, and books published or updated since 1990. Items of value, but not cited in this chapter, are marked in the references with an asterisk.

6.4 GENERIC RESEARCH QUESTIONS

6.4.1 UCGIS Spatial Analysis Research Agenda

The integration of spatial analysis with GIS has come a long way since the early call to action formulated by Goodchild (1987) in the late 1980s. This was followed in the 1990s with a flurry of activity in both the academic and commercial GIS communities dealing with three main lines of inquiry: (1) conceptual outlines for the integration of GIS and spatial analysis (*e.g.*, Openshaw, 1991; Anselin and Getis, 1992; Fischer and Nijkamp, 1993; Fotheringham and Rogerson, 1994; Fischer, Scholten, and Unwin, 1996; Fischer and Getis, 1997); (2) the types of techniques and issues that should be tackled in an integrated GIS-spatial analysis framework (*e.g.*, Goodchild *et al.*, 1992; Bailey, 1994; Haining, 1994; Openshaw and Fischer, 1995); and (3) the operational implementation of such linkages (*e.g.*, Anselin, 1991; Haining *et al.*, 1996; Anselin and Bao, 1997; Zhang and Griffith, 1997; Symanzik *et al.*, 1998). In addition, several commercial GIS software packages now include fairly advanced toolboxes for spatial analysis, the most widely known examples of which are perhaps the various "extensions" to the ArcView desktop GIS from Environmental Systems Research Institute (ESRI–spatial analyst extension, network analyst extension, three-dimensional analyst extension, S+ArcView extension, *etc.*).

While a lot of progress has been made, particularly from a technical viewpoint, as we enter the 21st century, a number of new research challenges are emerging that are not satisfactorily dealt with in the current state of the art. To some extent, these research challenges are qualitatively distinct from the challenges in the late 1980s in that the growing incorporation of GIS, spatial data and "spatial thinking" into the standard toolbox of the scientist has stimulated new theoretical, methodological and computational questions, to be addressed by the geographic information "science" research community. Interestingly, at least in the social sciences, where the adoption of spatial analysis has arguably seen the least advance, the quest for the next generation spatial analysis is rooted in some fundamental shifts in the emphasis towards a focus on spatial and social interaction. One could thus argue that an important aspect of the future demand for spatial analysis will be driven by new theoretical constructs that require alternative representation and measurement, different models, and new methods for their calibration, verification or falsification.

Before going into further specifics on the research challenges to spatial analysis in the three domains of *representation*, *modeling* and *methods*, it may be useful to elaborate on some of the broader theoretical bases for these challenges, with a special focus on the social sciences. For example, in economics, a major change in recent years has been a shift away from a conceptual framework built on an "atomistic" decision maker or agent towards an explicit incorporation of "social interaction" (Akerlof, 1997). This requires a formal treatment of location (in general attribute space as well as in geographic space), distance and spatial arrangement, concepts central to GIS and spatial analysis. However, this may well imply a need for a broadening of the "G" in GIS to deal with locations other than places on the earth, more specifically, to locations in abstract spaces, which provides interesting opportunities as well as challenges for the *representation* of space.

Some further examples of emerging theoretical constructs in economics, sociology and political science may help to illustrate the types of demands for advances in spatial analysis that are being generated by mainstream (non-geography) social theory. Models of interacting agents require the specification of the range of interaction, whether it is local (one-to-one) or global (one-to-many) in so-called mean-field interaction models that are becoming common in branches of the new macro economics (*e.g.*, Aoki, 1996). Growing interest in concepts, such as social capital and neighborhood externalities in sociology and political science, require the delineation of neighborhood boundaries and the measurement of the extent of interaction between "neighbors" (*e.g.*, Borjas, 1995; Sampson, Morenoff, and Earls, 1998). Similarly, emerging models of strategic interaction in the policy analysis arena formally incorporate the extent to which actors, such as local governments, take the positions and actions of others (their "neighbors") into account in their own decision making in terms of setting tax rates, adjusting welfare controls ("race to the bottom"), adopting growth controls, and the like (*e.g.*, Brueckner, 1998). Finally, perhaps the most familiar of these theoretical developments are the ones associated with the new economic geography that builds on the work of economists such as Krugman, and where notions of location, spatial externalities, agglomeration, and spillovers take a central role (*e.g.*, Krugman, 1991).

The common theme among these theory-driven demands is a need for new *models* that explicitly incorporate the role of space and spatial interaction, and of *methods* that allow for the estimation of such models. This has resulted in a tremendous growth in the use of spatial statistics, spatial econometric techniques and GIS in the social sciences. However, it has also highlighted the limitations in the current state of the art of those techniques with respect to dealing with changing definitions of neighborhood, space-time dynamics, latent variables and the nonlinearities required by theory (Anselin and Rey, 1997).

Paralleling these theoretical motivations are the tremendous amount of spatial (and space-time) data now available in geocoded form to both social and physical scientists. This has created very practical concerns of how to store, handle and manipulate these very large data sets (the traditional domain of spatial analysis). In addition, there are new demands on how to extract meaningful information from these data sets, as illustrated by the recently exploding interest in exploratory spatial data analysis and spatial data mining.

It is against this background of both theory-driven as well as data-driven demands for advances in spatial analysis that our discussion of the research agenda will be situated.

6.4.2 Representations of Geography in Spatial Analysis

At the heart of spatial analysis is measurement and representation of geography. Formal and computational models of geographic reality strongly condition the questions we can ask and the answers we can achieve from spatial analysis. Spatial analytic questions include questions of location and extent, distribution and pattern, spatial association, spatial interaction and spatial change (see Nyerges, 1991). This suggests three components to geographic measurement and representation, namely, *geographic space, geographic objects* and *geographic relationships*. We can measure all three components within a temporal context for analyzing change.

Traditional spatial analysis greatly simplifies measurement and representation of geographic space, geographic objects and geographic relationships. The "standard model" in spatial analysis is isolated planar space containing simple (and typically one type of) geometric objects. Geographic relationships among these objects often reduce to Euclidean distance-based functions. These limited representations are understandable from a historical perspective. Scientific understanding of geographic phenomena requires analytical tools. Although traditional spatial analytic techniques can accommodate selective relaxation of the standard model, they cannot handle a wholesale expansion of all three components.

Available technologies and the changing nature of scientific questions at the end of the 20th century suggest a comprehensive re-examination of geographic representation in spatial analysis. The rise of geographic information technologies and geographic information science have improved (to put it mildly) our ability to collect and process digital geographic data that accurately represents empirical geography. Environmental problems, equity and access issues and increasingly scarce resources require integrative, global-level perspectives and tighter linkages between the human and physical sciences. In addition, more sensitive analysis of human behavior at a variety of scales requires a richer representation of the Earth as perceived and used by individuals. Geographic information science can provide the geocomputational tools required for extending geographic representation in spatial analysis. Required is a conceptual and formal framework that can suggest the proper application of currently available tools and the research frontiers for developing new tools.

The quantitative geography/spatial analysis literature contains an undercurrent of rich conceptualizations of geographic space, geographic entities and geographic relationships. Some of these elements date back to the aftermath of the so-called "quantitative revolution" in geography of the late 1950s and 1960s when conceptualizations by spatial analysts were far ahead of the geocomputational tools available. In other cases, the conceptualizations and elements exist in other, non-geographic literatures that are (somewhat disturbingly) ahead of spatial analysis with respect to treatments of geography. This includes work by computer scientists on terrain navigation and statisticians on directional and spherical statistics. Integrating these existing elements into a coherent framework will identify the research frontiers for extending representations of geography in geographic information science.

In this section of the chapter, we discuss extended representations of geography in spatial analysis that can exploit the emerging data-rich and computation-rich environment of geographic information science. We discuss these extensions using as a framework a formal theory of geographic space by Begiun and Thisse (1979). This theory highlights the central role of measurable attributes in formal representations of geographic space. The framework will highlight the more limited representation in traditional spatial analysis. We will also discuss existing efforts in the literature for achieving fuller representations of geography. This will identify the spatial analytic theories and tools that should be integrated and extended using the geocomputational techniques of geographic information science.

6.4.3 A Framework for Expanded Geographic Representation in Spatial Analysis

6.4.3.1 Pre-geographic Space

Beguin and Thisse (1979) make a clear distinction between *pregeographic space* and *geographic space*. Pregeographic space refers to fundamental spatial properties devoid of any measured attributes, *i.e.*, "empty" space. The three basic components required are a *set of places X*, a *length metric d_L* and an *area-measure μ_A*. A place is an elementary spatial unit. The set of places must consist of at least two places that are distinguishable from each other. The set can be bounded (finite) or not bounded (infinite). These correspond to the common notions of *discrete space* and *continuous space* (respectively) in spatial analysis and GIS.

Length is the fundamental dimension of pre-geographical space. This consists of a length dimension over the set of non-negative real numbers and a length metric that obeys the metric space properties of non-negativity, identity, symmetry and triangle inequality (see Smith, 1989; Worboys, 1995). We typically interpret the length metric as the distance between two locations. An area-measure is defined for these subsets through an area dimension that is the product of the length dimension with itself. If the area-measure for a subset of places is positive and finite then that subset is dimensional (*e.g.*, lines, polygons). If the area measure for a subset of places is zero, then the subset is *adimensional* (*e.g.*, points).

The preceding concepts allow a formal definition of *pregeographic space*. Pregeographical space is defined by specifying a set of places, a length-metric and an area measure. This corresponds to the intuitive definitions of "absolute" or "isotropic" space. This is the standard representation of geographic space in spatial analysis, *i.e.*, a featureless plane with uniform transportation and resource characteristics. The latter term is a misnomer since "isotropic" means properties that do not vary with direction; this is a more limited definition than intended by spatial analysts.

6.4.3.2 Euclidean Space

More precisely, the so-called "isotropic plane" in spatial analysis is typically Euclidean space. Euclidean space derives from the assumptions that the set of places is unbounded and the distance between any two places is the length of the straight line segment connecting the two; this reflects the minimal or least-cost path between those locations. The well-known Euclidean distance function provides this length.

6.4.3.3 Metric Space

Euclidean space is only one type of metric space, that is, a space that obeys the metric space properties discussed above. An infinite number of metric spaces exist; these correspond to a generalization of the Euclidean function (see Love, Morris and Wesolowsky, 1988):

$$d(\mathbf{x_i}, \mathbf{x_j}) = \left[\sum_{k=1}^{n} \left| x_i^k - x_j^k \right|^p \right]^{\frac{1}{p}}$$

(6.1)

A metric space is realized when $p \geq 1$ Euclidean space is the special case where $p = 2$. Another important special case is *Manhattan space* where $p = 1$. Love and Morris (1972, 1979) demonstrate how to estimate the parameters of equation (6.1) from observed travel distances.

If $p < 1$, the resulting space is semi-metric since the triangle inequality property no longer holds but symmetry is still maintained (Smith, 1989). Mueller (1982) provides a method for estimating semi-metric spaces from observed travel distances or travel times. If both symmetry and triangular inequality are violated, the resulting space is known as *quasi-metric*. Modeling quasi-metric is possible using other distance functions (see Smith, 1989).

Current GIS software does not accommodate non-Euclidean spaces. Yet, as suggested above, these alternative spaces are easily formalized and therefore easily represented within a GIS. Useful GIS functionality would be a toolkit that allows the spatial analyst to read in a matrix of observed distances (*e.g.*, travel times, cognitive distances) and generate the space corresponding to the observed distances. This can support analysis and visualization of geography as perceived or experienced by individuals.

An alternative approach that fits well with the geographic data handling capabilities of GIS software is to introduce non-metric properties as *attributes*, i.e., as a feature of geographic space. We will discuss these methods below.

6.4.3.4 Spherical Space

Another important type of pregeographic space is *spherical space*. Most models of pregeographic space are planar representations of the Earth's surface. The widespread use of planar space results from the limited geographic domain of most spatial analyses. Extending spatial analysis to global-scale problems requires explicit treatment of space as a sphere (Raskin, 1994).

Similar to metric planar space, spherical space assumes an infinitely dense set of locations. However, these locations are restricted to the surface of a sphere with a known center and radius. The shortest path between any two locations on a sphere is the smaller arc of the *great circle* passing through points, where a "great circle" is a circle on the surface of a sphere whose center coincides with the center of the sphere. Spherical distance functions obey the metric space properties (see Raskin, 1994).

Some, although not all, of the basic building blocks for performing spatial analysis on the sphere exist in the literature. For example, measures of central tendency, dispersion and hypothesis tests for statistical distributions on the sphere are available (Raskin, 1994; Watson, 1983). Also available are methods for interpolation, line generalization and smoothing on the sphere (Wahba, 1981; Renka, 1984; Burt, 1989). Still required are methods for line and area processes and theory of spatial autocorrelation on the sphere (see Goodchild, 1988; Raskin, 1994; Watson, 1983). Location analysts have also examined optimal facility location on the sphere (*e.g.*, Drezner and Weslowsky, 1978; Wesolowsky, 1983; Drezner, 1985) although more complete statements of a spherical (that is, global-scale) location theory are not evident.

6.4.3.5 Geographic Space

Given the three components of pregeographic space, we can build *attributes* corresponding to geographic phenomena. A *simple attribute* is a mass defined using an arbitrary (user-defined) dimension on the measurable subset of places. Two types of simple attributes are *stock attributes* and *flow attributes*. The former is the mass available at one set of places. The latter is a mass associated with two disjoint sets of places and implies movement of the mass from an "origin" set to a "destination" set. *Composite attributes* are a function of some combination of length-metric, area-measure and simple attributes. These can include shortest-path distances within a network (see Beguin and Thisse [1979] for details).

The introduction of measured attributes allows formal definition of geographic space. This is defined by specifying a set of places, a distance-length metric, an area-measure and a set of measured attributes:

$$\left(X, d_L, \mu_A, \{\mu_h : h \in H\}\right) \tag{6.2}$$

This formal definition is consistent with Brian Berry's classic "geographic matrix" characterization of spatial analysis (Berry, 1964).

6.4.3.6 Shortest Paths in Geographic Space

As noted above, distance is a central property in pregeographic space. In geographic space, the corresponding property is shortest or least-cost paths. This occurs when some measured geographic attribute conditions movement or interaction. We interpret the attribute(s) as an *interaction cost*; this can be related to natural or human-made geographic features such as terrain or traffic congestion.

Since the interaction cost exists at each place, we treat it theoretically as a field. If the interaction cost at each place is adirectional, we treat this as *cost density field*. The minimum cost path between any two places can be stated as a special type of minimization program known as a variational problem (see Angel and Hyman, 1976; Goodchild, 1977). In other cases, interaction cost is also a function of direction from each place. Instead, we must characterize interaction cost at each place through vectors that show magnitude and direction. The result is a *cost vector field* (see Tobler, 1976, 1978).

An important research frontier is extending spatial analytic methodology to encompass geographical space. Since human and physical interaction occurs in (attributed) geographic and not in (sterile) pre-geographic space, extending spatial analytical methods to account for geographic space can greatly improve their realism. Examples include restating distance-based spatial analysis methods to account for the mitigating effects of geographic space on spatial interaction and spatial dependence. Some examples include spatial interaction models (see Fotheringham and O'Kelly, 1989), distance-based spatial autocorrelation measures (see Getis and Ord, 1992) and theories of accessibility and location rent (see Kellerman, 1989a, 1989b; Miller, 1999).

Geographic information science can contribute by providing a bridge between spatial analysis and the required geocomputational tools. Some of these techniques already exist and should be linked to the appropriate spatial analytic techniques. For example, solved special cases of the minimum path through cost fields include *cost polygons* (Werner, 1968; Smith, Peng, and Gahient 1989; Mitchell and Papadimitriou, 1991), *polyhedral surfaces* such as terrain (de Berg and van Kreveld, 1997; Mitchell, Mount, and Papadimitriou, 1987) and *cost lattices* (Goodchild, 1977; van Bemmelen *et al.*, 1993). Geocomputational versions of the minimum cost path through vector fields are still required, although the basic mathematics of this approach are well known from basic physics and vector calculus.

6.4.3.7 Morphology of Geographic Space

Pregeographic space allows definition of adimensional or dimensional subsets of places. However, it is more satisfying to treat these subsets as a feature of geographic space, *i.e.*, entities that emerge from measuring geographic attributes. Measuring geographic attributes can generate two major types of geometric forms, namely, *fields* or *objects*. Field representations treat attributes as collections of continuous spatial distributions often formalized through mathematical functions of location. Object-based representation treats attributes as discrete, identifiable, georeferenced entities, each homogeneous with respect to the measured attribute (see Couclelis, 1992; or Worboys, 1995 for discussions).

Treatment of geographic morphology in spatial analysis is much more advanced for field-based representations. These representations mostly correspond to physical applications. For example, analyzing and modeling terrain and landforms has a long history in spatial analysis (*e.g.*, King, 1969; Band, 1989). These methods have increased greatly in scope and sophistication due to GIS technology (see Band, 1989).

Modeling and analysis of geographic morphology is less advanced for object-based representations. These generally correspond to human applications. Yet, there is ample theoretic and empirical evidence that morphological characteristics can influence human spatial systems. For example, the *shape* of a city indicates influences on urban growth (Medda, Nijkamp, and Rietveld, 1998). Shape can also affect internal flow and interactions across boundaries (Griffith, 1982; Arlinghaus and Nystuen, 1990; Ferguson and Kanaroglou, 1998). Shape can also be an important property in facility site selection and configuration (Miller, 1996; Cova and Church, 1999). Shape indices have a long history in spatial analysis (*e.g.*, Boyce and Clark, 1964; Bunge, 1966; Austin, 1984; Griffith *et al.*, 1986), although most were developed for modest applications. Required are comprehensive shape analysis techniques that are scalable for very large geographic databases.

6.4.3.8 Geographic Relationships

Nystuen (1963) identified three fundamental relationships in spatial analysis: i) *distance*, ii) *connectivity*, and iii) *direction*. Traditional spatial analysis implements the first two relationships in very limited forms and virtually ignores the third. An expanded representation of geographic space means that geospatial analysis should capture a fuller spectrum of these relationships.

As noted above "distance" in spatial analysis usually refers to Euclidean distance although other metric spaces are sometimes used. Much of spatial analysis treats the "distance" between geographic entities as a single, scalar value. For example, a single centroid-to-centroid measurement typically summarizes the distance relationship between geographic entities. In fact, a *distribution* of distances exists between two entities if one is dimensional (Kuiper and Paelinck, 1982; ten Raa, 1983). A single, centroid-to-centroid distance measurement is often a poor surrogate for this distribution (Hillsman and Rhoda, 1978; Miller, 1996; Francis *et al.*, 1999).

Okabe and Miller (1996) develop tractable computational techniques for computing the minimum, maximum and average distances in Euclidean space between spatial objects stored in the vector GIS format. Algorithms for the minimum and maximum distances are scalable to large databases. The average distance algorithms are more computationally complex and are limited to moderate databases. GIScience research frontiers include developing and testing average distance heuristics against the exact methods of Okabe and Miller (1996), developing minimum, maximum and average distance measures for non-Euclidean metric spaces and (of course) incorporating these measures into spatial analytic techniques.

Another research frontier is developing scalable computational procedures for calculating *expected distances*. Average distance measures assume that inter-action is equally likely between all locations in the dimensional spatial objects. However, we often have good reason to believe that interaction will not be equally likely from all locations (*e.g.*, varying population densities). Traditional expected distance functions require a known or postulated probability density function describing the likelihood of interaction from each location in an object. Analytical solutions for these functions only exist for a small subset of simple geometric objects. GIScience can contribute by developing heuristic procedures that exploit empirical data (rather than density functions) on the spatial distri-butions of attributes within objects that affect the likelihood of interaction.

In traditional spatial analysis, the concept of "connectivity" is binary, *i.e.*, two geographic entities are either connected or not, although these relationships can also be weighted (*e.g.*, equal to the nearest neighbor distance, arc length or length of the shared boundary). Egenhofer and Herring (1994) analyze possible topological relationships between points, lines and areas and discover a much larger set than previously recognized in spatial analysis. They also develop a very tractable computational representation, supporting the integration of these rich topological relationships in spatial analysis.

Directional relationships among geographic entities have received only selective attention in spatial analysis. The importance of directional relationships is well recognized in physical spatial analysis but less common in human spatial analysis. However, directions can be important in human spatial behavior; for example, individuals have directional biases in their knowledge of and movement within an urban area (Adams, 1969) and at regional and continental scales (see Egenhofer and Mark, 1995). A large toolkit of *directional statistics* is available as a subset of spherical statistics (see Jupp and Mardia, 1989). However, directional statistics only concern point objects and therefore require extension to dimen-sional geographic objects. Papadias and Egenhofer (1996) and Peuquet and Ci-Xiang (1987) develop algorithms for determining directional relationships between dimensional geographic objects. These procedures could be used independently or in conjunction with extended directional statistics to incorporate directional relationships into spatial analysis.

6.5 AN EXAMPLE: A MODELING APPROACH TO A LOCATION PROBLEM SOLVED IN A GIS ENVIRONMENT

Compusearch Micromarketing Data and Systems

6.5.1 Overview

The intent of this section is to provide an example of the types of applied analyses that take place in a GIS environment in the private sector. The first application is a spatial interaction model for retail site evaluation and sales projection. The second is a location-allocation model for chain or branch network optimization. A novel feature of the second application is that it has the first application embedded within it and that the result has been operationalized within a state of the art GIS software environment. This application has been utilized by Compusearch clients principally in the banking sector such as Bank One Corporation in the United States and Toronto-Dominion Bank in Canada. In private sector applications, this methodology is best suited for retail delivery networks (*e.g.,* bank branches, shopping centers, drug stores, grocery stores, and car dealerships). Clients are in need of such models to implement conditional forecasting, or "what-if" scenario evaluation, and to quantify the impacts of adding, dropping or altering existing branches. The interest is in how business volumes respond to marginal changes to the facility system and in how to answer broader, network-wide questions.

What is motivating the use of these modeling approaches? On the demand side, bank clients are seeking to strike a balance between reducing their dependence on "bricks and mortar" facilities while maximizing convenience to customers. There has been a definite bias toward removing or "consolidating" existing facilities that are not strong performers. Typically, clients look to these models to provide systematic, unbiased guidance in restructuring large-scale networks on a market-by-market basis. We urge our clients to consider the results of these models not as the final word but rather as a strong basis from which informed and objective decision-making can emerge. One factor on the supply side is the fact that modern computing technology is facilitating large-scale spatial interaction modeling for multiple metropolitan markets and multiple product types.

Viewed in a top-down fashion, the key objective of this methodology is network optimization. In a given market, a network of branches and Automatic Teller Machines (ATMs) for a client bank and its competitors is conceptualized as an interdependent system where deletion, addition or alteration of facilities will impact the other branches in the network. Competitor branches and ATMs are also integral parts of the system but are both more difficult to model. While the engine of the network optimization package is clearly the spatial interaction model, it is the location-allocation model that "brings the model to life" and stimulates conceiving of the branch network as a system. Progressively better solutions literally emerge on the screen as the algorithm proceeds. Many location-allocation models are deterministic in the sense that all dollars from a demand point are allocated to the closest facility or supply point. Our application is novel, and substantially more complex, in that allocations are probabilistic as implied by the structure of the underlying spatial interaction model (Hodgson, 1978; O'Kelly, 1987; Lea and Simmons, 1995).

We are not able to discuss the results of application of an actual model for a specific market for obvious reasons of propriety. However, we do include a realistic example complete with maps. We discuss some of the practical issues surrounding these types of models as well as lessons we have learned and are still learning. We describe the types of data sets that we use in building these models, the limitations of the data and how this impacts the structure of the model. We note the balance that we try to strike between theoretical rigor and producing reasonable results in time-limited circumstances. We discuss the importance of software in the process. In the private sector, it is extremely important that the client have a software tool on their desk that they can use. Software interface design is much more important than in academia.

6.5.2 Data Requirements

Many data sources are needed to build a spatial interaction model and network optimization capability for a bank. Some databases originate from the client and others from third party commercial sources and the census. Data are generally assembled and stored a market at a time.

One fundamental database is a comprehensive list of all branch facilities, including those of the competitors as well as the client. At a minimum, these data must include the full address and key attributes. We then geocode these facilities, using high quality street address based geocoding. The bank must provide the data necessary for Compusearch to construct a "dollar flow" origin-destination matrix. This database includes all household or account level dollar balances and the branches at which the accounts are domiciled. These data are then aggregated for each unique combination of demand area and branch to form product-specific flow matrices expressed both in dollar terms and number of households. Average travel speed on network links, if available, permits travel effort to be measured by travel time as opposed to drive distance and provides the best proxy for effort.

For both bank and competitor locations, high quality site and situation variables are important (see table 6.1). In our models these are considered as attractiveness attributes of the destinations or choice units. Site variables refer to characteristics of the immediate facility location. Useful site variables include the average number of tellers, ATMs and parking spaces as well as age of facility, hours of service and facility square footage. If the client can develop subjective but reasonably standardized measures of facility appearance (*e.g.*, visibility, signage) and immediate accessibility, then these are useful members of a site variable data set. Measures of the management and staff quality and experience at each branch, although difficult to obtain, are very helpful in explaining business volumes. Site variables are typically not available for competitors since a comprehensive survey is needed.

Table 6.1 Typical explanatory variables used in site selection

Category	Examples
Site variables	Number of square feet
	Number of ATMs
	Number of parking spaces
	Site visibility rating
	Years facility at present location
Situation variables	Locational context (freestanding, strip plaza, etc.
	Nearby traffic volumes
	Grocery store "potential"
	Shopping center "potential"
Socioeconomic and demographic	Average household income
	Percent population ages 65+
	Percent university educated persons
	Average number of children per household
Competition	Number of significant competitors within 2 miles
	Competition "potential"
Other	Measures of management quality

Situation variables measure the characteristics of the surrounding neighborhood or environment. Proximity of the bank facility to grocery stores, shopping centers and universities are often important determinants of a branch's success. Estimates of daytime population and the total number of employees within each block group address the issue of not having census data for daytime population that drives business volumes at many branches in downtown and industrial areas of cities. Counts of vehicular and pedestrian flows in the vicinity of a branch can also help. Finally the locations and numbers of "visitors" at recreational sports entertainment and cultural attractions can be helpful in capturing variance relating to a situation that would otherwise be elusive. The task of developing aggregate measures of situation and accessibility is a modeling process in itself and is described below.

The conceptual framework of a Spatial Interaction Model (SIM) involves allocation of dollars from demand areas (block groups in the United States and Enumeration Areas in Canada are best) to branches. The total dollars associated with these demand areas must be estimated. It is possible to use a geodemographic cluster system such as PRIZM or PSYTE to obtain the estimates. Third party data suppliers provide data on the average holdings of households by type of account in each of a set of neighborhood level geodemographic market segments. For example, all block groups belonging to cluster 17 might be assigned an average of $752 per household in savings accounts. If such demand data are unavailable, it is necessary to undertake a survey to obtain the values of dependent variables for households in order to run a regression on demographics. (Note that the client's database on its own will not constitute a representative sample.) The set of explanatory variables for such demand models is composed largely of socioeconomic and demographic census variables. The key variables for each small demand area are the number of households, income, education levels, dwelling values and tenure, ages of maintainers, number of children, employment variables and car ownership statistics. Variables that capture population density and membership of the demand area in the urban/rural hierarchy are very useful. Other variables relating to race, ethnicity, mother tongue and mobility can also help.

In order to ensure high-level consistency, a good estimate of the total dollars by account type for all facilities (including competitors) in the market is required. Even if the demand estimates are of high quality, the sum of the demands is unlikely to equal the sum of these supplies since there will be net inflows or outflows of banking dollars external to the market. In both Canada and the United States, "semi-official" data are available which accurately represent market-specific banking dollars. With a consistent measure of total market dollars, an estimate of the total draw and market share of competitors and client is possible. With all relevant dollar data, appropriate modeling and rescaling processes, and the inevitable informed assumptions, a consistent dollar data set is derived which allows the supply and demand sides to balance.

6.5.3 Structure of the Predictive Model

Most of the SIMs designed and used by Compusearch are based on a discrete choice conceptualization of the process. In particular, the multinomial logit model is utilized. Anas (1983), for example, has shown the link between logit modeling in the spatial context and the traditional types of "Wilson" models with which most geographers are familiar. The bank models discussed below are viewed as production-constrained SIM models whereby known dollar volumes associated with polygon entities representing demand areas are allocated among a series of point entities which are bank branches or perhaps ATMs. We will refer below to "dollar flows" or "dollar volumes" which in the case of total volume bank models are aggregated quantitative connections between account holders and bank branches.

The objective of the SIM is accurate characterization of the determinants of the allocation process so that observed dollar flow patterns are well reproduced and so that the model is as well behaved as possible when changes (for example, branch additions or subtractions) are made to the system. The main reason that clients choose a modeling approach is to provide a framework through which the effects of such changes can be assessed objectively. They are not typically very interested in how successful we are in explaining the process to be modeled.

While the multinomial logit model is derived from first principles of utility maximization at the individual choice-maker level, we treat the small area origins as the observational units in the model. The assumption is made that we are seeking to characterize and project the behavior of the average individual resident in a block group or enumeration area. While the assumption is not ideal, since individuals within a block group will clearly have their own set of behaviors, the results are much more realistic and robust than those that would be provided by the typical regression site evaluation model where all origins are aggregated together. Block groups or enumeration areas are the chosen demand units for these models because these are the smallest spatial units for which most census variables are available. It is possible to use demand areas that are at higher levels of census geography such as the census tract but the origins become quite large. Since all dollars are assigned to origin centroids (and not to real residential locations), the potential for error can be quite large given that a substantial proportion of bank dollar volumes often "travel" less than 1 or 2 kilometers only.

The structure of the logit model is very much affected by the absence of competitor-specific dollar data. This is a pervasive theme in the private sector. The researcher will typically have superb data on the client network but little more than locations for the competitors that typically constitute the majority of facilities in the network system. A detailed observed flow matrix can be generated for the client but at best a set of facility dollar totals, probably estimated, is available for the competitors. Competitor attractiveness attributes may be available. If not, we urge our clients to agree to a rigorous process of data collection or to purchase this service. One solution to the problem of having little data on competitors is to develop, for each origin in the model, an estimate of the "expected maximum utility" (Ben-Akiva and Lerman, 1985) provided by the competition as a group. The composite utility approach, widely used in sequential estimation of the nested logit model, is employed but it is not based on a calibration since there are no competitive flow data. Instead, plausible parameters are judiciously selected and evaluated through a heuristic algorithm that ensures the marginal facility volumes (if available) are accurately reproduced. The final composite utility term for the competition acts as a variable in the model just as it would in the nested logit model. Its associated parameter is typically estimated in the unit range (0–1) as is consistent with theory. In this conceptualization, we do not need to model the origin-level flows associated with individual competitors since it is the aggregate effect of the competition on an origin that is considered most important. In this sense, the absence of detailed competitor attribute data is not critical. If it is available or collected, then a stronger case can be made for including competitors as explicit destination choices. Since competitive flow data are absent and our origin-destination flow matrix is partially vacant, it follows that to infer total origin dollars, a separate demand modeling procedure of the type outlined earlier must be undertaken.

A particularly challenging issue in banking applications of the SIM is that many branches exist to serve workplaces. Since people often commute large distances to work, a residential address is often a poor indicator of where people choose to bank. Perhaps up to 40 percent of banking trips can in some way be related to workplaces. Unfortunately, banks do not typically gather or retain workplace addresses of their customers in their databases so we are forced to model all banking activity as if it were originating from the place of residence. One approach is to build an uncalibrated "pre-model" that judiciously re-allocates a percentage of demand areas' dollars to workplace origins and then do the calibration on the basis of the modified data set. The other is to undertake calibrations on unadjusted flow matrices. In the latter case, specification techniques such as facility-specific distance decay parameters and the inclusion of variables that capture the access to workplaces are useful. We typically choose the latter.

We generally calibrate a separate SIM for each type of financial product such as checking accounts, mutual funds or mortgages that the client wants modeled. In the ideal implementation of this model, the products would be modeled jointly but this is not considered feasible at this stage. Sometimes, we calibrate a single spatial interaction model that considers all products as a group. When we calibrate multiple models and not just the model of total dollars, each must be embedded within the location-allocation network optimization model so that they contribute jointly to the final facility pattern, even though the models were not jointly estimated.

6.5.4 Specification of the Predictive Model

Specification of these models is partially a variable creation process and partially the creation of design matrices set up for parameter estimation. With SIMs, the variable creation process is extremely important. The reason is that so many important variables must capture the spatial positioning of various entities in the system with respect to other entities. A GIS is very useful here in understanding the implications for modeling. Three branches of a certain bank can all be equal distances from a given origin but two of the banks may be extremely close together while the other is off by itself. This situation is an approximate spatial analogy of the red bus/blue bus example that is often mentioned in the discrete choice modeling literature. The typical solution in theory is the application of a nested logit model or probit model to account for the natural groupings of alternatives that occur. Given the large number of branches and origins to be modeled, neither option is particularly appealing in private sector spatial interaction modeling contexts where results must be produced in a reasonable time frame (see Figure 6.1).

Instead we have turned to the development of powerful situational variables. Situational variables typically should not be specified as simple counts (*e.g.*, number of grocery stores within x kilometers). Rather, we place considerable emphasis on variables known by the term "potentials" (akin to classical mathematical potentials) which seek to capture many of the accessibility and situational relationships associated with a particular set of point entities. For a given point in space, the aggregate potential or accessibility for the particular entity is basically a distance-discounted average of attractiveness scores for all the nearby occurrences of that entity. We will typically calculate the potentials using various assumptions about the associated parameters and then simply use the few potentials out of the larger set that work best in the calibration of the SIM.

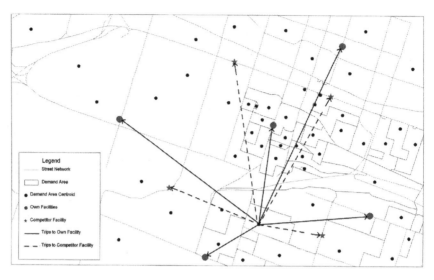

Figure 6.1 SIM model in an urban market

Banks can often increase business volumes by locating near grocery stores, for example, implying that the grocery potential score for a bank might be a useful variable. The effects of cannibilization can be assessed through an "own-banner" potential which measures the extent to which branches of one bank compete with one another. Traffic volume potentials, shopping center potentials and employment potentials are all very important. The employment potential for a branch would be inversely related to the distance of the branch from the various significant places of work and positively related to the attractiveness of these places of work in terms of the number of employees. Employment potentials are critical given our efforts to capture the workplace banking phenomenon.

In theoretical terms we are in essence using a competing destinations model (Fotheringham, 1983; Fotheringham and O'Kelly, 1989). Specialized variables are created to account systematically for the natural groupings of alternatives. In so doing it is expected that the random error components across alternatives are essentially independent as they should be in any application of the multinomial logit model. If we are successful in meeting this condition then we are justified in not using methods such as multinomial probit or nested logit models that have more theoretical appeal but less practical appeal.

Other variables are specified in a more straightforward manner. The most important variable in any spatial interaction analysis is distance (or travel time) as our experience has confirmed. It is preferable that distances be measured over the road network, or at least around barriers, as opposed to being measured in a straight line. Drive distances themselves are no small matter to create for large markets and are one of the major reasons that assembling and cleaning all the inputs for a SIM is almost as involved as *developing* the model. Typically, a non-linear transformation of distance is used which will have one or two associated parameters to be estimated. Branch attributes such as hours of operation, size, age of facility enter the specification as standard attributes of alternatives. Since the number of alternatives is typically large in these models, there is a lot of flexibility in replacing a full set of alternative-specific constants with the facility attributes instead. This is desirable from the point of view of parsimony and generating explanation. The parameters on distance can be specified in an alternative-specific fashion also with varying decays for individual branches. While this may improve fit for particular branches and address the workplace banking issue, the overall improvement in model fit may well not justify the number of extra parameters. In these situations, many distance parameters can and should be constrained to equality across subsets of branches. Various criteria can be used to specify subset membership such as location, type of clientele or nature of branch.

The third and final variable group to be considered is socio-economic and demographic characteristics of origins. While demographics strongly underpin the total demand associated with an origin, these same variables can assist in explaining how dollars are allocated across the branch network. Mostly, we use these variables to assess how demographics affect the likelihood of choosing a certain type of bank "banner" since certain banks appeal to different segments of the population. The question of how branches are chosen given that the bank has been decided upon (note that we are not trying to imply a sequential decision process here) is really represented in the models as a matter of spatial relationships and facility attributes. The choice of banks is modeled only as well as our characterization of "observed" competitor dollar volumes permits. It also suffers from aggregation. For example, average ages of household maintainers at the block group level will have much smaller variance than at the household level.

At this stage, the reader may wonder how ATMs enter into our models. Since dollars cannot be domiciled at ATMs, they must enter a "dollar balance" model as a situational attribute of a branch or as a branch-specific ATM potential. Branches that are surrounded by several of the banks' ATMs are expected to be more attractive in that they receive greater "support." In reality, ATMs are best viewed as "transactions machines." Models of transactions are considerably more complex than models of dollar balances and model performance is generally not as high. When ATMs are included as potential destinations for transactions, the number of alternatives in the models becomes so huge that alternative methodologies like regression are probably better suited. One can argue that a facility-type such as an ATM is too commoditized to justify the extensive treatment offered by a SIM.

6.5.5 Estimation and Validation of the Predictive Model

Our use of the multinomial logit model stems from the belief that the mathematical properties of the model are well understood, that specification of the models is very flexible and that estimation methodologies are well established.

McFadden (1974) had shown many years ago that the multinomial logit model in theory has a unique maximum likelihood solution. In addition, highly informative applied texts such as that of Ben-Akiva and Lerman (1985) have made a major impact. On the other hand, rigorous maximum likelihood solutions for the gravity conceptualization of the spatial interaction model have been much slower in development (Sen and Smith, 1997). There is no question that rigorous estimation methodologies have yet to make the rounds of the private sector. Even recently published books on retail site evaluation do not discuss estimation or calibration of SIMs but instead speak of simply "balancing" such models to come up with a reasonable set of parameters (Buckner, 1998, Ch. 9).

Our calibration procedure is a GAUSS implementation of the standard maximum likelihood approach to estimating a multinomial logit model (Ben-Akiva and Lerman, 1985, Ch. 5). The analytical results with regard to the gradient vector and the Hessian matrix have been coded to maximize the efficiency of the iterative procedure. A Newton-Raphson procedure is implemented to arrive at the optimal set of parameters. A useful outcome of the maximum likelihood approach is information on the statistical significance of the variables entering the spatial interaction model. Hence it is possible to do repeated estimations to refine the specification if the results seem theoretically perverse (although calibration is a computationally intensive process). The fact that GAUSS is used in the implementation is important. Large matrices are handled very efficiently and the code itself is quite compact. The inefficient looping characterizing traditional computer languages is bypassed. In our implementation we can calibrate a 30-parameter model with 1,000 origins and thirty facilities in about 5 minutes. That would be considered a relatively small market. For large metropolitan markets, the models are much larger and take longer to calibrate.

Our process to validate the SIM is not painstaking. We make sure that the signs of parameters are consistent with expectations and that the best balance between parsimony and fit is achieved. We check the percentage errors associated with the various modeled facilities to ensure that none are too far out of line. We spend time studying the GIS generated map pattern of residuals. The problem is that in a SIM, there are often several thousand residual maps that *could* be done but there simply is no time to produce and act on such maps. The biggest concern theoretically is that a model will be in violation of the Independence from Irrelevant Alternatives (IIA) property. This will manifest itself in the results of the location-allocation through "optimal" patterns that excessively cluster facilities. Rather than implementing some of the more exotic tests that have been suggested to test for violation of the IIA property (McFadden *et al.*, 1977), we have preferred to rely on experience and making sure that we do not make the same sorts of specification errors in subsequent models. Any SIM that yields perverse results is respecified and recalibrated.

In general, we find that as researchers, we tend to fret over the rigor of a particular implementation of the model more than the clients do. Not surprisingly, clients tend to be overly concerned with "bottom-line" forecasted business volumes being as close as possible to actual volumes as if this were in itself some guarantee of a good model. There is less concern among clients about accurate replication of individual dollar flows from origins to branches although this pattern of flows has much more to do with how the model is *actually* working, and indeed is the primary reason that we are using a spatial interaction model instead of a much simpler regression model.

6.5.6 The Location-Allocation Model

Although most day to day use of our models is on incremental branch-specific decisions, there is also a demand for branch network optimization writ large. The question posed is: what is the optimal number, locational configuration and set of branch attributes in a network of branches in a market? Uses of the answer to this question vary from intrinsic interest, to long-range facility system planning, to knowing whether one should be adding or reducing the number of facilities in a market. We do not recommend taking model solutions as targets in the short run.

The problem of finding the optimal branch network is known in the operations research literature as a plant location problem. In this formulation, a fixed cost is associated with establishing a facility at each location. It is this fixed cost that allows determination of the cost minimizing number of branches. There is also a linear (or concave) variable cost that relates to the processing of demand. Our clients tend to find this conceptual framework too demanding, preferring instead to make relatively simple assumptions about facility costs. Typically, clients are asked to provide data on facility construction and operating costs so that we can build a more robust model. We have yet to have a client who wanted to provide these data. Clients tend to specify the number of facilities (often by size or type), see where "the algorithm" puts them and how much total demand is attracted. A relationship our clients have found intriguing is that between market share and the number of branches in the market. Figure 6.2 shows a typical curve of this type where the marginal impacts on market share are seen to increase as the number of facilities decreases. Faced with a curve of this type, clients then brood about costs.

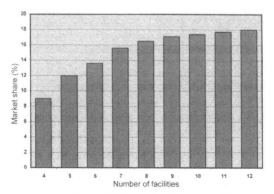

Figure 6.2 The market share trade-off

This problem's conceptualization, at its simplest, is known as the p-median problem. Typically, demand is assigned to the closest (cheapest) supply facilities but in the case of retail site evaluation, it would be extremely crude to follow this rule. Indeed, as we have already spent a good deal of effort to build a SIM-based allocation model, it is only natural to include this as the allocation model. With SIM allocation processes, location-allocation models, which are already combinatorially explosive, become even more highly non-linear. Such a model would not normally be considered a candidate for true mixed integer programming optimization. Rather, we use some powerful and robust heuristic procedures to solve these problems. We have generally used a GAUSS implementation of the now widely used Teitz and Bart (1968) heuristic algorithm. See also Densham and Rushton (1992).

This is a discrete space conceptualization wherein the analyst specifies in advance the possible locations of facilities (typically including existing locations) on the network. Starting with an initial solution of p facilities, each free location is tried as the location for an existing facility and the best such substitution that reduces the objective is made. The algorithm proceeds to try each free location in sequence and then repeats this process until no improvement can be found which reduces the objective. Although an optimal solution is not guaranteed, many researchers have reported results indicating that very good, if not optimal, solutions are routine (Church and Sorensen, 1997).

A major challenge in making these models truly realistic is properly capturing facility attractiveness. In almost all applications of p-median optimization, facilities are implicitly equally attractive; they simply process all the demand that falls upon them. With the presence of the SIM allocation model, there is a true conceptual challenge in optimization. Clearly an optimal solution features not just the best locational configuration but also the best set of facility types and attributes. The attractiveness of a facility is influenced profoundly by its size among other factors. We have tried several different approaches to this interesting specification question, but will not report details here.

As noted above, facility attractiveness is comprised of attributes that are facility and site specific (hours, size, parking, and visibility) and those that are situation specific (access to grocery stores or employment). While all attributes are taken as exogenous in the case of the SIM, facility and site specific attributes could be viewed as endogenous in the optimization, staying with the facilities as they are swapped by the heuristic. Situation attributes clearly are exogenously set and fixed to the ground.

Despite the obvious appeal of site attributes "floating" with the facility, we prefer fixing them exogenously along with the situation attributes. Approximate facility sizes or hours of operation, for example, are associated in advance with appropriate areas of a market. There are conceptual reasons for this approach and also reasons of computational tractability. Since our location-allocation model is focused on the revenue side, there are no constraints on how attributes are swapped. It is hard, for example, to keep branches with very late hours out of daytime workplace locations without taking costs into account. Given that we are utilizing a straightforward, revenue-based approach, exogenous specification of attributes makes the most sense.

6.5.7 Using the Location-Allocation Model

Two objectives are served by the network optimization modeling process. The first is to find the best locational configuration of a specified number of facilities. If there are 15 facilities currently, this might involve locating 13 or 17 in separate runs, while keeping competitor facilities fixed. The second objective is studying the effects on levels of realized demand and associated market share of quite different numbers of optimally located facilities. Both should be considered long range planning problems to distinguish them from simpler incremental what-if scenario runs of the SIM models.

It is common that the client is interested in optimizing the locations of a subset of facilities given fixed locations for several highly profitable, existing facilities. For example, of 15 current locations, 5 might be held fixed while the algorithm works on the best locations for the other 10. When 13 or 17 are attempted, the same 5 would be held fixed to obtain a consistent picture. This problem is frequently encountered because clients rarely want to "wipe the slate clean" and disregard existing facilities that they have no intention of altering. Our implementation makes it easy to hold certain locations fixed while allowing others to "float".

It is natural and desirable from the client's point of view to expect some locational "persistence" of facilities as networks expand or contract. There is no necessity for this to be the case in unconstrained formulations but we have tended to find locational persistence when the "best" network solution is judiciously selected from the candidates as N varies. Since facility age (as a surrogate for familiarity and branch maturity) often enters our SIM models positively, locational persistence may be encouraged but not determined in the optimal set of solutions. We are not yet to the point where we solve dynamic location-allocation models with SIM allocation models and non-linear constraints to show optimal system evolution.

Figures 6.3 and 6.4 show location-allocation "before and after" GIS mapped results that are typical of what we encounter, although the majority of the markets that we model in reality will have a larger number of branches. Nine optimization runs containing the full range from 4 to 12 client facilities would actually be carried out (Figure 6.2) but the given maps refer to the original and the optimal scenarios. Note the pair of branches (8 and 9) in the south of the city. These branches perform poorly and are in downscale areas. The southern area is characterized by a relatively low daytime population, as few people are employed nearby, and a large proportion of residents commute either downtown or to the workplace oriented branch (7) on the industrial western side of the city. In the optimal solution, one branch (8A) replaces the two in the south. One of the downtown branches (2) is absent in the optimal solution because the facility is located in a declining inner city neighborhood. Suburbanization has eroded much of the branch's business over the years. Some of the facilities (1, 5, and 7) remain fixed at the client's request since they are happy with those branches, while other facilities (3, 6, and 4) drift a bit from their original locations. Such drifting would not be considered as grounds for an actual change to the network. Given that the city is not growing, the

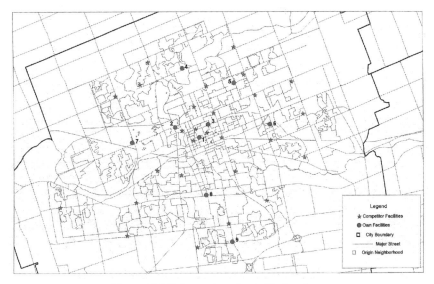

Figure 6.3 Initial scenario—nine facility locations

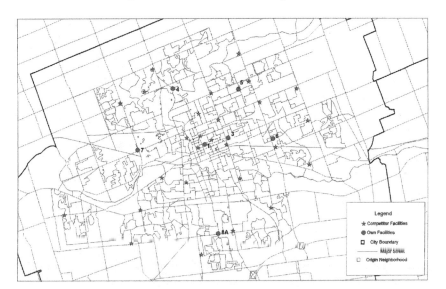

Figure 6.4 Optimal scenario—seven facility locations

accepts the idea of reducing the number of branches, especially when they consider the relationship between number of facilities and market share in Figure 6.2 and feel comfortable with the projected trade-off. The client subsequently goes on to do their own evaluations, particularly on the cost side, about the implications of removing the two facilities.

We provide other reports to assist in the process. One illustrates for each facility: the size, several other key components of attraction, the aggregate attraction score and the final realized demand. A second report shows the percentage distribution of each facility's business volumes across demand areas. This report is helpful in assessing how "predatory" each branch in the system is.

The use of the network optimization capability can be in the hands of the client but is more likely to be run by the experts at Compusearch. In the latter case, we will deliver a set of database and mapped results of what we consider to be some of the best solutions. If the client requires their own optimization capability, then the issues of software user interface come to the fore once again; having GIS generated maps of each solution is crucial.

6.5.8 Designing the Software Environment

Delivering a complex SIM application in a form that the client can efficiently utilize is an arduous task. User-friendly interfaces and programs without error messages, requirements from the client's point of view, are extremely difficult to implement and effectively double the amount of work compared to a "researcher's" version. A less taxing approach would require that we deliver hard copy reports on a particular scenario or scenarios that the client has described in advance. Although this is a straightforward approach, it has a number of restrictions. First, the client must describe, in advance, all of the scenarios to be run. Second, the results are static. Once delivered there is no way, other than redoing part of or the entire project, to modify the analysis. Finally, hardcopy reports have limited functionality to the client. In many cases these reports end up gathering dust on a shelf.

A more flexible approach, and the one we have chosen, is to develop a software system that fulfils the role of "what if" scenario tester while allowing the client to interactively manipulate the environment under study and implement marginal changes to the network. The client, in this case, has a dynamic system that can be used whenever necessary. The basic functionality includes the adding, dropping and moving of facilities and the altering of facility attributes. The package utilizes Visual Basic for all aspects of interface and Map Objects for mapping. All intensive computation is done within GAUSS as it runs in the background while working seamlessly with Visual Basic. The net result is a package that leverages the strongest points of all the components involved. The software is designed to provide access to the SIMs for several markets at once. It is possible to move quickly from market-to-market, undertaking unique scenarios in each even though the markets are not linked in any mathematical sense.

A number of design principles were decided upon early in the project and were used to shape the development of the system. The first and most important was that the software package delivered to the client would be driven by the same SIM model that also forms the basis of the optimization process. This measure would ensure that the client was provided with consistent and reliable results. Several weeks of effort were put into integrating the SIM with the "what if" software so that the model was invisible to the end user. The second principle was to develop a very focused application. This software is highly efficient at assisting the client in answering "what if" questions but does not have a great deal of extraneous functionality. For example, mapping functionality is very limited because it was assumed that if the client wanted to do maps for presentation purposes, they would use a desktop GIS package. The role of the map here is for setting up problems and understanding spatial context. A third principle was to rely heavily on the concept of the "benchmark". The software is designed so that after doing a change to the network, the client can compare the current facility allocation, both in mapped and quantitative terms, with a base allocation that acts as a reference. Generated reports, done in Crystal Reports, show clearly and precisely how the facility allocation has been altered at any given time. Ultimately, we wanted to ensure that the needs of the client were clearly understood and fulfilled. This was accomplished by involving the client in the development process. User needs analyses were performed so that both parties understood all requirements in terms of the types of analyses, user interface and reporting.

6.6 CONCLUSIONS

We have succeeded in developing a valuable set of models and software to assist analysts in designing better retail networks. Spatial interaction models, when calibrated on real-world flow data, are generally regarded as the state-of-the-art in retail site modeling. Our discrete choice implementation of this model consistently generates high quality results that serve our clients' daily need to ask and answer "what-if" type scenario questions. While we use the SIM model to answer marginal system change questions, we use a location-allocation model, with the SIM embedded as the allocation model, for a true network optimization capability. The two modeling frameworks have separate software implementations that include GAUSS as the computational engine, ESRI's Map Objects for mapping, and Visual Basic to integrate. The SIM based "what-if" software called "Branch Manager" is very user friendly and is designed to be on the desks of numerous branch network analysts. The location-allocation software called "Branch Optimizer", for which we are currently developing an improved interface, is designed for a smaller number of long-range network planners.

It is no surprise that we have encountered challenges in design and implementation. In SIM modeling, we have come up with good solutions to characterize locational context by using a system of carefully created potentials. We have devised reasonable solutions to deal with the absence of flow and attribute data for competitors. Although we envision continued progress on these fronts, our principal remaining challenge is modeling the reality of work based banking behavior in the absence of work-based addresses and demand estimates. Despite any shortcomings of our system, our clients find the functionality we have provided to be a huge leap forward, in the sense that it provides them with a solid competitive edge over other banks that are using relatively crude methods and tools.

GIS enthusiasts will have noticed that we make few explicit references to GIS functionality or technology in our description. We have concluded that it is extremely inefficient, if not impossible, to implement large-scale, complex SIM and location-allocation models using standard GIS software. Far too much time must be spent writing large volumes of code and the resulting programs are far too slow. The number of occasions in our procedure when it is efficient to undertake explicitly map-based computations (*e.g.*, in the computation of potentials) are few and far between. The best approach is to use maps for (1) user friendly visualization of the current facility system and its geographical context, (2) allowing the user a graphical and map interface for altering specific elements of the system, including locations and attributes of facilities, and (3) showing the user the results of runs. We have found that ESRI's Map Objects is ideally suited for the embedded mapping functionality in our software.

Some might think that we are not in fact using GIS to model branch networks, but we disagree; what we are doing is clearly GIS. The problems being modeled are intrinsically geographical, the data are stored in spatial databases, capturing complex spatial relationships is critical and maps are used throughout. Ours is a process that very much celebrates geography. Applying GIS clearly does not have to be done in a conventional GIS software system. In the present case, taking such an approach would be substantially inferior.

Applying our knowledge of the earth's environment can aid society in resolving increasingly difficult problems involving the efficient and equitable distribution and use of earth's resources. Geographic phenomena are related in subtle and multidimensional ways. Knowledge of spatial analytic methods and the solutions to the types of problems mentioned at the chapter's outset are required to avoid the failures that often result from uninformed research, and the mishandling and processing of georeferenced information. We firmly believe that geographic information science can fulfill its promise to help us better understand and solve the types of spatial problems outlined in this chapter.

6.7 REFERENCES (Items not cited in text are indicated by *)

Adams, J.S., 1969, Directional bias in intra-urban migration. *Economic Geography*, 45: 302–323.

Akerlof, George A., 1997, Social distance and social decisions. *Econometrica*, vol. 65: 1,005–1,027.

*Amrhein, C.G. and Reynolds, H., 1996, Using spatial statistics to assess aggregation effects. *Geographical Systems*, 3: 143–158.

Anas, A., 1983, Discrete choice theory, information theory and the multinomial logit and gravity models, *Transportation Research B*, 17: 13–23.

Angel, S. and Hyman, G.M., 1976, *Urban Fields: A Geometry of Movement for Regional Science*, London: Pion.

*Anselin, L., 1988, *Spatial Econometrics: Methods and Models*, Kluwer, Dordrecht.

*Anselin, L., 1990, What is special about spatial data? Alternative perspectives on spatial data analysis. In Griffith, D.A. (ed.), *Statistics, Past, Present and Future*, Institute of Mathematical Geography, Ann Arbor, MI: pp. 63–77.

Anselin, L., 1991, *SpaceStat: A program for the analysis of spatial data*, Department of Geography, Santa Barbara, University of California, Santa Barbara.

*Anselin, L., 1994, Local indicators of spatial association—LISA. *Geographical Analysis*, 27: 93–115.

*Anselin, L., 1998, Exploratory spatial data analysis in a geocomputational environment. In Longley, P., Brooks, S., McDonnell, R. and Macmillan, B. (eds.), *Geocomputation, A Primer*, New York: Wiley, pp. 77–94.

Anselin, L. and Bao, S., 1997, Exploratory spatial data analysis linking SpaceStat and ArcView. In Fischer, M. and Getis, A. (eds.), *Recent Developments in Spatial Analysis*, Springer-Verlag, pp. 35–59.

Anselin, L. and Getis, A., 1992, Spatial statistical analysis and geographic information systems. *Annals of Regional Science,* 26: 19–33.

*Anselin, L. and Griffith, D.A., 1988, Do spatial effects really matter in regression analysis. *Papers of the Regional Science Association*, 65, 11–34.

Anselin, L. and Rey, S., 1991, Properties of tests for spatial dependence in linear regression models. *Geographical Analysis,* 23: 112–31.

Anselin, L. and Rey, S., 1997, Introduction to the special issue on spatial econometrics. *International Regional Science Review,* 20: 1–7.

Aoki, Masanao, 1996, *New Approaches to Macroeconomic Modeling*, Cambridge: Cambridge University Press.

*Arbia, G., Benedetti, R. and Espa, G., 1996, Effects of the MAUP on Image classification. *Geographical Systems*, 3: 159–180.

*Arbia, G., 1989, *Spatial Data Configuration in Statistical Analysis of Regional Economic and Related Problems*, Kluwer, Dordrecht.

ARC/INFO, Environmental Systems Research Institute, Redlands, CA.

ArcView, Environmental Systems Research Institute, Redlands, CA.

Arlinghaus, S.L. and Nystuen, J.D., 1990, Geometry of boundary exchanges. *Geographical Review*, 80: 21–31.

Austin, R.F., 1984, Measuring and representing two-dimensional shapes. In Gaile, G.L. and Wilmott, C.J. (eds.), *Spatial Statistics and Models*, Dordrecht: D. Reidel, pp. 293–312.

Bailey, T.C., 1994, A review of statistical spatial analysis in Geographical Information Systems. In Fotheringham, S. and Rogerson, P. (eds.), *Spatial Analysis and GIS*, London: Taylor & Francis, pp. 13–44.

*Bailey, T.C. and Gatrell, A.C., 1995, *Interactive Spatial Data Analysis*, Essex: Longman Scientific and Technical.

Band, L.E., 1989, Spatial aggregation of complex terrain. *Geographical Analysis*, 21: 279–293.

*Bao, S. and Henry, M., 1996, Heterogeneity issues in local measurements of spatial association. *Geographical Systems*, 3: 1–13.

*Barnsley, M.J. and Barr, S.I., 1996, Inferring urban land-use from satellite sensor images using kernel-based spatial reclassification. *Photogrammetric Engineering and Remote Sensing*, 62: 949–58.

*Batty, M. and Longley, P., 1994, *Fractal Cities*, London: Academic Press.

Beguin, H. and Thisse, J.F., 1979, An axiomatic approach to geographical space. *Geographical Analysis*, 11: 325–341.

Ben-Akiva, M.E. and Lerman, S.R., 1985, *Discrete Choice Analysis: Theory and Application to Travel Demand*, Cambridge, Mass.: MIT Press.

Berry, B.J.L., 1964, Approaches to regional analysis: A synthesis. *Annals of the Association of American Geographers*, 54: 2–11; reprinted in Berry and Marble (1968).

Berry, B.J.L. and Marble, D.F, 1968, *Spatial Analysis: A Reader in Statistical Geography*, Englewood Cliffs, N.J.: Prentice Hall.

Besag, J., 1977, Discussion following Ripley. *Journal of the Royal Statistical Society*, B 39: 193–5.

Boots, B. and Kanaroglou, P.S., 1988, Incorporating the effect of spatial structure in discrete choice models of migration. *Journal of Regional Science*, 28: 495–507.

*Boots, B. and Getis, A., 1988, *Point Pattern Analysis*, Newbury Park, CA: Sage.

Borjas, G., 1995, Ethnicity, neighborhoods and human-capital externalities. *American Economic Review*, 85: 584–604.

Boyce, R. and Clark, W., 1964, The concept of shape in geography. *Geographical Review*, 54: 561–572.

Brueckner, J., 1998, Testing for strategic interaction among local governments: The case of growth controls. *Journal of Urban Economics*, 44: 438–67.

Buckner, R.W., 1998, *Site Selection: New Advancements in Methods and Technology*, New York: Chain Store Publishing Corp.

Bunge, W., 1966, *Theoretical Geography, Lund Studies in Geography, Series C, General and Mathematical Geography number 1*, 2nd edition.

Burt, J.E., 1989, Line generalization on the sphere, *Geographical Analysis*, 21: 68–74.

*Casetti, E., 1972, Generating models by the expansion method: applications to geographic research. *Geographical Analysis*, 4: 81–91.

Church, R.L. and Sorensen, P., 1997, Integrating mormative location models into GIS: Problems and prospects with the p-median model. In Longley, P. and Batty, M. (eds.), *Spatial Analysis: Modeling in a GIS Environment*. Cambridge: GeoInformation International.

*Clark, W.A.V., 1969, Applications of spacing models in intra-city studies. *Geographical Analysis*, 1: 391–9.

*Cliff, A.D. and Ord, J.K., 1973, *Spatial Autocorrelation*, London: Pion.

*Cliff, A.D. and Ord, J.K., 1981, *Spatial Process: Models and Applications*, London: Pion.

Couclelis, H., 1992, People manipulate objects (but cultivate fields): Beyond the raster-vector debate in GIS. In Frank, A.U., Campara, I. and Formentini, U. (eds.), *Theories and Methods of Spatio-Temporal Reasoning in Geographic Space*, Berlin: Springer, pp. 65–77.

Cova, T.J. and Church, R.L., 1999, Exploratory spatial optimization in site search: a neighborhood operator approach. *Proceedings Geocomputation '99*, Arlington, Virginia.

*Cressie, N., 1996, Change of support and the modifiable areal unit problem. *Geographical Systems*, 3: 159–80.

*Cressie, N., 1991, *Statistics for Spatial Data*, Chichester: John Wiley.

*Dacey, M.F., 1965, A review of measures of contiguity for two and k-color maps. In Berry, B.J.L and Marble, D.F (eds.), *Spatial Analyses: A Reader in Statistical Geography,* Englewood Cliffs, NJ: Prentice Hall: pp. 479–95.

de Berg, M. and van Kreveld, M., 1997, Trekking in the Alps without freezing or getting tired. *Algorithmica,* 18: 306–323.

Densham, Paul J., and Rushton, Gerard, 1992, A more efficient heuristic for solving large p-median problems. *Papers in Regional Science,* 71(3): 307–329.

Diggle, P.J., 1979, Statistical methods for spatial point patterns in ecology. In Cormack, R.M. and Ord, J.K. (eds.), *Spatial and Temporal Analysis in Ecology,* Fairland, MD: International Co-operative Publishing House: pp. 95–150.

*Diggle, P.J., 1983, *Statistical Analysis of Spatial Point Patterns,* London: Academic Press.

Donnelly, K.P., 1978, Simulations to determine the variance and edge effect of total nearest neighbor distance. In Hodder, I. (ed.), *Simulation Methods in Archaeology,* Cambridge: Cambridge University Press: pp. 91–5.

Drezner, Z., 1985, A solution to the Weber location problem on a sphere. *Journal of the Operational Research Society,* 36: 333–334.

Drezner, Z. and Wesolowksy, G.O., 1978, Facility location on a sphere. *Journal of the Operational Research Society,* 29: 997–1004.

Dubin, R., 1995, Estimating logit models with spatial dependence. In Anselin, L. and Florax, R.J.G.M. (eds.), *New Directions in Spatial Econometrics,* Berlin: Springer: pp. 229–42.

*Eastman, J.R., 1993, *IDRISI: A Geographical Information System,* Worcester, MA: Clark University.

Egenhofer, M.J. and Herring, J.R., 1994, Categorizing binary topological relations between regions, lines and points in geographic databases. In Egenhofer, M., Mark, D.M. and Herring, J.R., (eds.), *The 9-intersection: Formalism and its Use for Natural-language Spatial Predicates,* National Center for Geographic Information and Analysis Technical Report 94-1, pp. 1–28.

Egenhofer, M.J. and Mark, D.M., 1995, Naïve geography, National Center for Geographic Information and Analysis Technical Report 95-8.

Faulkenberry, G.D. and Garoui, A., 1991, Estimating a population total using an area frame. *Journal of the American Statistical Association,* 86: 445–9.

Ferguson, M.R. and Kanaroglou, P.S., 1998, Representing the shape and orientation of destinations in spatial choice models. *Geographical Analysis,* 30: 119–137.

*Fischer, M.M., Gopal. S., Staufer, P. and Steinnocher, K., 1997, Evaluation of neural pattern classifiers for a remote sensing application. *Geographical Systems 4.*

*Fischer, M.M. and Getis, A. (eds.), 1997, *Recent Developments in Spatial Analysis: Spatial Statistics, Behavioural Modelling, and Computational Intelligence*. Heidelberg: Springer.

Fischer, M. and Getis, A. 1997, *Recent Developments in Spatial Analysis*, Berlin: Springer-Verlag,.

Fischer, M. and Nijkamp, P., 1993, *Geographic Information Systems, Spatial Modelling and Policy Evaluation*, Berlin: Springer-Verlag.

Fischer, M., Scholten, H. and Unwin, D., 1996, *Spatial Analytical Perspectives on GIS in Environmental and Socio-Economic Sciences*, London: Taylor & Francis.

*Foster, S.A., Gorr, W.L., 1986, An adaptive filter for estimating spatially varying parameters: application to modeling police hours in response to calls for service. *Management Science,* 32: 878–89.

Fotheringham, A.S., 1983, A new set of spatial interaction models: The theory of competing destinations. *Environment and Planning A*, 17: 213–30.

*Fotheringham, A.S., Densham, P.J. and Curtis, A., 1995, The zone definition problem in location-allocation modeling. *Geographical Analysis*, 27: 60–77.

Fotheringham, A.S. and O'Kelly, M.B., 1989, *Spatial Interaction Models: Formulations and Applications*, Dordrecht: Kluwer Academic.

Fotheringham, A.S. and Rogerson, P., 1994, *Spatial Analysis and GIS*, London: Taylor & Francis.

*Fotheringham, A.S. and Wong, D.W.S., 1991, The modifiable areal unit problem in multivariate statistical analysis. *Environment and Planning A,* 23: 1,025–1,044.

Francis, R.L., Lowe, T.J., Rushton, G. and Rayco, M.B., 1999, A synthesis of aggregation methods for mulifacility location problems: Strategies for containing error. *Geographical Analysis*, 31: 67–87.

*Gamma Design Software, 1995, *GS+ Geostatistics for the Environmental Sciences*, Plainwell, MI: Gamma Design Software.

*Gatrell, A.C., Bailey T.C., Diggle, P.J. and Rowlingson, B.S., 1996, Spatial point pattern analysis and its application in geographical epidemiology. *Transactions Institute of British Geographers*, 21: 256–74.

*Getis, A., 1995, Spatial filtering in a regression framework: experiments on regional inequality, government expenditures, and urban crime. In Anselin, L. and Florax, R.J.G.M. (eds.), *New Directions in Spatial Econometrics*. Berlin: Springer: pp. 172–188.

*Getis, A. and Boots, B., 1978, *Models of Spatial Processes*. Cambridge: Cambridge University Press.

*Getis, A. and Franklin, J., 1987, Second-order neighborhood analysis of mapped point patterns. *Ecology*, 68: 473–7.

*Getis, A. and Ord, J.K., 1992, The analysis of spatial association by use of distance statistics. *Geographical Analysis*, 24, 189–206.

*Getis,. A, Ord, J.K., 1996, Local spatial statistics: An overview. In Longley, P. and Batty, M. (eds.), *Spatial Analysis Modelling in a GIS Environment*, Cambridge: GeoInformation International, pp. 269–285.

*Gong, P. and Howarth, P.J, 1990, An assessment of some factors influencing multispectral land cover classification. *Photogrammetric Engineering and Remote Sensing*, 56: 597–603.

Goodchild, M.F., 1977, An evaluation of lattice solutions to the problem of corridor location. *Environment and Planning A*, 9: 727–738.

Goodchild, M.F., 1987, A spatial analytical perspective on Geographical Information Systems, *International Journal Systems,* 1: 327–334.

Goodchild, M.F., 1988, The issue of accuracy in global databases. In Mounsey, H. and Tomlinson, R. (eds.), *Building Databases for Global Science*, London: Taylor & Francis.

*Goodchild, M.F, and Gopal, S., 1989, *Accuracy of Spatial Databases*. London: Taylor & Francis.

Goodchild, M.F., Haining, R. and Wise S., 1992, Integrating GIS and spatial analysis—Problems and possibilities. *International Journal of Geographical Information Systems,* 6. 407–423.

*Green, M, Flowerdew, R., 1996, New evidence on the modifiable areal unit problem. In Longley, P. and Batty, M. (eds.), *Spatial Analysis Modelling in a GIS Environment*, Cambridge: GeoInformation International.

*Griffith, D.A., 1996, Computational simplifications for space-time forecasting within GIS: the neighbourhood spatial forecasting model. In Longley, P. and Batty, M. (eds.), *Spatial Analysis Modelling in a GIS Environment*, Cambridge: GeoInformation International.

*Griffith, D.A., Haining, R. and Arbia, G., 1994, Hetrogeneity of attribute sampling error in spatial data sets. *Geographical Analysis*, 26: 300–320.

*Haining R.P., 1990, *Spatial Data Analysis in the Social and Environmental Sciences*. Cambridge: Cambridge University Press.

Haining, R., 1994, Designing spatial data analysis modules for Geographical Information Systems. In Fotheringham, S. and Rogerson, P. (eds.), *Spatial Analysis and GIS*, London: Taylor & Francis, pp. 45–63.

*Haining, R.P., 1995, Data problems in spatial econometric modeling. In Anselin, L. and Florax, R.J.G.M. (eds.), *New Directions in Spatial Econometrics*, Berlin: Springer: pp. 156–71.

Haining, R., Ma, J. and Wise, S. 1996, Design of a software system for interactive spatial statistical analysis linked to a GIS. *Computational Statistics*, 11: 449–66.

*Haslett, J., Bradley, R., Craig, P., Unwin, A. and Wills, G., 1991, Dynamic graphics for exploring spatial data with application to locating global and local anomalies. *The American Statistician*, 45: 234–42.

*Haslett, J., Wills, G. and Unwin, A., 1990, SPIDER—An interactive statistical tool for the analysis of spatially distributed data. *International Journal of Geographic Information Systems*, 4: 285–96.

Hillsman, E.L. and Rhoda, R., 1978, Errors in measuring distances from populations to service centers. *Annals of Regional Science*, 12(3): 74–88.

Hodgson, M.J., 1978, Toward more realistic allocation in location allocation models. *Environment and Planning*, 10: 1,273–1,280.

*Holt, D., Steel, D.G., and Tranmer, M., 1996, Area homogeneity and the modifiable areal unit problem. *Geographical Systems 3*, pp. 181–200.

*Hunt, L. and Boots, B., 1996, MAUP effects in the principal axis factoring technique. *Geographical Systems,* 3: 101–122.

Jupp, P.E. and Mardia, K.V., 1989, A unified view of the theory of directional statistics, 1975-1988. *International Statistical Review*, 57: 61–294.

Kellerman, A., 1989a, Agricultural location theory, 1: Basic models. *Environment and Planning A*, 21: 1,381–1,396.

Kellerman, A., 1989b, Agricultural location theory, 2: Relaxation of assumptions and applications. *Environment and Planning A*, 21: 1,427–1,446.

*Knox, E.G., 1964, The detection of space-time interactions. *Applied Statistics*, 13: 25–9.

Krugman, P., 1991, Increasing returns and economic geography. *Journal of Political Economy*, 99: 483–99.

Kuiper, J.H. and Paelinck, J.H.P., 1982, Frequency distribution of distances and related concepts. *Geographical Analysis*, 14: 253–259.

Lea, A.C., and Simmons, J., 1995, *Location-Allocation Models for Retail Site Selection: The N Best Sites in the Toronto Region. Research Report Number 1,* Centre for the Study of Commercial Activity, Ryerson Polytechnic University, Toronto, Canada.

*Longley, P. and Batty, M., 1996, Analytical GIS: The future. In Longley, P. and Batty, M. (eds.), *Spatial Analysis Modelling in a GIS Environment*, Cambridge: GeoInformation International.

Love, R.F. and Morris, J.G., 1972, Modelling inter-city road distances by mathematical functions. *Operational Research Quarterly*, 23: 61–71.

Love, R.F. and Morris, J.G., 1979, Mathematical models of road distances. *Management Science*, 25: 130–139.

Love, R F., Morris, J.G. and Wesolowsky, G.O., 1988, *Facility Location: Models and Methods*, New York: North-Holland.

*Marceau, D.J., Gratton, D.J., Fournier, R.A. and Fortin, J.P., 1994, Remote sensing and the measurement of geographical entities in a forest environment: The optimal spatial resolution. *Remote Sensing of the Environment*, 49: 105–117.

*Marshall. R.J., 1991, A review of methods for the statistical analysis of spatial patterns of disease. *Journal of the Royal Statistical Society A*, 154: 421–441.

*McMillen, D.P., 1992, Probit with spatial autocorrelation. *Journal of Regional Science*, 32: 335–48.

Medda, F., Nijkamp, P. and Rietveld, P., 1998, Recognition and classification of urban shapes. *Geographical Analysis*, 30: 304–314.

*Mertes, L.K., Daniel, D.L., Melack, J.M., Nelson, B., Martinelli, L.A. and Forsberg, B.R., 1995, Spatial patterns of hydrology, geomorphology, and vegetation on the floodplain of the Amazon River in Brazil from a remote sensing perspective. *Geomorphology*, 13: 215–32.

*Michaelsen, J., Schimel, D.S., Friedl, M.A., Davis, F.W.and Dubayah, R.C., 1996, Regression tree analysis of satellite and terrain data to guide vegetation sampling and surveys. *Journal of Vegetation Science*, 5: 673–686.

Miller, H.J., 1996, GIS and geometric representation in facility location problem. *International Journal of Geographical Information Systems*, 10: 791–816.

Miller, H.J., 1999, Measuring space-time accessibility benefits within transportation networks: Basic theory and computational methods. *Geographical Analysis*.

Mitchell, J.S.B., Mount, D.M. and Papadimitriou, C.H., 1987, The discrete geodesic problem. *SIAM Journal of Computing*, 16: 647–668.

Mitchell, J.S.B. and Papadimitriou, C.H., 1991, The weighted region problem: Finding shortest paths through a weighted planar subdivision. *Journal of the Association for Computing Machinery*, 38: 18–73.

*Morrison, A.C., Getis, A., Santiago, M., Rigau-Peres, J.G. and Reiter, P., 1996, *Exploratory Space-time Analysis of Reported Dengue Cases during an Outbreak in Florida, Puerto Rico, 1991–1992.* Manuscript, Dengue Branch, Centers for Disease Control and Prevention, San Juan, PR.

Mueller, J.-C., 1982, Non-Euclidean geographic spaces: Mapping functional distances, *Geographical Analysis*, 14: 189–203.

Nyerges, T.L., 1991, Analytical map use. *Cartography and Geographic Information Systems*, 18: 11–22.

Nystuen, J.D., 1963, Identification of some fundamental spatial concepts. *Papers of the Michigan Academy of Science, Arts and Letters*, 48, 373-384; reprinted in Berry and Marble, 1968.

*O'Loughlin, J. and Anselin, L., 1991, Bringing geography back to the study of international relations: Dependence and regional context in Africa, 1966-1978. *International Interactions*, 17: 29–61.

*Okabe, A., Boots, B. and Sugihara, K., 1992, *Spatial Tesselations: Concepts and Applications of Voronoi Diagrams*, New York: John Wiley.

Okabe, A. and Miller, H.J., 1996, Exact computational methods for calculating distances between objects in a cartographic database. *Cartography and Geographic Information Systems*, 23: 180–195.

*Okabe, A. and Tagashira, N. 1996, Spatial aggregation bias in a regression model containing a distance variable. *Geographical Systems*, 3: 77–100.

O'Kelly, M.B., 1987, Spatial interaction based location-allocation models. In Ghosh, A. and Rushton, G. (eds.), *Spatial Analysis and Location-Allocation Models*, New York: Van Nostrand Reinhold, pp. 302–326.

*Oliver, M.A. and Webster, R., 1990, Kriging: A method of interpolation for geographical information systems. *International Journal of Geographic Information Systems* 4: 313–32.

Openshaw, S., 1991, Developing appropriate spatial analysis methods for GIS. In Maguire, D., Goodchild, M.F. and Rhind, D. (eds.), *Geographical Information Systems: Principles and Applications*, Vol 1, London: Longman, pp. 389–402, 1991.

*Openshaw, S. 1996, Developing GIS-relevant zone-based spatial analysis methods. In Longley, P. and Batty, M. (eds.), *Spatial Analysis Modelling in a GIS Environment*, Cambridge: GeoInformation International

Openshaw, S. and Fischer, M., 1995, A framework for research on spatial analysis relevant to geo-statistical information systems in Europe, *Geographical Systems*, 2: 325–337.

*Openshaw, S. and Taylor, P., 1979, A million or so correlation coefficients: Three experiments on the modifiable areal unit problem. In Wrigley, N. and Bennett, R. J. (eds.), *Statistical Applications in the Spatial Sciences*. London: Pion.

*Ord, J.K., 1975, Estimation methods for models of spatial interaction. *Journal of the American Statistical Association*, 70: 120–6.

*Ord, J.K. and Getis, A., 1993, Local spatial autocorrelation statistics: Distributional issues and an application. *Geographical Analysis*, 27: 286–306.

*Paelinck, J. and Klaassen, L. 1979, *Spatial Econometrics*, Farnborough: Saxon Hill.

Papadias, D. and Egenhofer, M., 1996, *Algorithms for Hierarchical Spatial Reasoning*, National Center for Geographic Information and Analysis Technical Report 96-2.

Peuquet, D.J. and Ci-Xiang, Z., 1987, An algorithm to determine the directional relationship between arbitrarily-shaped polygons in the plane. *Pattern Recognition*, 20: 65–74.

Pielou, E.C., 1977, *Mathematical Ecology*, New York: Wiley-Interscience.

Raskin, R.G., 1994, *Spatial Analysis on a Sphere: A Review*, National Center for Geographic Information and Analysis Technical Report 94-7.

Renka, R.J., 1984, Interpolation of data on the surface of a sphere. *Association for Computing Machinery Transactions on Mathematical Software*, 10: 417–436.

*Ripley, B.D., 1977, Modeling spatial patterns. *Journal of the Royal Statistical Society B*, 39: 172–94.

*Ripley, B.D., 1981, *Spatial Statistics*, New York: John Wiley.

*S+SpatialStats, 1996, Seattle: MathSoft, Inc.

Sampson, R.J., Morenoff, J. and Earls F., 1998, *Beyond social capital: neighborhood mechanisms and structural sources of collective efficacy for children.* Working Paper, Department of Sociology, University of Chicago.

Sen, A. and T.E. Smith, 1995, *Gravity Models of Spatial Interaction Behavior*, New York: Springer-Verlag.

Smith, T.E., 1989, Shortest-path distances: An axiomatic approach. *Geographic Analysis*, 21: 1–31.

Smith, T.R., Peng, G. and Gahinet, P., 1989, Asynchronous, iterative, and parallel procedures for solving the weighted-region least cost path problem. *Geographical Analysis*, 21: 147–166.

Symanzik, J., Kötter, T., Schmelzer, S., Klinke, S., Cook, D. and Swayne, D. F., 1998, Spatial data analysis in the dynamically linked ArcView/XGobi/XploRe environment. *Computing Science and Statistics*, Vol. 29.

Teitz, Michael B. and Bart, P., 1968, Heuristic methods for estimating the generalized vertex median of a weighted graph, *Operations Research*, 16: 955–961.

ten Raa, T., 1983, Distance distributions. *Geographical Analysis*, 15: 164–169.

Thompson, S.K., 1992, *Sampling*, New York: Wiley.

*Tiefelsdorf, M. and Boots, B., 1994, The exact distribution of Moran's I. *Environment and Planning A*.

Tobler, W.R., 1976, Spatial interaction patterns. *Journal of Environmental Systems*, 6: 271–301.

Tobler, W.R. 1978, Migration fields. In Clark, W.A.V. and Moore, E. G. (eds.), *Population Mobility and Residential Change*, Northwestern University Studies in Geography Number 25, pp. 215–232.

*Tobler, W.R., 1979, Smooth pycnophylactic interpolation for geographic regions. *Journal of the American Statistical Association*, 74: 519–36.

*Tobler, W.R., 1989, Frame independent spatial analysis. In Goodchild, M.G. and Gopal, S. (eds.), *The Accuracy of Spatial Databases*. London: Taylor & Francis: pp. 115–22.

*Upton, G.J. and Fingleton, B., 1985, *Spatial Statistics by Example, Vol. 1: Point Pattern and Quantitative Data*, Chichester: John Wiley.

van Bemmelen, J., Quak, W., van Hekken, M. and van Oosderom, P., 1993, Vector vs. raster-based algorithms for cross-country movement planning. *Proceedings, Auto-Carto 11*, pp. 304–317.

Wahba, G., 1981, Spline interpolation and smoothing on a sphere. *SIAM Journal of Scientific Statistical Computation*, 2: 5–16.

*Wartenberg, D., 1985, Multivariate spatial correlation: A method for exploratory geographical analysis. *Geographical Analysis*, 17: 263–83.

Watson, G.S., 1983, *Statistics on Spheres*, New York: John Wiley and Sons.

Werner, C., 1968, The law of refraction in transportation geography: Its multi-variate extension. *Canadian Geographer*, 12: 28–40.

Wesolowsky, G.O., 1983, Location problems on a sphere. *Regional Science and Urban Economics*, 12: 495–508.

Worboys, M.F., 1995, *GIS: A Computing Perspective*, London: Taylor & Francis.

Wrigley, N., Holt, T., Steel, and Tranner, M., 1996, Analysing, modelling, and resolving the ecological fallacy. In Longley, P. and Batty, M. (eds.), *Spatial Analysis Modelling in a GIS Environment*, Cambridge: GeoInformation.

Zhang, Z. and Griffith, D., 1997, Developing user-friendly spatial statistical analysis modules for GIS: An example using ArcView, *Computers, Environment and Urban Systems*, 21: 5–29.

CHAPTER SEVEN

Research Issues on Uncertainty in Geographic Data and GIS-Based Analysis

A-Xing Zhu, University of Wisconsin–Madison

7.1 INTRODUCTION

Increasingly, policy decision-making involves the use of geographic data and geographic information systems (GIS) techniques. For example, making detailed policy decisions on how to preserve the Florida Everglades calls for detailed analyses of the environment using state-of-the-art GIS technology and geographic data about the Everglades. Location and allocation of urban resources (such as transportation planning, fire station location and fire truck routing, school zoning, *etc.*) often employ the use of GIS techniques and geographic data. The reliability of the resulting policy decisions very much depends on the quality of geographic data used for reaching these decisions since the quality of data affects the quality of decisions and the evaluation of decision alternatives.

Geographic data are often used under the assumption that they are free of errors. The beguiling attractiveness, the high aesthetic quality of cartographic products from GIS, the analytical capability and the high precision that GIS can provide further contribute to an undue credibility, at times, of these products (Abler, 1987, p. 305). However, undeserved and inappropriate acceptance of the accuracy of these data is often not warranted due to the complex nature of geographic data and the inability to adequately sample geographic features and phenomena (Goodchild and Gopal, 1989, pp. xii-xiii). Error-laden data, used without consideration of its intrinsic uncertainty, has a high probability of leading to inappropriate decisions.

Although quality of geographic data in GIS and the propagation of errors through GIS analyses have received much attention recently (Goodchild and Gopal, 1989; NCGIA, 1989; Lunetta *et al.*, 1991; Guptill and Morrison, 1995; Lowell and Jaton, 1999; Mowrer and Congalton, 2000; Hunsaker *et al.*, 2001), the current state of GIS technology in dealing with these issues falls far short of the goal described by Goodchild (1993, p. 98) in which: i) each object in a GIS database would carry information describing its accuracy; ii) every operation or process within a GIS would track and report error; and iii) accuracy measures would be a standard feature of every product generated by a GIS. In addition, little has been done to examine the effect of accuracy of geographic data and errors in GIS analyses on the outcome of decision making process involving geographic data and GIS analyses.

The term data quality encompasses two aspects. The first is data accuracy that is related to data production. The other is fitness-for-use that is the suitability of the data for a particular application. The two aspects are closely related but also different. For instance, the poor accuracy of a digital elevation model (DEM) over a study area would cause the DEM to be unsuitable for flow routing analysis in a hydrological application. Meanwhile, an accurate yield map would not be suitable for determining direction flow due to its little relevance to the application. On the other hand, improper use of the geographic data can introduce error into the results derived from GIS analyses. For example, the use of a vegetation data layer, designed to portray spatial vegetation patterns over a large area, for determining the habitats of certain grassland birds over a small area (large scale) would lead to a wrong conclusion about the relationship between vegetation types and the bird habitats, due to the scale incompatibility between this coarse vegetation data layer and other detailed data layers (such as terrain attributes derived from high resolution DEM).

This chapter highlights the research issues to be addressed on data accuracy in GIS. The fitness-for-use will only be discussed when it is related to the issue of data accuracy. The chapter first clarifies terminology used in this discussion and provides a general framework under which to discuss the research issues in data accuracy. Sections 7.3 through 7.4 discuss the research issues.

7.2 BACKGROUND

7.2.1 Nature of Geographic Data

Geographic data are unique. A datum about a geographic feature contains three different kinds of attributes: thematic attributes (describing the thematic properties of a geographic feature), spatial attributes (including the location and its spatial relationship with other features), and temporal attributes (the temporal coordinates associated with the status of the feature). For example, a datum about a land parcel can be the crop yield of the parcel (as thematic attributes), the location and spatial extent of the parcel (the location attributes) and its relationships with its surrounding parcels (spatial relationships), and the date of creation of this parcel and the season for which a particular yield was observed (temporal attributes).

Geographic data are observations of geographic features or phenomena (referred to as geographic reality). Geographic reality often cannot be exhaustively measured since it is next to impossible to obtain measurements for every point over a landscape. It is also difficult to measure geographic reality accurately due to its continuous (slow or rapid) variation over time and due to the limitations of instruments, financial budgets, and human capacity. In these respects, geographic data at creation are only approximations of the geographic reality. Furthermore, the basic schemes (Couclelis, 1992) used to represent geographic data in GIS are not dynamic and deal only with a static, invariable world. They do not deal with complex objects that may consist of interacting parts, or display variation at many different levels of details over space and over time. A discrepancy therefore exists between geographic data and the reality these data are intended to represent. This discrepancy propagates through, and may be further amplified by, spatial data management and analyses in a GIS environment. Thus uncertainty is an integral part of results from GIS analyses.

7.2.2 Elements of Data Accuracy

The following terms are used in the context of accuracy of geographic data: accuracy, precision, resolution, consistency, and completeness. Each of these terms captures different aspects of data accuracy (Veregin, 1999; Mowrer and Congalton, 2000). The term "accuracy" is the closeness of the attribute value in GIS database to the actual attribute value for a given geographical entity/ phenomena. Error is the inverse of accuracy and defined as the discrepancy between the attribute value in the database and the actual attribute value. The respective discrepancies in thematic, spatial, and temporal attributes can be expressed as errors in each of the attribute types (Veregin, 1999). Thematic and spatial errors have been extensively discussed by many authors (*e.g.*, Blakemore, 1983; Chrisman, 1991; Goodchild, 1991; Leung and Yan, 1998; Shi and Liu, 2000; Seong and Usery, 2001), but errors in temporal attribute receive much less attention and deserve some elaboration here. According to Veregin (1999), temporal accuracy is distinct from "currentness". The former refers to the degree of agreement between the recorded and the actual temporal coordinates while the latter refers to the degree to which the database is up to date. A perfect temporal accuracy does not necessarily mean the database is "current" or "up to date". An out of date database may not necessarily have a poor temporal accuracy. For example, suppose a land parcel database records, among other things, the ownership of land parcels and ownership changes over time. Suppose a given land parcel changed ownership on May 28, 1994, and the database recorded the change as such, so the temporal accuracy for this change is perfect. If all changes prior to 1996 were recorded accurately, we would think the database is temporally accurate. Suppose that the ownership of the land parcel changed again on June 20, 1996. The database has not been updated to include this change. The database is out of date but still accurate in terms of temporal coordinates it contains. If this database is used to answer the query of land ownership changes between 1994 and 1995, we would obtain an accurate answer from the information in the database. However, if the query is changed to "land ownership changes since 1994", we would consider the database is unsuitable for this query since the database is not current. Thus, we consider "currentness" is more of an issue of fitness-for-use, rather than an issue of accuracy.

Precision refers to the degree of detail that can be recorded for the value of a given attribute in database. For example, the elevation of a point is one-third meter. At low precision it might be represented as 0.3 while at high precision it might be represented as 0.3333. However, high precision does not mean high accuracy since we may be able to represent an attribute value at very high precision but the value itself can be inaccurate compared to what actually is.

Resolution, particularly spatial resolution, refers to the level of spatial detail that can be represented in database. It is often used in the context of raster representation and remote sensing imagery. Spatial resolution is inversely related to the size of pixel. The level of spatial details that can be discerned increases as the spatial resolution increases (as the pixel size decreases). In the vector domain, the term "scale" is used to imply the level of spatial details which are retained in the database although the concept of scale is much more related to cartographic generalization needed to present geographic information on a piece of paper or similar media. In fact, the concept of "scale" does not apply when it comes to the representation of geographic entities in the digital world since adjacency on digital storage media has no meaning to adjacency in physical space. So, spatial resolution is more appropriate term to use when referring to the level of spatial details the database is representing. For example, in representing a coastline in a digital database, it is much more meaningful to indicate the minimum length of line segment at which the coastline is represented. Normally, high spatial resolution would result in high accuracy. However, it is not always true. For example, if one re-samples a remote image acquired at 1 kilometer by 1 kilometer resolution (such as imagery produced by the National Oceanic and Atmospheric Administration's [NOAA's] Advanced Very High Resolution Radiometer [AVHRR] sensor) to produce an image at 30 meter by 30 meter resolution, one would not increase the spatial accuracy since the result image inherits the spatial resolution of the original image.

Consistency refers to the conformance of representation of geographic features with a prescribed set of rules or guidelines across the area covered by the database. For example, it is a common practice, in remote sensing, to mosaic images of different dates to provide a spatially complete coverage of an area due to the limited cloud free days. This practice will work okay for phenomena that are not dynamic. It would make the mosaic inconsistent if the features to be represented are highly dynamic since the images at different dates capture the different status of the features at different times. For example, if an early spring image of an area is mosaic with a late spring of image of a neighboring area in the mid latitude for monitoring crop yield, then the information on the two images are inconsistent since one was taken at early spring when the crop was just starting to sprout while the other image is showing the crop at the time when it has grown quite a bit. Another example is topological consistency where certain topological rules should be consistently applied across the entire representation domain (Kainz, 1995; Veregin, 1999). However, high consistency does not mean high accuracy but databases of high accuracy must maintain a high consistency.

Completeness refers to the degree at which the database coverage meets the set of content guidelines (specification) prescribed for the database. Data completeness includes the following aspects: entity completeness, attribute completeness, and value completeness (Brassel *et al.*, 1995). Entity completeness refers to the level at which the entities specified in the guidelines have been included. This encompasses two aspects. First is the degree to which different kinds of entities specified in the guidelines (specification) are included. Second is the level at which entities of same kind are included. The attribute completeness refers to the degree to which relevant attributes for each type of entities are included, and value completeness means the level at which attribute values are present for all attributes (Veregin, 1995). Data completeness includes both data omission and data commission (redundancy). Data completeness is not only a data accuracy issue but also a fitness-for-use issue. For example, a highway network data layer may be considered complete based on some prescribed guidelines by the transportation department but may not be suitable to be used as a street network layer for conducting population census.

These different elements of spatial data accuracy are not independent of each other but interconnected. Change in one element could also lead to change in another, which makes spatial data accuracy a very complex issue. Some elements of accuracy can be easily detected and corrected while others are difficult to detect and measure, and very expensive to correct.

7.2.3 Uncertainty as an Expression of Data Accuracy

Often, accuracy of a geographic data layer is reported as a single value that is computed from the discrepancies between the recorded and observed values at few locations over the map area and used to summarize the overall accuracy over the map area. What we need to know is the accuracy of every feature at every location. It is difficult, if not impossible, to report the accuracy of every feature at every location since the true value for every geographic feature or phenomenon represented in a geographic data set is rarely determinable and the exact value of error for every feature or phenomenon cannot be obtained in most cases. Furthermore, if the true value for every feature at every location were known, there would not be a need to compute the error value since the data layer is perfectly accurate. Thus the term "error" is misleading when it is used in the context of data accuracy for geographic data. Instead, the term "uncertainty" is used to express the degree, not the actual value, of discrepancy between geographic data in GIS and the geographic reality these data are intended to represent. The difference between uncertainty and error is that uncertainty is a relative measure of the discrepancy while error tends to measure the actual value of this discrepancy (Goodchild *et al.*, 1994, p.142).

Data accuracy information associated with a geographic data set should be perceived as a map depicting varying degrees of uncertainty associated with each of the features or phenomena represented in the data set. The discussion on the elements of data accuracy also applies to uncertainty.

7.2.4 Data Lifecycle and Uncertainty

A piece of geographic data typically goes through the following lifecycles (Figure 7.1). The data is first either created through measurement of geographic features/phenomena (such as field observation and remote sensing techniques) or derived from existing geographic data layer(s) through spatial analysis (such as calculating slope gradient data layer from a DEM). The data could be further transformed (such as coordinate transformation and spatial interpolation) so that it is in compliance with other data. The data or its transformed version could be directly used in the decision-making process or be used as part of the inputs for deriving a new data layer. Uncertainty in geographic data arises during the process of data creation and data transformation as it is inevitable that the process introduces errors due to the limitation of the model and techniques used.

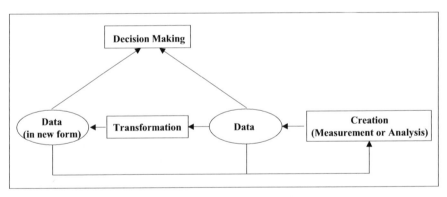

Figure 7.1 Lifecycle of geographic data

This lifecycle of geographic data can be simplified if we focus on data flow in GIS-based analyses (Figure 7.2). A typical GIS-based geographic analysis operates on a set of inputs and produces an output, and the output is then used to generate and evaluate decision alternatives in a decision making process. Uncertainty associated with geographic data also changes along this data flow path. Uncertainty in each of the input data gets passed along into the analysis. In addition the incompatibility (such as spatial and attribute incompatibility) among the different input data also adds onto the uncertainty. Furthermore, the model on which the spatial analysis is based may not be perfectly accurate for the process to be modeled. This model imperfection will compound with the aforementioned uncertainty and the uncertainty propagates through the analysis into the output. The uncertainty in the output will then impact the quality of the decision made based on the output.

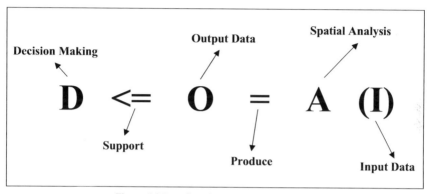

Figure 7.2 Data flow through GIS based analyses

7.3 RESEARCH AGENDA ON UNCERTAINITY IN GIS DATA AND GIS-BASED ANALYSES

The research agenda outlined here calls for a systematic effort to advance the understanding of uncertainty associated with geographic data and the propagation of this uncertainty through GIS based analyses. This understanding is required for developing strategies for managing uncertainty in decision-making involving geographic data. Strategies are needed for identifying, quantifying, tracking, reducing and reporting (including visualizing) uncertainty in geographic data and GIS-based analyses, and for development of standardized means by which uncertainty can be addressed in daily GIS applications. The discussion on the research agenda in this chapter will follow the data flow in Figure 7.2.

7.3.1 Representation of Uncertainty Information

For comprehensive and accurate assessment of uncertainty in GIS-based analyses it is necessary to have every input data layer accompanied by information on spatial variation of uncertainty associated with that data layer. To achieve this, a representation scheme needs to be developed so that spatial variation of different components of uncertainty can be represented together with the data layer. For example, object-based data layers (Goodchild, 1992), that are often represented using vector data model, should have a scheme to represent the different uncertainty elements associated with geographic entities in the each data layer. One of the questions would be "how are the different elements of uncertainty attached?" Should they be attached to a feature (entity) or to parts (such as points, arcs, polygons) of an entity? Concepts in object-oriented data modeling would be useful in addressing this issue since each object can be encapsulated and yet interconnected with other objects (Rumbaugh *et al.*, 1991).

With the field-based data layers, that are often represented using raster data model, it is possible to represent the uncertainty information as separate layers with the value of pixel representing the uncertainty at the corresponding pixel in the data layer. For field-based data layers with categorical attribute values (such as forest classes, soil types), Burt and Zhu (2001) suggest a packed decimal approach to representing data and uncertainty in one layer. The idea is to use the integer part of a float number to represent the class (category) the pixel belongs to and use the decimal part of the number to represent the uncertainty of the pixel to be in that class. However, this approach would not be applicable to data layers whose attribute values take the form of floating point number themselves.

7.3.2 Detection of Errors

Detection of errors is the first step in the process of spatial data accuracy management. Many errors in spatial data can be corrected rather easily after their detection, such as stripes in DEM, inconsistencies in topology, incomplete areal coverage, and omission of certainty features from a database. What is missing is a set of automated procedures that can detect errors or inconsistencies in spatial data by examining the sources (rules) from which the data is created since manual detection of errors is time consuming and error-prone. The quality of spatial data will certainly be much improved if every data layer created goes through a consistent and efficient error detection process.

Examination of errors and inconsistency should not only be based on the traditional measures of geometry and proximity, but also on the geographic knowledge in the datasets. For example, De Groeve and Lowell (2001) argue that boundary uncertainty can be assessed based on the nature of boundary existing in the real world. Thus, knowledge on the characteristics of the natural boundary can be used to check the accuracy of the boundary represented in database. Much of the work conducted in the field of spatial data conflation is applicable to the detection of errors and inconsistencies in spatial databases (Lynch and Saalfeld, 1985; Chung *et al.*, 1995; Foley *et al.*, 1997). Cobb and her colleagues (Cobb *et al.*, 1998) developed a hierarchical rule-based approach augmented with capabilities of reasoning under uncertainty for conflating attributed vector data. The use of knowledge-based approaches coupled with fuzzy logic concepts in detecting errors or inconsistencies definitely deserves more attention than it has received in the past (Robinson, 1988; Robinson and Miller, 1989). The utility of these techniques can be augmented with graphic display methods (Beard and Buttenfield, 1999) for human intervention to efficiently detect errors and inconsistencies in spatial databases.

7.3.3 Quantification of Uncertainty

It is inevitable that some errors in spatial databases cannot be detected due to the nature of spatial data and the method used to capture and the model used to represent the data. Thus, methods or models for quantifying the uncertainty associated with spatial data must be developed. There are two basic theoretic frameworks under which uncertainty is characterized (quantified). One is the probability theory based on crisp (Boolean) logic and the other is the possibility theory based on fuzzy logic. Each of them is applied to the quantification of uncertainty in both representation forms of geographical data: object-based (mostly using a vector data model) and field-based (mostly using a raster data model) (Goodchild, 1992).

7.3.3.1 Probability Theory for Uncertainty Quantification

The application of probability theory in the quantification uncertainty is based on the assumption that the representation of a geographic feature in the database is a result of one of multiple realizations of the geographic feature (Blakemore, 1983; Chrisman and Lester, 1991; Goodchild *et al.*, 1992). The accuracy of this representation depends on the process which generates these realizations. For example, a digitization of a point is a process of realizing the point. How accurate the point is digitized depends on the process (the accuracy of the digitizer, the experience and dedication of the operator). One way to assess the accuracy of the digitized point is to have the operator digitize the same point a number of times (realizations) and then examine the distribution of these realizations. If these realizations of the point are very dispersed, then the accuracy is low. Often, the standard deviation of realizations is used to characterize the uncertainty associated with the representation of the point in the databases.

There are two major challenging issues associated with the probability approach to uncertainty quantification. The first challenge is the difficulty in obtaining the standard deviation without conducting multiple realizations. For example, in quantifying the uncertainty associated with the coordinates of a point due to digitization, one has to repeat the digitization of the point a number of times. This would be prohibitively expensive and time consuming if one wants to quantify the spatial uncertainty associated with every point in a data layer created through digitization. If we assume that for a given person digitizing a map using a given digitizer the uncertainty across a digitized map is the same, then we can have the operator conduct multiple digitizations at each of few selected points. The uncertainty associated with these points then can be used as a measure of uncertainty across the entire map. This also assumes that the variation of density of features on the map does not impact the variation of uncertainty across the map area.

The probability approach to quantify uncertainty might be applicable if the data layer is created directly through digital means such as global positioning systems (GPS), spatial interpolation, and softcopy photogrammetry, since repetition can be easily done with most of the digital measurement techniques. For example, the common approach to obtain the coordinates of a point using GPS is to set the GPS receiver at the location for a certain period of time and let it take many readings. Often, the mean coordinates of these readings are used as the coordinates of the point. The standard deviation of these readings can be used to quantify the uncertainty associated with the coordinates obtained for this point. Spatial interpolation techniques such as Kriging produce estimation variance which can be used to quantify the uncertainty associated with the interpolated attribute (Isaaks and Srivastava, 1989; Deutsch and Journel, 1992; Goovaerts, 1997). However, the validity of these uncertainty measures still needs to be studied.

The other challenge is the characterization of spatial autocorrelation of uncertainty in a given database. For example, let us assume that the positional error of a point takes on a circular normal distribution. A line segment is made of two points each of them with a positional error of circular normal distribution. The question then is "How are these two circular normal distributions related to each other?" In other words, what would be the realizations of the line segment (Caspary and Scheuring, 1993)? What would be the realizations of a polygon constructed by line segments? Should we use a bottom up approach to the derivation of uncertainty of complex features, that is, using the uncertainty associated with points to derive the uncertainty associated with the more complicated geometric features (such as line and polygon) as described by Leung and Yan (1998)? Or should we use a top down approach, that is, using the more complicated geometric features (such as polygon) to infer the uncertainty associated with simpler geometric features (such as line and point)? Or should we take the middle ground by using the uncertainty associated with arcs as the building blocks? Which approach to take may very much depend on how the representation of geographic features is created. For example, when the spatial extent of spatial feature is delineated using GPS, the bottom up approach might be more appropriate since multiple observations are made at a vertex before the operator moves onto the next vertex. But if the spatial extent is digitized off a map, the middle ground approach might be appropriate since the common practice in digitizing is to decompose the spatial extents of features into arcs and create the representation of each arc by digitizing vertices along the arc sequentially.

We often assume that spatial autocorrelation of uncertainty in attribute values exists. This assumption may hold for spatial data represented using the field-based approach (mostly in the form of raster data model). For example, the uncertainty associated with a groundwater contamination map created through Kriging may exhibit spatial autocorrelation since the samples used to make the estimation of contamination at a given point overlap significantly with those used to estimate the contamination at a neighboring point. Thus, the uncertainty associated with the estimation of contamination at the two points is expected to be correlated. However, the assumption may not hold for uncertainty associated with attribute values represented using the object-based approach (mostly vector data model). For example, the uncertainty associated with the attribute value of a soil polygon may not relate to that of a neighboring soil polygon.

7.3.3.2 Possibility Theory for Uncertainty Quantification

The appropriateness of using the probability approach to measure the uncertainty is obvious if the uncertainty is due to the randomness during the process of measurement such as measurement of temperature at a given location and the digitization of a point. However, the appropriateness becomes questionable conceptually when the approach is applied to measure the uncertainty caused by fuzziness (vagueness) (Fisher, 1999). For example, when remote sensing classification techniques are used to create a forest cover type map based on spectral signatures the assignment of a given pixel to a particular class is not a random process. Rather the assignment is based on the similarity of the spectral signature of the pixel to the prescribed spectral signature of the given class. The uncertainty associated with the assignment is due to the difference between the two signatures so the concept of multiple realizations really does not apply here.

The possibility theory is more appropriate to describe the uncertainty due to fuzziness. Under fuzzy logic an object can bear partial membership in multiple classes and allows the expression of different levels of similarity to a set of prescribed classes. The difference between probability theory and possibility theory can also be illustrated by the following example. Suppose we need to classify the soil at a particular location and there are two candidate soil classes (A and B) we can use for the assignment. We further assume that the soil at a location is intermediate to the two candidate soil classes and fit into the concept of class A at 60 percent and to B at 40 percent. A probability statement about this situation is "there is a 60 percent chance that the soil is Class A and 40 percent chance that it is Class B". The possibility statement is "the soil is 60 percent similar to the typical case of Class A and 40 percent similar to the typical case of Class B". The first statement expresses the likelihood that Class A or Class B soil occurs at that location while the second statement expresses the similarity of the soil to each of the two soil classes in terms of soil properties. Since the soil is already there, the question is not about whether the particular soil type occurs or not, rather it is the status (property values) of the local soil which is under concern. Thus the second statement is more suitable conceptually and meaningful in terms of interpretation of soil properties.

The possibility theory is especially useful in measuring uncertainty associated with the classification of spatial features based on their attribute values (such as classification of remotely sensed images (Lillesand and Kiefer, 2000) and automated soil mapping (Zhu *et al.*, 1996; Zhu, 1997a; and Zhu *et al.*, 2001).

Zhu (1997b) and Zhu (2001) used a similarity model to represent the spatial variation of geographic features. Under the similarity model, a geographic feature at a given location is represented by a vector of membership values with each membership value representing the similarity of the local feature to the typical case of a given class. Coupled with a raster data model spatial variation of geographic features can be represented as continuum in both geographic and attribute spaces. When information represented under this model is used to produce a conventional categorical map showing the spatial distribution of geographic features of various classes uncertainty are introduced. To quantify the uncertainty associated with the generation of the conventional categorical map he developed indices measuring membership exaggeration and membership ignorance. Case studies (Zhu, 1997b; Zhu, 2001) have shown that these uncertainty measures are effective in measuring the uncertainty associated with the assignment of individual pixels to categories in the production of conventional categorical resource maps. However, research in this field needs to be expanded and implemented in the production of categorical maps using remote sensing and other automated classification techniques.

The possibility theory would also be useful to characterize the uncertainty in boundary definition due to the fuzziness of boundary between two geographic features/phenomena. De Groeve and Lowell (2001) argue that attribute differences, geometry, and geometry differences between neighboring polygons influence the existence and the width of a boundary. The relationships between boundary conditions and feature attributes can be used to build a rule base that can then be used to assess the uncertainty of boundary between two geographic features. For example, they suggested "the positional uncertainty of a forest-stand boundary is proportional to the difference in attributes of the forest on the left and the right side of the boundary." This line of research in quantifying spatial uncertainty definitely deserves more attention.

7.3.4 Research Issues on Visualization of Uncertainty

Visualization techniques not only allow us to detect errors or inconsistencies in the spatial data but also allow users to gain an appreciation of the quality of a given spatial data set. Tools for visualizing uncertainty in GIS data layers have been explored by many authors (*e.g.*, Hancock, 1993; McGranaghan, 1993; Fisher, 1994; Goodchild *et al.*, 1994; MacEachren, 1994; Paradis and Beard, 1994; Burt and Zhu, 2001). The approaches used for visualizing uncertainty can largely be grouped into two categories: those "blurring" the map and those "focusing" on the data elements of high quality.

The "blurring" approaches (such as Fisher, 1994; Goodchild *et al.*, 1994; MacEachren, 1994; Mitasova *et al.*, 1995; Burt and Zhu, 2001) convey uncertainty by adding one or more visual variables representing the uncertainty associated with the data. The basic idea is to introduce "blurring" proportional to the uncertainty when the data is displayed. The blurring can be introduced in a number of forms (such as unsaturated colors, duration a category is displayed, shifting location of line, and defocusing of the object or its boundaries). For example, Fisher (1994) used error animation to view the reliability of classified imagery. The length of time a class label is displayed at a given pixel is proportional to the accuracy of that class. The frequency of different classes being displayed at a given pixel conveys the uncertainty inherent in the assignment of pixel to a class.

The "blurring" approaches sometimes "annoy" the viewers too much and as a result users may not use these approaches. There are two issues concerning these type of approaches. One is how to introduce uncertainty without annoying the viewers too much. The question is what level of "blurring" is visually acceptable for reviewers to have the "blurring" on while viewing a data layer. The other issue is the precision at which the uncertainty is conveyed to viewers. In other words, do we need to have reviewers quantitatively differentiate uncertainty when the data are displayed. For example, Burt and Zhu (2001) combine color hue and saturation to display a raster categorical map and its associated uncertainty. They use color hues for different classes but the purity of the color (saturation) being displayed at a given pixel is inversely proportional to the uncertainty associated with the assignment of the class to the local pixel. However, for a given saturation but with different hues the perceived amount of whiteness (*chroma*) is different. For example, suppose that blue is to represent Class A and yellow to represent Class B, and Pixel 1 is assigned class A with uncertainty of 0.6 and Pixel 2 is assigned Class B with the same amount of uncertainty. When the two pixels are displayed Pixel 2 appears to be whiter than Pixel 1 due to the inherent difference in chroma between blue and yellow. As a result, Pixel 2 is perceived to be less certain to be Class B. This type of issues needs to be addressed if uncertainty is to be accurately portrayed to viewers.

The "focusing" approaches (such as Paradis and Beard, 1994) convey uncertainty by "filtering" out the items which do not meet a given certainty requirement. For example, Paradis and Beard (1994) developed a data quality filter which allows a viewer to specify the data quality component to be viewed (such as positional accuracy), a quality measure (such as root mean square error [RMSE]), and a quality threshold. The filter only displays data that meet or exceed this threshold. It also allows the viewer to toggle the display to see data that do not meet the threshold value. While this approach certainly overcomes the "annoyance" associated with the "blurring" approaches and reduces the visual complexity of displaying uncertainty, it has its own problems. First, it does not differentiate the different levels of uncertainty associated with features once these features pass through the filter and makes them equal in terms of data quality. Second, it could give viewers the sense that the features are more accurate once they pass through the filter. It might be worth the efforts to explore the combination of the "blurring" approaches with the "focusing" approaches to overcome the "annoyance" associated with the "blurring" approaches and the "indifference" associated with "focusing" approaches so that visualizing uncertainty is more effective.

The development of virtual reality has brought new dimensions for geo-visualization (Bishop and Karadaglis, 1994; Fairbairn and Parsley, 1997; MacEachren *et al.*, 1998). The development in this research area will certainly bring new opportunities and challenges for uncertainty visualization.

7.3.5 Research Issues on Propagation of Uncertainty through GIS-Based Analysis

The uncertainty in the output of a GIS-based analysis contains two parts: one due to the propagation of uncertainty from the input data and the other due to the error in the model on which the analysis is based. Although analysis of uncertainty in the output should consider the uncertainty due to model error, for simplicity of this discussion we assume that there is no error in the model used and uncertainty in the output is propagated from the uncertainty in the input data.

GIS-based analyses can be divided into two basic types: those operating on a single input data layer and those operating on multiple input data layers (Bonham-Carter, 1994, pp. 177–334). For the second set of operators the propagation of uncertainty should also include the component due to the incompatibility among the input data layers in addition to the propagation of uncertainty in each of the input data layers.

7.3.5.1 Scale Incompatibility Issue

The incompatibility can be caused by differences in spatial resolution (scale), differences in attribute precision, and differences in currentness of data among the input data layers, but uncertainty due to scale incompatibility among input data layers is a major concern. While scale and its effects have been studied extensively (Band, 1993; Ehleringer and Field, 1993; Foody and Curran, 1994; Schneider, 1994; Band and Moore, 1995; Quattrochi and Goodchild, 1997; Van Gardingen *et al.*, 1997; Walsh *et al.*, 1997), but impacts of scale incompatibility among input data layers on the output of geographic analyses have not received much attention. To overcome the scale incompatibility one would need to perform geographic analyses at the scale at which the process operates. However, it is perceived to be difficult, if not impossible, when the analyses involve multiple data layers which may describe different processes. The first reason for this difficulty is that some, if not all, geographic processes operate at multiple scales and it would be challenging to determine an optimal scale to perform the analysis. Second, it will require us to find the scale at which the process described by each input data operates. Third, we need to acquire the input data represented at that scale. One important issue facing us is to determine the scale at which geographic analyses should be performed for a given set of input data so that uncertainty caused by scale incompatibility is minimized. Zhu (2002) examined this issue by comparing simulated hydro-ecological responses produced using two different soil datasets: one from the conventional soil map and the other from an automated soil inference approach (Zhu *et al.*, 2001). The resolution of the soil data set produced from the automated approach is much higher than that from the conventional soil map (Zhu and Mackay, 2001). He found that the difference between the simulated responses produced from the two soil datasets is at its lowest when the spatial partitioning of the landscape approaches the modal size of the soil polygons in the conventional soil map. This indicates that there might exist some scale (level of spatial partitioning of landscape) at which the scale incompatibility among the input data might be minimal for the given set of input data. However, research is needed to confirm and to generalize this finding.

7.3.5.2 Methods of Analyzing Propagation of Uncertainty and Research Issues

The propagation of uncertainty in the input data through GIS-based analyses to the output has received much attention (*e.g.*, Fisher, 1991; Stoms *et al.*, 1992; Heuvelink and Burrough, 1993; Veregin, 1995; Davis and Keller, 1997; Arbia *et al.*, 1998; Burrough and McDonnell, 1998, pp. 241–264; Heuvelink, 1999; Crosetto and Tarantola, 2001). The basic form of a GIS-based operation can be expressed as Equation 7.1

$$o = g(I) \tag{7.1}$$

where *I* represents a set of *m* inputs $(I_1, I_2, ..., I_i, ..., I_m)$, $g(.)$ is a GIS analysis which operates on the set of inputs, and *o* represents the output from the GIS analysis. The study of uncertainty propagation focuses on two parts (Burrough and McDonnell, 1998, pp. 241–264; Crosetto and Tarantola, 2001). The first is to examine the uncertainty in *o* given *I* with known uncertainty. The second part is to study the sensitivity of $g(.)$ to uncertainty associated with each input so that error reduction efforts can be efficiently directed toward the input to which $g(.)$ is most sensitive (referred to as balance of error (Heuvelink, 1999)). Basically there are two major types of approaches to modeling of the propagation of uncertainty through GIS-based analyses: Monte Carlo simulation and analytical methods.

7.3.5.3 Monte Carlo Method and Issues

The Monte Carlo method to analyze the propagation of uncertainty is based on the concept of realizations (Hammersely and Handscomb, 1979). The idea is to provide a set of realizations of input *I* based on the uncertainty associated with each input and the co-variation among the inputs. For each of the realizations, compute *o*, the result of $g(I)$. The results from this process form a distribution of *o*. The parameters (such as mean and variance) of the distribution can be estimated from the set of results. The mean and variance can be then used to assess the expected result and its associated uncertainty from $g(.)$ given I with known uncertainty. The method consists of the following steps (Heuvelink, 1999):

1) Generate a realization I_i, $i = 1, \ldots, m$

2) For this realization, compute and store the output o

3) Repeat Steps 1) and 2) for N times

4) Compute and store statistics from the N outputs o

The most important part of the Monte Carlo method is the generation of each realization. Let's start at the simplest case, that is, a point analysis requiring one input, I. A realization of I can be created by randomly sampling a value from a distribution defined for I. The distribution for I is often assumed to be Gaussian with mean to be the recorded value of I and the standard deviation to be the uncertainty value associated with I. The creation of realizations becomes a bit complicated when a GIS analysis involves two or more inputs (still for a single location) since we now need to sample from the joint distribution of these inputs to produce a realization which now includes a value for each of the inputs. To define a joint distribution among inputs, we need to know the cross-correlation among uncertainty associated with these inputs. To further complicate the generation of realizations, let's now extend the process to an area which is made of many spatial units (whether as polygons or pixel) and each realization consists of a set of inputs values for each of spatial units. In other words each realization is a set of maps with each map representing the spatial variation of one input. For such a realization to be meaningful, we need to know two different correlations: spatial autocorrelation of uncertainty associated with each input and spatial cross-correlation among uncertainties so that the spatial structures existing in the inputs are not totally destroyed by the random sampling process.

Most of the current efforts in generating realizations for uncertainty analysis are based on the stationarity assumption that the uncertainty associated with a given input is the same over space. To create a realization, a random error field is generated based on the uncertainty (often the RMSE value over the entire map area) and the error field is then added to the input data layer. A prescribed spatial autocorrelation and spatial cross-correlation may also be included during the creation of the error field to maintain the spatial cohesiveness of the error field. There are two major challenges associated with this approach to generate realizations. First, the assumption of uniform uncertainty within an input data layer is highly inappropriate since clearly uncertainty in an input data layer often varies from one location to another. Although this assumption is not native to the Monte Carlo approach and rather it is the result of lack of information on the spatial variation of uncertainty in an input data layer, it does reiterate the importance of quantifying and representing spatial variation of uncertainty as outlined in previous sections.

The second challenge is that the current way of creating realizations for capturing the uncertainty associated with input data completes ignore the process by which uncertainty associated with an input data layer is created. In particular, the validity of the use of prescribed spatial auto/cross correlations is very questionable. The issue of how to use uncertainty maps, which contains the spatial structure (spatial autocorrelation and spatial cross-correlation) of uncertainty, to create realizations must be addressed before Monte Carlo analysis of uncertainty propagation becomes meaningful.

The Monte Carlo method is conceptually simple but computationally expensive. The application of the Monte Carlo method does not require an analytical understanding of the model. What is involved is the generation of realizations and re-running the analysis using the realization. The accuracy of the Monte Carlo method is inversely related to the square root of the number of runs N (Heuvelink, 1999) given that the realizations are generated meaningfully. To achieve high accuracy a huge number of simulations (in the thousands) may be required. This puts a huge demand on computing resources even with the computing power offered by today's computers. Thus, sensitivity analysis of output to the uncertainty associated with inputs using the Monte Carlo method is very time consuming and computationally expensive.

7.3.5.4 Analytical Approaches and Issues

There are many analytical approaches to the analysis of uncertainty propagation (Burrough and McDonnell, 1998, pp. 241–264). The Taylor series method is among the ones commonly used (Taylor, 1982). The assumption of Taylor series method is that the result of a model, such as $g(.)$, can be expressed as a Taylor series expansion about the means of the model inputs, $\bar{I} = (\bar{I}_1, \bar{I}_2, ..., \bar{I}_i, ..., \bar{I}_m)$, as shown in Equation (7.2).

$$o = g(\bar{I}) + \sum_{i=1}^{m} \frac{\partial g(\bar{I})}{\partial I_i}(I_i - \bar{I}_i) + \frac{1}{2} \sum_{i=1}^{m} \sum_{j=1}^{m} \frac{\partial^2 g(\bar{I})}{\partial I_i I_j}(I_i - \bar{I}_i)(I_j - \bar{I}_j) + ... \quad (7.2)$$

The first term in the right hand side of Equation 7.2 gives the result of using the mean values of the model inputs and the next two terms add the deviations to the result caused by the deviation of each model input from its mean value and that caused by the covariance between the model inputs. If the model is not highly non-linear over the domain of I, the result, o, can then be approximated using the first order of the Taylor series by ignoring the higher order terms in Equation (7.2), which gives us:

$$o = g(\bar{I}) + \sum_{i=1}^{m} \frac{\partial g(\bar{I})}{\partial I_i}(I_i - \bar{I}_i) \quad (7.3)$$

The mean and variance of *o* are given as (Heuvelink *et al.*, 1989):

$$\xi \approx g(\bar{I}) \tag{7.4}$$

$$\tau^2 \approx \sum_{i=1}^{m} \sum_{j=1}^{m} \rho_{ij} \sigma_i \sigma_j \, \frac{\partial g(\bar{I})}{\partial I_i} \, \frac{\partial g(\bar{I})}{\partial I_j} \tag{7.5}$$

where ρ_{ij} is the spatial correlation between the uncertainty associated with input *i* and that associated with input *j* (cross correlation when $i \neq j$ and autocorrelation when $i = j$) and σ_i describes the uncertainty (variation) associated with input *i*.

The Taylor series method to the analysis of uncertainty propagation takes an analytical form. It can easily accommodate uncertainty represented as maps due to its explicit treatment of spatial correlation of uncertainty. Sensitivity analysis is easily accomplished since the uncertainty propagation is expressed in an analytical form. However, the method does require an analytical understanding of the model and the results are only approximate. Sometimes it is difficult to determine whether a particular level of approximation is acceptable due to the non-linearity of the model over the domains of its inputs.

Both the Monte Carlo method and the Taylor series method do not meet all the needs in the analysis of uncertainty propagation. They do provide us the starting point. However, at this point, all that we have are illustrations of how these methods work. An important issue facing us is to determine how useful these techniques are in assessment of uncertainty in outputs of GIS-based analyses. One specific issue related to the examination of the usefulness of these methods is the collection of empirical data for validation of these methods.

7.3.5.5 Research Issues on Impact of Uncertainty on Decision Making

The ultimate goal of data quality analysis and management is to help end users of geographic data to understand and manage the risk in making a particular deci-sion involving the use of these data. Yet, little research has been done on how to use the uncertainty information associated with geographic data in reaching policy decisions (Agumya and Huner, 1999). This is partly due to the lack of infor-mation on the uncertainty associated with geographic data, particularly informa-tion on the spatial variation of uncertainty in a data set. The other reason is the disjoint among the data producers, data end users (the people who use the data to assist their decision making), and specialists in uncertainty analysis. As the popu-larity of GIS analyses increases and as more and more geographic data in the form of GIS data layers are used for provide interpretation in decision making, the issue of uncertainty in the form of reliability will rise. Research on the impacts of uncertainty on decision making is urgently needed. An important issue to address is how to perform risk assessment of a decision based on imperfect data.

7.3.5.6 Metadata Issues

Data quality is one of the most important parts of Metadata. Data users use the data quality information together with other information documented in the metadata to assess the fitness-for-use of the given data layer. The current Metadata standard only facilitates the documentation of an overall accuracy measure over the entire map area. For example, the attribute accuracy component in the Metadata standard contains only three pieces of information: an explanation of the accuracy of the identification of the entities, a value assigned to summarize the accuracy of the identification of the entities and assignments of values in the data set, and the identification of the test that yielded the attribute accuracy value (FGDC, 1998). It does not provide means to document the spatial variation of uncertainty associated with the given data layer.

The other issue is the update of metadata when a data layer is modified or created with the use of GIS-based analysis (such as coordinate transformation, spatial interpolation, and overlay analysis). Most of the current GIS systems do not provide means to automatically update (create) the metadata once the data layer is modified (created). The updating of metadata is the sole responsibility of the user. This makes the updating very time-consuming and tedious and as a result updating is often not done at all (Goodchild, 1995). Some pilot work on automated updating metadata as part of the GIS-based analysis is underway (Lanter, 1991; Veregin, 1991; Lanter and Surbey, 1994), but collaboration from major GIS vendors is needed if this is to become the reality of every day GIS operations.

7.4 A COORDINATED EFFORT NEEDED

As shown through the above discussions uncertainty exists in every phase of geographic data life cycle transcending boundaries of disciplines and organizations. Yet, most of the work on uncertainty in spatial data was conducted in research laboratories as academic research exercises. Very few of these research findings were transferred to the geospatial industry (including both the private and public sectors). For example, the quantification of uncertainty due to classification has been well studied and some of the measures for quantifying uncertainty associated with the creation of categorical resources maps are quite robust (Wang, 1990; Goodchild *et al.*, 1994; Zhu, 1997b). Yet, no agencies or private sectors use any of these measures to quantify the uncertainty in the categorical resource maps they produce. This is largely due to lack of communication among the research community and the geospatial data production industry. Thus, an inter-institutional and inter-disciplinary team, consisting of domain experts, GIS experts including spatial statisticians, application users including decision makers, data producers, and GIS software vendors, is needed to facilitate the communication among the different parties dealing with spatial data, to identify needs in research on uncertainty, and to plan uncertainty management strategies. In this way, new needs and demands on uncertainty management can be promptly identified and advancements in the field can be quickly transferred into daily operations of GIS.

Such a team should also participate in promoting and advising a test institute which will conduct applied research on linking research findings on uncertainty from various stages of data creation and data analyses via case studies so that isolated research findings can be transferred out of laboratory settings into geospatial industry. Some of the specific tasks for this test institute are:

- Testing new methods of managing uncertainty in geographic data and GIS analyses;

- Implementing strategies and methodologies for reducing, quantifying, tracking, and reporting uncertainty in GIS implementation, in geographic data collection and generation, and in spatial data standards and decision making processes; and

- Streamlining research findings on uncertainty through complete GIS applications from data collection through decision making processes.

7.5 SUMMARY

This paper provides an overview of the issues related to uncertainty in geographic data and GIS-based analyses from the perspective of geographic data life cycle. While many issues related to uncertainty in geographic data and GIS-based analyses were discussed, the author wishes to highlight the following three areas as the priorities of research. First is the quantification of spatial variation of uncertainty in geographic data. As it can be seen from the above discussions data quality cannot be addressed if spatial variation of uncertainty is not quantified first. It should be quantified for every geographic feature at every location. Second, the impact of uncertainty in geographic data and results of GIS-based analyses on decision making virtually received no attention in the past. Research on how to incorporate uncertainty associated with geographic data and GIS analyses in the decision making process is rarely done. Third, little effort is made to link research findings on uncertainty to geographic data production. A coordinated effort in the form of multi-institutional and multi-disciplinary team and a test institute is needed to conduct applied research to transfer research findings into the geospatial industry.

7.6 ACKNOWLEDGEMENTS

Much of the materials presented in this paper were drawn from the 1996 UCGIS white paper on research priorities in uncertainty and GIS-based analyses. When drafting the white paper this author received help from many people. William Reiners and Michael Goodchild were the editors for the white paper. They provided this author much needed help during the process of writing the white paper. Carolyn Hunsaker also contributed to the initial draft of the white paper. The white paper was presented at the 1998 UCGIS summer assembly in Park City, Utah and commented on by many participants prior to and at the summer assembly. In particular, Nicholas Chrisman, Kate Beard, and Stephen Guptill were the panelists on the topic and provided constructive comments on the white paper. The author also wishes to acknowledge the support received from the Graduate School at University of Wisconsin-Madison for attending the UCGIS summer assemblies. The support made it possible for the author to be part of this exciting intellectual discussion on the topic.

7.7 REFERENCES

Abler, R.F., 1987, The National Science Foundation National Center for Geographic Information and Analysis. *International Journal of Geographical Information Systems*, Vol. 1, pp. 303–326.

Agumya, A. and Hunter, G.J., 1999, A risk-based approach to assessing the "Fitness for Use" of spatial data, *URISA Journal*, Vol. 11, pp. 33–44.

Arbia, G., Griffith, D. and Haining, R., 1998, Error propagation modeling in raster GIS: overlay operations. *International Journal of Geographical Information Science*, Vol. 12, pp. 145–167.

Band, L.F., 1993, Effect of land surface representation on forest water and carbon budgets. *Journal of Hydrology*, Vol. 150, pp. 749–772.

Band, L.E. and Moore, I.D., 1995, Scale: landscape attributes and geographical information systems. *Hydrological Processes*, Vol. 9, pp. 402–422.

Beard, M.K. and Buttenfield, B.P., 1999, Detecting and evaluating errors by graphical methods. In Longley, P.A., Goodchild, M.F., Maguire, D.J. and Rhind, D.W. (eds.), *Geographical Information Systems: Principles and Technical Issues*, John Wiley & Sons, Inc.: New York, pp. 219–233.

Bishop, I.D. and Karadaglis, C., 1994, Use of interactive immersive visualization techniques for natural resources management. *SPIE*, 2656, pp. 128–139.

Blakemore, M. 1983, Generalization and error in spatial databases. *Cartographica*, Vol. 21, pp. 131–139.

Bonham-Carter, G.F., 1994, *Geographic Information Systems for Geoscientists*, Pergamon: Tarrytown, New York, 398 p.

Brassel, K., Bucher, F., Stephan, E.M. and Vckovski, A., 1995. Completeness. In Guptill, S.C. and Morrison, J.L. (eds.), *Elements of Spatial Data Quality*, Elsevier Science: Oxford, pp. 81–108.

Burrough, P.A. and McDonnell, R.A., 1998, *Principles of Geographical Information Systems*, Oxford University Press: New York, 333 p.

Burt, J.E. and Zhu, A.X., 2001, *Depicting uncertainty in fuzzy-classified images*, (poster), Annual Meeting of American Association of Geographers, February 27–March 3, 2001, New York City, N.Y.

Caspary, W. and Scheuring, R., 1993, Positional accuracy in spatial databases. Computers, *Environment and Urban Systems*, Vol. 17, pp. 103–110.

Chrisman, N.R., 1991, The error component in spatial data. In Maguire, D.J., Goodchild, M.F. and Rhind, D.W. (eds.), *Geographical Information Systems: Principles and Applications, Vol. 1*, John Wiley & Sons Inc.: New York, pp. 165–174.

Chrisman, N.R. and Lester, M.K., 1991, A diagnostic test for error in categorical maps. *Auto-Carto 10: Technical Papers of the 1991 ACSM-ASPRS Annual Convention*, Baltimore, Maryland, Vol. 6, pp. 330–348.

Chung, M., Cobb, M., Arctur, D.K. and Shaw, K.B., 1995, An object-oriented approach for handling topology in VPF products. *Proceedings of GIS/LIS'95*, Nashville, TN, pp. 163–174.

Cobb, M.A., Chung, M.J., Foley III, H., Petry, F.E.and Shaw, K.B., 1998, A rule-based approach for the conflation of attributed vector data. *GeoInformatica*, Vol. 2, pp. 7–35.

Couclelis, H., 1992, People manipulate objects (but cultivate fields): beyond the raster-vector debate. In Frank, U., Campari, I. and Formentini, U. (eds.), *Theories and Methods of Spatio-temporal Reasoning in Geographic Space, Lecture Notes in Computer Science*, Springer-Verlag: Berlin-Heidelberg, Vol. 639, pp. 65–77.

Crosetto, M and Tarantola, S., 2001, Uncertainty and sensitivity analysis: tools for GIS-based model implementation. *International Journal of Geographical Information Science*, Vol. 15, pp. 415–437.

Davis, T.J. and Keller, C.P., 1997, Modeling uncertainty in natural resource analysis using fuzzy sets and Monte Carlo simulation: slope stability prediction. *International Journal of Geographical Information Science*, Vol. 11, pp. 409–434.

De Groeve, T. and Lowell, K., 2001, Boundary uncertainty assessment from a single forest-type map. *Photogrammetric Engineering and Remote Sensing*, Vol. 67, pp. 717–726.

Deutsch, C.V. and Journel, A.G., 1992, GSLIB—geostatistical software library and user's guide, Oxford University Press: New York, [computer file]

Ehleringer, J.R. and Field, C.B. (eds.), 1993, *Scaling Physiological Processes: Leaf to Globe*, Academic Press: San Diego, 388 p.

Fairbairn, D. and Parsley, S., 1997, The use of VRML for cartographic presentation. *Computers & Geosciences*, Vol. 23, pp. 475–482.

FGDC (Federal Geographic Data Committee), 1998, *Content Standard for Geospatial Metadata*, Washington D.C., 78 p., *(http://www.fgdc.gov/metadata/contstan.html)*.

Fisher, P.F., 1991, Modeling soil map-unit inclusion by Monte Carlo simulation. *International Journal of Geographical Information Systems*, Vol. 5, pp. 193–208.

Fisher, P.F., 1994, Visualization of the reliability in classified remotely sensed images. *Photogrammetric Engineering and Remote Sensing*, Vol. 60, pp. 905–910.

Fisher, P.F., 1999, Models of uncertainty in spatial data, In Longley, P.A., Goodchild, M.F., Maguire, D.J. and Rhind, D.W. (eds.), *Geographical Information Systems: Principles and Technical Issues*, John Wiley & Sons, Inc: New York, pp. 191–205.

Foley, H., Petry, F., Cobb, M. and Shaw, K.B., 1997, Utilization of an expert system for the analysis of semantic characteristics for improved conflation in geographic information systems. *Proceedings of the 10th International Conference on Industrial and Engineering Applications of AI*, Atlanta, GA, pp. 267–275.

Foody, G.M. and Curran, P.J. (eds.), 1994, *Environmental Remote Sensing from Regional to Global Scales*, John Wiley & Sons: New York, 238 p.

Goodchild, M.F., 1991, Issues of quality and uncertainty. In Muller, J.C. (ed.), *Advances in Cartography*, Elsevier Science: Oxford, pp. 111–139.

Goodchild, M.F., 1992, Geographical data modeling. *Computers and Geosciences*, Vol. 18, pp. 401–408.

Goodchild, M.F., 1993, Data models and data quality: problems and prospects. In Goodchild, M.F., Parks, B.O. and Steyaert, L.T. (eds.), *Environmental Modeling With GIS*, Oxford University Press: New York, pp. 94–103.

Goodchild, M.F., 1995, Sharing imperfect data. In Onsrud, H.J. and Rushton, G. (eds.), *Sharing Geographic Information*, Center for Urban Policy Research: New Brunswick, pp. 413–425.

Goodchild, M.F. and Gopal, S. (eds.), 1989, *Accuracy of Spatial Databases*, Taylor and Francis: New York, 290 pp.

Goodchild, M.F., Sun, G. and Yang, S., 1992, Development and test of an error model for categorical data. *International Journal of Geographical Information Systems*, Vol. 6, pp. 87–104.

Goodchild, M.F., Chin-Chang, L. and Leung, Y., 1994, Visualizing fuzzy maps. In Hearnshaw, H.M. and Unwin, D.J. (eds.), *Visualization in Geographical Information Systems*, John Wiley & Sons: New York, pp.158–167.

Goovaerts, P., 1997, *Geostatistics for Natural Resources Evaluation*, Oxford University Press: New York, 483 p.

Guptill, S.C. and Morrison, J.L. (eds.), 1995, *Elements of Spatial Data Quality*, Elsevier Science: Tarrytown, New York, 202 p.

Hammersley, J.M. and Handscomb, D.C., 1979, *Monte Carlo Methods*, Chapman & Hall: London, 178 p.

Hancock, J.R., 1993, Multivariate regionalization: an approach using interactive statistical visualization. *Proceedings of Auto Carto 11*, Minneapolis, pp. 218–227.

Heuvelink, G.B.M., 1999, Propagation of error in spatial modeling with GIS. In Longley, P.A., Goodchild, M.F., Maguire, D.J. and Rhind, D.W. (eds.), *Geographical Information Systems: Principles and Technical Issues*, John Wiley & Sons, Inc: New York, pp. 207–217.

Heuvelink, G.B.M. and Burrough, P.A., 1993, Error propagation in cartographic modeling using Boolean logic and continuous classification. *International Journal of Geographical Information Systems*, Vol. 7, pp. 231–246.

Heuvelink, G.B.M., Burrough, P.A. and Stein, A., 1989, Propagation of errors in spatial modeling with GIS. *International Journal of Geographical Information Systems*, Vol. 3, pp. 303–322.

Hunsaker, C.T., Goodchild, M.F., Friedl, M.A. and Case, T.J. (eds.), 2001, *Spatial Uncertainty in Ecology*, Springer: New York, 402 p.

Isaaks, E.H. and Srivastava, R.M., 1989, *An Introduction to Applied Geostatistics*, Oxford University Press: New York, 561 p.

Kainz, W., 1995, Logical Consistency. In Guptill, S.C. and Morrison, J.L. (eds.), *Elements of Spatial Data Quality*, Elsevier Science: Oxford, pp. 109–137.

Lanter, D., 1991, Design of a lineage-based meta-database for GIS. *Cartography and Geographic Information Systems*, Vol. 18, pp. 255–261.

Lanter, D. and Surbey, C., 1994, Metadata analysis of GIS data processing: a case study. In Waugh, T.C. and Healey, R.G. (eds.), *Advances in GIS Research*, Taylor and Francis: London, pp. 314–324.

Leung, Y. and Yan, J., 1998, A locational error model for spatial features. *International Journal of Geographical Information Science*, Vol. 12, pp. 607–620.

Lillesand, T.M. and Kiefer, R.W., 2000, *Remote Sensing and Interpretation*, John Wiley & Sons, Inc.: New York, 724 p.

Lowell, K. and Jaton, A., (eds.), 1999, *Spatial Accuracy Assessment: Land Information Uncertainty in Natural Resources*, Ann Arbor Press: Chelsea, Michigan, 455 p.

Lunetta, R.S, Congalton, R.G., Fenstermaker, L.K., Jensen, J.R., McGwire, K.C. and Tinney, L.R., 1992, Remote sensing and geographic information system data integration: error sources and research issues. *Photogrammetric Engineering and Remote Sensing*, Vol. 57, pp. 677–687.

Lynch, M.P. and Saalfeld, A.J., 1985, Conflation: automated map compilation – a video game approach. *Proceedings of Auto-Carto 7*, Falls Church, VA.

MacEachren, A.M., 1994, Some truth with maps: a primer on symbolization and design. *Association of American Geographers*, Washington, D.C.

MacEachren, A.M., Haug, D., Quian, L., Otto, G., Edsall, R. and Harrower, M., 1998, *Geographic visualization in immersive environments*, GeoVISTA Center, Pennsylvania State University.

McGranaghan, M., 1993, A cartographic view of data quality. *Cartographica*, Vol. 30, pp. 8–19.

Mitasova, H., L. Mitas, W. Brown, D.P. Gerdes, I. Kosinovsky, and Baker, T., 1995, Modeling spatially and temporally distributed phenomena: new methods and tools for GRASS GIS. *International Journal of Geographical Information Systems*, Vol. 9, pp. 433–446.

Mowrer, H.T. and Congalton, R.G., 2000, *Quantifying Spatial Uncertainty in Natural Resources: Theory and Applications for GIS and Remote Sensing*, Ann Arbor Press: Chelsea, Michigan, 244 p.

NCGIA, 1989, The research plan for the National Center for Geographic Information and Analysis. *International Journal of Geographical Information Systems*. Vol. 3, No. 2, pp. 117–136.

Paradis, J. and Beard, K., 1994, Visualization of data quality for the decision-maker: a data quality filter. *Journal of the Urban and Regional Information Systems Association*, Vol. 6, pp. 25–34.

Quattrochi, D.A. and Goodchild, M.F. (eds.), 1997, *Scale in Remote Sensing and GIS*, Lewis Publishers: New York, 406 p.

Robinson, V.B., 1988, Some implications of fuzzy set theory applied to geographic databases. *Computers, Environment and Urban Systems*, Vol. 12, pp. 89–98.

Robinson, V.B. and Miller, R., 1989, Intelligent polygon overlay processing for natural resource inventory maintenance: prototyping a knowledge-based spurious polygon processor. *AI Applications*, Vol. 3, pp. 23–34.

Rumbaugh, J., Blaha, M., Premerlani, W., Eddy, F. and Lorensen, W., 1991, *Object-Oriented Modeling and Design*, Prentice Hall: Englewood Cliffs, New Jersey, 500 p.

Schneider, D.C., 1994, *Quantitative Ecology: Spatial and Temporal Scaling*, Academic Press: New York, 395 p.

Seong, J.C. and Usery, E.L., 2001, Assessing raster representation accuracy using a scale factor model. *Photogrammetric Engineering & Remote Sensing*, Vol. 67, pp. 1,185–1,191.

Shi, W.Z. and Liu, W.B., 2000, A stochastic process-based model for the positional error of line segments in GIS. *International Journal of Geographical Information Science*, Vol. 14, pp. 51–66.

Stoms, D.M., Davis, F.W. and Cogan, C.B., 1992, Sensitivity of wildlife habitat models to uncertainties in GIS data. *Photogrammetric Engineering & Remote Sensing*, Vol. 58, pp. 843–850.

Taylor, J.R., 1982, *An Introduction to Error Analysis: The Study of Uncertainties in Physical Measurement*, Oxford University Press: Oxford.

Van Gardingen, P.R., Foody, G.M. and Curran, P.J. (eds.), 1997, *Scaling-up: From Cell to Landscape*, Cambridge University Press: New York, 386 p.

Veregin, H., 1991, *GIS Data Quality Evaluation for Coverage Documentation Systems*, Environmental Monitoring Systems Laboratory, US Environmental Protection Agency: Las Vegas.

Veregin, H., 1995, Developing and testing of an error propagation model for GIS overlay operations. *International Journal of Geographical Information Systems*, Vol. 9, pp. 595–619.

Veregin, H., 1999, Data quality parameters. In Longley, P.A., Goodchild, M.F., Maguire, D.J. and Rhind. D.W. (eds.), *Geographical Information Systems: Principles and Technical Issues*, John Wiley & Sons, Inc.: New York, pp. 177–189.

Walsh, S.J., Moody, A., Allen, T.R. and Brown, D.G., 1997, Scale dependence of NDVI and its relationship to mountainous terrain. In Quattrochi, D.A. and Goodchild, M.F. (eds.), *Scale in Remote Sensing and GIS*, Lewis Publishers: New York, pp. 27–55.

Wang, F., 1990, Improving remote sensing image analysis through fuzzy information representation. *Photogrammetric Engineering and Remote Sensing*, Vol: 56, pp. 1,163–1,169.

Zhu, A.X., 1997a, A similarity model for representing soil spatial information. *Geoderma*, Vol. 77, pp. 217–242.

Zhu, A.X., 1997b, Measuring uncertainty in class assignment for natural resource maps using a similarity model. *Photogrammetric Engineering and Remote Sensing*, Vol 63, pp 1,195–1,202.

Zhu, A.X., 2001, Modeling spatial variation of classification accuracy under fuzzy logic. In Hunsaker, C.T., Goodchild, M.F., Friedl, M.A. and Case, T.J. (eds.), *Spatial Uncertainty in Ecology*, Springer: New York, pp. 330–350.

Zhu, A.X., 2002, Effect of soil landscape parameterization on simulation of watershed hydro-ecological responses with change of scale. *Proceedings of Geoinformatics'2002: GIS and Remote Sensing for Global Change Studies and Sustainable Development*, Nanjing, China.

Zhu, A.X., Band, L.E., Dutton, B. and Nimlos, T., 1996, Automated soil inference under fuzzy logic. *Ecological Modeling*, Vol. 90, pp. 123–145.

Zhu, A.X., Hudson, B., Burt, J., Lubich, K. and Simonson, D., 2001, Soil mapping using GIS, expert knowledge, and fuzzy logic. *Soil Science Society of America Journal*, Vol. 65, pp. 1,463–1,472.

Zhu, A.X. and Mackay, D. S., 2001, Effects of spatial detail of soil information on watershed modeling. *Journal of Hydrology*, Vol. 248, pp. 54–77.

CHAPTER EIGHT

The Future of the Spatial Information Infrastructure

Harlan Onsrud, University of Maine
Barbara Poore, U.S. Geological Survey
Robert Rugg, Virginia Commonwealth University
Richard Taupier, University of Massachusetts
Lyna Wiggins, Rutgers University

8.1 INTRODUCTION

Information about human and economic activities relative to the location and character of natural and cultural resources is essential to making decisions in every day life. Geographic information systems (GIS) and associated technologies respond directly to the need to relate activities to resources. Government agencies, the commercial sector, the scientific community, community groups, and individual citizens are using digital geographic data and technologies for wide-ranging purposes and their uses for many applications have become commonplace. Geographic digital technologies are now in common use to:

- update and reproduce government tax maps, zoning maps, and land use maps;

- create maps as needed that overlay multiple sets of information such as aerial images, zoning, tax parcels, topography, sewer and water lines, and wet lands;

- select optimal sites for locating businesses and other facilities;

- track and manage urban growth;

- dispatch and route emergency vehicles;

- map and analyze crime patterns;

- farm more effectively through precision application of fertilizers and pesticides;
- optimize delivery of rural health and medical services;
- evaluate sites for waste disposal;
- inventory and manage roads, utilities, and other physical facilities;
- advance economic development through provision of detailed asset information;
- track and model the spread of pollutants or destructive biological agents;
- evenly distribute student populations among schools and route school buses efficiently;
- identify hazardous waste sites and map brown fields;
- provide citizens with remote access to local government information;
- optimize preservation of farmlands;
- track power outage locations;
- provide detailed planning for efficient and sound land development;
- track and model the quality of ground and surface water;
- forecast weather;
- map the territories of animal and plant species;
- navigate aircraft and ships;
- track depletion and recovery patterns of fisheries, forests, and soil erosion;
- enable citizens to access and evaluate local government records for themselves;
- locate sites for telecommunication towers and cell phone facilities; and
- a host of similar activities.

Government has been the dominant force in the initial development of the spatial data sets used throughout U.S. society. Such data developed at all levels of government is used for myriad government and private sector purposes. An Office of Management and Budget Report (OMB, 1994) indicates that the federal government alone accounts for over $4 billion each year in geographic information-based activity while activity at state and local government levels approaches a similar magnitude. Much of the data that agencies at all levels collect for governmental purposes is made available widely in order to encourage public education and enlightenment and to assist economic development within the commercial sector. Among the economic values at work is that individuals and commercial businesses ought to be able to derive economic benefit from public goods (such as public information) (U.S. Congress, 1986).

In addition to government spending on geographic information-based activity, a National Academy of Public Administration (NAPA) report indicates that the estimated 1998 commercial market for geographic data sales and services involves an additional $4.2 billion (NAPA, 1998, 298). Conservatively one can estimate that well over $10 billion of direct geographic information activity occurs annually in the U.S. and the effects of this investment in facilities and data play an increasingly important role in the national and global economy.

The *spatial information infrastructure* is an institutional concept being advanced in order to better respond to needs for spatially referenced information in various problem-solving domains. As indicated, these diverse needs are spread widely across government, industry, academic, community organization, and public interest sectors.

The term "infrastructure" typically brings to mind public facilities such as roads, sewer lines, electric lines, airports, and similar physical structures or networks in which government has played a major role in their construction or ongoing support. Thus, terms such as the National Information Infrastructure (NII) and National Spatial Data Infrastructure (NSDI) typically bring to mind the facilities being financed through tax dollars in order to allow the more efficient transmission and communication of information in support of the general and widespread interests of broad sectors of society. Information infrastructure also brings to mind the facilities, processes, and standards by which information essential to the operation of government is made available to governmental agencies and bodies. Government involvement in information infrastructure development is critical to advancing the economic and social well being of our nation's citizens.

The NSDI has been defined as "the technology, policies, standards, human resources, and related activities necessary to acquire, process, distribute, use, maintain, and archive spatial data" (OMB, 2001). Alternatively, NSDI has been defined as "the means to assemble geographic information that describes the arrangement and attributes of features and phenomena on the earth. The infrastructure includes the materials, technology, and people necessary to acquire, process, store, and distribute such information to meet a wide variety of needs" (National Research Council–NRC, 1993).

In contrast to the term "infrastructure", the term "library" brings to mind the image of an institution in which the works of government, the private sector, and individuals are made available to the public through a decentralized yet networked national or global system. The "library" as an institution may be defined as a distributed societal arrangement for systematically indexing, storing, and providing access to the intellectual works and information resources contributed by millions of individuals from all walks of life. A "geolibrary" may be defined as a digital library filled with works or information that can be associated with a geographic location, such as by footprint on the Earth's surface or by place name, and for which a primary search mechanism is *place* (NRC/PDG, 1999). Its users, services, metadata, and information assets are assumed to be distributed among many distinct locations but also are integrated.

It might be argued that the metaphor of "infrastructure" is more appropriate for use when focusing on the role of government in advancing technologies, policies, standards, and archives for spatial data whereas the metaphor of "library" is more appropriate when thinking in terms of a broader milieu of potential data, resources, contributors, users and services as well as when placing a heavier emphasis on the public goods aspects of an institutional arrangement. We make no such distinction. Similarly, terms such as Global Spatial Data Infrastructure (GSDI), Global Map, and Digital Earth are used to distinguish particular perspectives or to highlight specific aspects of advancement that users of those terms may be pursuing. Our intent is to encompass these and similar concepts within an overall concept we refer to as the "spatial information infrastructure." Thus, in subsequent use of terms such as "geolibrary" or "spatial data infrastructure", the subject matter addressed under one term should be broadly interpreted to include subject matter germane to the others.

Facilitating the advancement of the spatial information infrastructure requires focused research in several technical areas as well as new knowledge regarding policy, institutional, and legal alternatives that might best meet the needs of widely diverse sectors of society. Because the technical issues are addressed elsewhere in the UCGIS research agenda, we address in this chapter priority social and institutional research issues germane to the development of spatial information infrastructure concepts.

Each of the priority research topics described is directed at advancing knowledge that will help decision-makers evaluate and understand the likely consequences of choices they might make among competing policy alternatives. The consequences of information policy choices are intertwined with issues such as the ownership and control of geographic information, economics of spatial information production and dissemination, protection of personal information privacy, access to the spatial data compiled and held by government agencies, liability for spatial information products and services, and overall effects on the quality of life of our nation's people. In order to facilitate the growth and utilization of spatial information resources toward meeting societal needs, priority research topics should be directed at knowledge advancements that help policy-makers, scientists, business leaders, and citizen groups better understand the relationships between government information policies and spatial information resources, products, and services. Both reflective studies as well as prospective models are needed. Technological solutions alone will not suffice. Neither will legal or policy solutions. Technological, legal, and policy solutions must be addressed in the context of each other and concurrently. The following documentation and research agenda was prepared in 2002.

8.1.1 Background

In the early 1990s, the NRC's Mapping Science Committee articulated how spatial information handling might best be approached from an organizational perspective (NRC, 1993). This led to a plan for the creation of a NSDI, recognized as critical to national priorities (Office of President, 1994). In addition, many executive science and technology priorities, such as homeland security, science education, technology transfer, high-performance computing and networking, digital libraries, emergency and disaster response, global environmental change, and international competitiveness, have significant geographic information components, as do traditional land management activities. These priorities are mirrored at state and local levels of government. There is, however, a growing need for increased coordination between programs and for making the outcomes of these activities both appropriate and available to address social needs.

Despite large investments in geographic data development by government and the private sector, there is often a lack of knowledge of the complex policy-related issues arising from the community-wide creation, compilation, exchange, and archiving of large and small spatial data sets. Technical, legal, and public policy uncertainties interact, making it difficult to utilize information resources fully in order to pursue social goals. The ownership of digital geographic data, protection of privacy, access rights to the geographic data compiled and held by governments, and information liability are concepts that require greater clarity in the new, automated context. Observation of the social and economic ramifications of following different policy choices are needed to help guide future choices.

The government sector plays an important role in developing the fundamental spatial information infrastructure because of its activities in the systematic collection, maintenance, and dissemination of geographic data. These resources have significant uses beyond their governmental purposes. For example, subsequent use of geographic information by organizations can stimulate the growth and diversity of the information services market. At the same time, public access to government information remains essential to ensuring government accountability and democratic decision-making. Reconciliation of the tensions inherent in these and other policies is increasingly more important as we move toward global economies and international networked environments. Rigorous, impartial analysis is needed to inform decision makers on the economic, legal, and political ramifications of choosing one policy over another. Further, prospective models should be developed and offered that build on past lessons and that incorporate technological, legal, and policy considerations in the context of each other.

8.1.2 Spatial Information Infrastructure Research Tenets

The research agenda for the spatial information infrastructure as articulated below has arisen from three foundation tenets, which involve a mix of natural and social science perspectives:

- *Underlying social principles, institutions, and traditions matter:* Democratic governments must develop in ways that enhance public participation while supporting responsible use of science and technology.

- *Technology matters:* The policy issues associated with the development of spatial information infrastructures cannot be fully understood without an understanding of the technical aspects and technological possibilities for handling spatial information.

- *Information policy issues arise at all levels, from local to global and from public to private sources:* Each jurisdiction, whatever its size, has its own culture and set of practices. In the modern, automated communications environment, these jurisdictions are less independent and influence one another in new ways.

Research activities focusing on the future of the spatial information infrastructure should draw upon specialists from various academic disciplines including information science, planning, law, economics, geography, political science, and engineering as contributors to project work. In addition, perspectives and insights from experts, users, and data subjects in government, industry, and academia and citizen's groups should be drawn upon. Rigorous analysis of local, state, national, and international initiatives should be undertaken from independent, multidisciplinary perspectives. Researchers should employ various methodologies (*e.g.*, surveys, case studies, impact assessment, comparative analysis) to evaluate the ramifications of alternative legal, economic, and information policies. Multidisciplinary participation should be supported in order to deliver a comprehensive analysis that would not be possible otherwise.

8.1.3 Benefits for the Nation

The advance of electronic networks (the Internet, the World Wide Web [WWW], intranets, mobile technologies, and emerging location based services) has made it practical to share data among many organizations at all levels and over great distances. The social and economic benefits of sharing spatial data with public, private, and other government sectors are substantial and increased benefits will accrue to the nation by expanding research on priority spatial information infrastructure topics. Increasing the capabilities and usability of the spatial information infrastructure will:

- improve the efficiency and effectiveness of investments in spatial information across all major economic sectors;

- promote the continued growth of the domestic geographic information industry and stimulate economic growth generally;

- enhance public access and participation in decisions relying on spatial data; and

- contribute to the improved functioning of democratic processes generally.

Because government institutions are the single largest producers of spatial information, they can serve as model developers of a spatial information infrastructure that promotes community-wide sharing and use of spatial data and spatial technologies.

8.2 ACCESS AND LEGAL POLICIES

One aspect of *access* related to spatial information infrastructure developments involves the ways by which individuals and groups of individuals are able to respond to the constraints of effort, time, and cost in order to gain access to work, shopping, recreation, and other desirable human activities. The nature of *accessibility* is changing as many goods and services may be accessed without recourse to physical movement (Janelle and Hodge, 1998). For instance, the expansion of networks and the spread of the general information infrastructure have opened new opportunities for individuals to work and shop at a distance. Thus, information, in combination with the infrastructure that carries it, is a new and expanding resource that often replaces labor, capital, and physical resources. The evolving spatial information infrastructure offers new access opportunities for businesses, government, and individuals to accomplish effectively and efficiently activities that depend on spatial data and resources.

A second theme of access in regard to spatial information infrastructure developments concerns access as a basis of wealth and power in society. This theme is concerned with societal issues such as equity, ownership, and control. It is concerned with freedom of citizens to access government information, use and protect intellectual property, and protect their personal information privacy.

The legal right to access data is a necessary but insufficient condition for gaining and sustaining practical access to the evolving spatial information infrastructure. Practical access also requires actual physical connectivity to the infrastructure as well as to the technical tools and knowledge that make access meaningful. Regardless, legal access remains critical. While in many ways we may be drowning in data and offered services, legal scholars argue that the legal foundations of citizen rights in data and intellectual works are being undermined rapidly as society becomes more dependent on networked digital environments (NRC, 2000). The ability to control information about oneself is also being challenged severely (Agre and Rotenburg, 1998).

In the design, development, and implementation of spatial information infrastructure technological and institutional approaches, it is important to identify the processes by which diminishment in legal access is occurring, explore alternatives for halting or reversing the diminishment, and investigate models for expanding access or providing more equitable access. How the full range of human rights and corporate interests might best be protected or balanced against each other as the spatial information infrastructure expands and evolves are important research topics.

8.2.1 Access to Government Data

The U.S. Freedom of Information Act (USCS, Title 5 § 55) and the open records laws of the individual states create a balance between the right of citizens to be informed about government activities and the need to maintain confidentiality of some government information. As compared to other nations, the existence of these laws in the U.S. has increased greatly the ability of citizens and businesses to access and copy geographic data maintained or used by government agencies. Further, federal agencies are obliged to disseminate their information in a proactive manner in conformance with the provisions of OMB Circular A-130. They are particularly encouraged to disseminate freely that raw data upon which additional value-added products and services may be built by other private, government, and non-profit interests. The increased use of WWW servers by government agencies has resulted in large numbers of geographic data sets being offered freely to anyone with the ability to access them over the internet. The general deference in U.S. policies towards openness in government agency records has allowed private sector businesses and the economy to thrive (Pira International, 2000).

However, the government data policy situation is not lacking in problems and challenges in the U.S. The laws and policies affecting access to geographic data that apply to federal agencies vary greatly from those that apply to state and local government agencies. Conflicts among data policies at different levels of government have been a major impediment to the widespread sharing of geographic data in the context of an integrated national spatial information infrastructure. Means and methods for facilitating increased spatial data sharing through the resolution of competing information policies remains one of the most daunting research and development challenges. Models incorporating incentives and technological, legal and policy approaches that would allow local governments and the private sector to better contribute to development and maintenance of the spatial information infrastructure are sorely needed.

8.2.2 Licensing

The shift from the sale and purchase of books, maps, and other works in paper form to the widespread licensing of equivalent works through electronic transactions has resulted in a major shift in the practical legal rights of consumers in such works. Those utilizing licenses in digital environments argue that license provisions should and do control over conflicting copyright law provisions. Under licensing, publishers argue that the equivalent right to that of the right of a consumer to buy a book and then give it away, lend it, or sell it, is in direct opposition to a copyright holder's desires fails to exist (*i.e.* negation of the "first sale doctrine"). Similarly, the general legal principle of "fair use" in support of public interest objectives under copyright law no longer exists by default if a transaction involves a license rather than a sale. The shift to licensing has caused a major shift in the balance between public versus private rights in the data and intellectual works created within society and is greatly affecting the ability to access, use, and archive data.

Problems surrounding the licensing of geographic data and information services by government are of particular concern and will affect the overall utility and sustainability of a practical and useful spatial information infrastructure. Crafting mutually beneficial government agency, commercial sector, and citizen access arrangements requires addressing in concert with each other multiple policy, legal, and regulatory issues that affect cooperation in licensing spatial data and services. Studies need to document the ramifications of government agencies choosing to license rather than purchase geographic data and services from the private sector. Similarly, observations need to be made of the ramifications of government licensing data and services to private firms and to the public as compared to continuing open access approaches. Prospective models should be developed from the lessons learned.

8.2.3 Copyright and Public Commons Approaches

A primary objective of copyright law is to encourage the expression of ideas in tangible form so they become accessible to others. The ideas made accessible are able thereby to benefit society generally. Copyright provides an incentive to authors to make their ideas, knowledge, artistic expression, and information known so that others might use the expression for economic or social gain.

Copyright is limited both justifiably and logically from a societal perspective. As expressed in the 1986 Berne Convention (U.S. Treaty Doc. 99-27, KAV 2245) to which the U.S. is a signatory, copyright protects only expression and not facts. This has raised the specter of potential additional legislative protections for data collections. Although some works, including spatial data sets, are not protected to the extent some compilers might desire, many scholars argue that such works already are protected sufficiently by alternative laws or practical protections. Contract law, misappropriation law, and trade secret law as well as technological and code protections provide substantial protection for many data sets lacking the creativity requisite for protection under copyright. The commercial community in the U.S. is very much split on whether additional protections would be advisable. Some are arguing for much stronger protections due to the ease of copying and distributing digital data. However, preliminary empirical studies indicate that database legislation as implemented in Europe to date may actually be detrimental to the economic vitality of nations as well as damaging to the advancement of science (Mauer *et al.*, 2001). Thus new laws alone are not likely the best answer in realigning the balance of ownership rights among various commercial sector interests, private individuals, and government sector interests. Regardless, detailed empirical studies on the growth of the availability of geographic databases in Europe before and after database legislation as well as comparisons with progress in other nations should be performed.

Many national governments throughout the world are involved in developing spatial information infrastructures that will better facilitate the availability and access to spatial data for all levels of government, the commercial sector, the non-profit sector, academia, and citizens in general. A key premise in most of the initiatives is that national governments will be unable to gather and maintain more than a small percentage of the geographic data that users in their nations want and desire. Thus, the national initiatives are depending typically on the cooperation of those already gathering spatial data and using GIS to help construct, populate with data, and maintain the spatial information infrastructures for their nations. Some of the impediments to widespread spatial data sharing are well known from asking geographic data set creators why they are not currently creating metadata or making their data sets more readily available to others. Most of these impediments are unrelated to a need for increased funds. For many organizations, even if their budgets were doubled they still would not use the increased funds to make their geographic data sets more accessible to their own communities or the rest of the world. They are inhibited by further impediments that money alone is unable to address.

Common wisdom suggests that intellectual property laws and the markets they protect create the only practical environment for producing and sharing useful information. That is, profit motivations drive all major resource development. Yet the history of the web shows otherwise and provides numerous examples of massive voluntary resource production and sharing. In some instances, tens of thousands of individuals have worked collaboratively or as independent contributors in creating new knowledge resources or producing new software. In many instances these collaborative voluntary efforts have resulted in far superior information products and services than traditional commercial economies have been able to achieve (*e.g.*, Linux). Such approaches need to be explored in depth in the context of creating the incentives that will allow the spatial information infrastructure to expand and maintain itself over time.

The "public commons" may be defined as data, information, creative works, and even services whose use by others is unimpeded by intellectual property regimes. Works within the public commons may be used by others without express authorization. While works within the "public domain" are completely free of any intellectual property restrictions, we define works within the "public commons" as being available for anyone to freely use, copy, and develop derivative products. Public commons works may invoke copyright law and may have licensing or contractual restrictions imposed upon them to help ensure that they remain freely available. Thus, spatial data files distributed under an open source or open access license contribute to the "public commons" but are not by legal definition within the "public domain."

A general incentive premise of emerging spatial data sharing models is that, as individuals, most of our conduct in daily life is not driven by profit motives. For instance, many creators of spatial data have indicated they would be more than willing to share their data sets if, among other reasons, sharing was much easier to do, creators could reliably retain credit and recognition for their contributions to the public commons, and creators could acquire substantially increased liability protection from use of the data they make available to the public. Combined technological and legal models utilizing open access licensing approaches and automated detection and enforcement of creator licensing preferences show substantial promise for overcoming these and similar impediments to sharing spatial data.

Well-established mechanisms for preserving the public commons exist. These include public archives and data centers along with ever increasing web sites. Within the spatial information infrastructure development community, publicly accessible archives are already maintained by government agencies, universities, the non-profit sector, and even some private firms. Outside of the spatial data community, very innovative legal and institutional models are now being developed in the scientific, legal, and library communities for preserving and expanding the public commons in information. The potential of these new models to enhance expansion of spatial information infrastructure institutional developments should be explored fully.

In summary, pursuit of explicit conceptual models with substantial potential for providing incentives and overcoming impediments in wide-scale spatial data sharing environments should be a major thrust of the research community.

8.2.4 Personal Information Privacy

The expansion and continued development of the spatial information infrastructure will increase the ability to amass detailed information by its location and will allow such data to become much more widely available. This will have many positive benefits. Yet the ability to amass and merge information from many spatial technology applications in networked environments will also increase the potential for misuse of location information about individuals.

The legal right to privacy is essentially the right to be left alone. Developing further laws to protect individuals from location privacy intrusions by commercial or government interests may be a component but is not a complete or even realistic sole solution for protecting the location privacy interests of citizens. Similarly, a code of conduct for professionals by itself will not suffice in controlling privacy abuses in a future information environment where all members of society have ready access to track and merge location information about others. In order that new spatial technologies may be more readily adaptable for use in e-commerce and in society in general, privacy considerations need to be accounted for from the outset in spatial information infrastructure developments and in the design and coding of intelligent spatial technologies. One area ripe for investigation is the development of protection approaches that may be embedded in the code of location based services to support the privacy preferences of users of such systems.

The spatial information infrastructure as an overall multi-level and multi-participant system will need to delicately balance the privacy interests of individuals, the security interests of the nation, the trade secret interests of private companies, and the rights of citizens in democratic societies to access and scrutinize the databases compiled and used by their governments. Therefore another area ripe for investigation is development and exploration of models that account for the complex interplay among technological, legal, economic, and institutional issues in achieving such balances.

In each of the policy issue areas raised in the above section, the National Research Council, through various assembled expert committees, has already accomplished numerous studies and issued recommendations concerning legal and policy principles that should be adhered to in building and evolving the nation's information infrastructure (*e.g.*, NRC 1998, NRC/CSPA 1999, NRC 2000, NRC/CGED 2001). Whether these recommendations are being followed in the context of building the nation's spatial information infrastructure should be thoroughly assessed, particularly in regard to the intellectual property, personal information privacy, and security approaches recommended.

8.3 ECONOMICS OF INFORMATION

8.3.1 Use and Value of Geographic Information Systems

The importance of being able to effectively measure or estimate the benefits of GIS can be discussed from several perspectives. The most pressing of these may be that many public and private organizations are considering substantial new investments in GIS and the data that are necessary to their successful implementation. These investments are often on the order of tens of millions of dollars. When aggregated at the national level those investments easily amount to hundreds of millions.

Yet, the efforts of those agencies that view geographic data and technologies as vital to their interests are hampered by their inability to provide clear answers to questions concerning the value of the benefits that could be realized. The lack of clear benefit valuation methods means that in some cases we may forego investments in GIS initiatives in lieu of other initiatives simply because the benefits of GIS initiatives may be less clearly defined and understood.

Because geographic information has some of the characteristics of a public good (Didier, 1990; Taupier, 1995; NAPA, 1998; Onsrud, 1998), all the problems of estimating the demand for public goods come into play when attempting to estimate its value. It is a classic economic problem and one with many unsolved elements. One of the potential results of failure to address the problem is that we may be under or over producing geographic information due to the lack of proven methods for effectively measuring demand.

The report, *Geographic Information for the 21st Century* (NAPA, 1998) by the National Academy of Public Administrators, is primarily concerned with building a national strategy for spatial data development and dissemination and an examination of the organization of the federal mapping agency roles in support of that goal. The study is useful in illustrating the role that geographic information plays in many sectors of the economy and argues persuasively that geographic information is important to the strategic interests and competitiveness of national economies. The study includes an examination of the national economy in an effort to describe the extent to which geographic information contributes to success in eleven economic sectors, building a case that geographic information has a high strategic national value and that geographic information plays a role in about half of the economic activities of the United States. The report offers eight elements for an ideal NSDI. Among those is one that states that

> "The costs of generating, maintaining, and distributing such data are justified in terms of public benefits and/or private gains; overlap and duplication among participating organizations is avoided wherever possible" (NAPA, 1998, page xiii).

The report makes it fairly clear that such a goal is yet rather distant and summarizes the economic issues surrounding GIS with the statement that "The full importance of GI systems within global, national, regional, local, and personal contexts, however, cannot be easily described or summarized. The GIS field is still too immature, the tool kits too experimental, and the value too imprecise to make a universal assessment" (NAPA, 1998, page 10).

There is a statement in the Preface to the edited rendering into English of Michel Didier's *Utility and Value of Geographic Information* that the literature on the evaluation of the costs and benefits of geographic information and GIS is "very small and often methodologically weak." During the early part of this decade there was a minor flurry of studies dealing with the measurement of benefits from the use of GIS (Didier, 1990; Gillespie, 1992; Smith and Tomlinson, 1992). These studies occurred within the same time frame as a report by the General Accounting Office (GAO, 1990b) that criticized the efforts of the USDA Forest Service to quantify the benefits of a proposed 12-year, $1.2 billion GIS initiative. Those concerns were later resolved (GAO, 1992) and the procurement allowed to go forward. Yet, nearly 8 years later in the NAPA report (1998), concern is still expressed about lack of success in measuring benefits and the distribution of benefits from GIS.

It would be more accurate to conclude not that past studies are poorly documented but rather that very few useful studies exist. Those that do exist tend to be incomplete and often inconclusive. While these studies all lay important groundwork, most acknowledge that they have been successful in measuring only some of the assumed benefits and that other apparent benefits continue to elude efforts at quantification. Those benefits that have been addressed with the least success include positive externalities and the economic value and social benefit of new or improved products developed through the use of GIS. The methods recommended to address these shortcomings appear not to have been applied yet.

Other studies addressing the benefits of GIS have concerned themselves with approaches that have to do with measures of user satisfaction (Torkzadeh and Doll, 1988; DeLone and McLean, 1992), increased accuracy of data (Budic,

1994), gains in productivity (Tveitdal and Hesjedal, 1989; Brown, 1990; Gillespie, 1992), and other non-economic measures of system performance. All of those studies make important contributions to the means by which we evaluate the effectiveness of GIS technology.

8.3.2 Literature on the Economics of Geographic Information Systems

A review of the literature on GIS over the past two decades reveals that there have been scarcely more than a dozen articles and papers dealing with the analysis of costs and benefits. These are supplemented by a small number of studies done for U.S. and other national government agencies as well as documents developed for submission with budget requests for U.S. federal agencies. These latter documents are not widely circulated and are very difficult to identify and locate.

The 1990 French publication, l'Utilite et Valeur de l'Information Geographique, by Didier (1990) provides the most extensive treatment of the benefits and value of geographic information. This study, commissioned by the French National Council for Geographic Information, may be one of only two significant studies on the economics of geographic information systems undertaken by an economist. Its focus is on the need for a formal economic evaluation of geographic information and it "makes a proposal for doing this by applying the methodology of cost-benefit analysis (CBA) in a decision-making framework." Chapter 4 of the Didier report begins the discussion of the economics of GIS through a discussion of the economic characteristics of geospatial data and the dominance of public institutions in its generation. It continues with an examination of the microeconomics faced by producers of geographic information.

Geographic information is evaluated from the point of view of a mixed public good and it is asserted that the value of private goods generated from public geographic information can be evaluated by the market but that substantial non-market benefits exist that can only be evaluated through the use of CBA. Didier defines mixed public goods as having at least one of three conditions: the impossibility of exclusion of many users, a global obligation to use them, and the absence of a congestion effect. The first two of his conditions are encompassed by the concept of *nonexcludable* while the third is synonymous with the concept of *nonrival* (Comes and Sandler, 1986). If goods are both nonexcludable and nonrival, many economic theorists argue that such goods must be produced by government otherwise they will not be produced in society. Alternatively, the marketplace may be subsidized in order to provide an incentive to the marketplace to produce such public goods.

For instance, copyright law may be viewed as a subsidy to the marketplace in which costs and burdens of law enforcement and the courts are placed on society in order to restrict access to a good (intellectual work) that would otherwise be free for individuals to use without dispossessing others of the good. By further example, technological protections such as encryption are now being applied to many commercial information products to make them excludable. At least some geographic data and information products are already protected in this manner. Although highly controversial in its potential for effectively eliminating legal "fair use" of works, the Digital Millennium Copyright Act makes it a crime to distribute code to break through such technological protections. Thus, geographic data may be far more excludable in the future as encryption and similar technological protection techniques come into greater use. Similar to copyright law overall, the Digital Millennium Copyright Act is a subsidy to the marketplace by nature of the economic burden the law places on the courts and law enforcement. It further burdens society by potentially eliminating the economic stimulation and new innovations that "fair uses" often bring.

While some geographic information will be removed from the realm of "public goods", much will remain. For instance, public agencies will often desire or need to make their data sets widely available to the general public regardless of the state of laws protecting intellectual property or technological protections available.

The value of the products of a GIS in a commercial market is "the aggregate of what users are willing to pay." In the absence of a satisfactory solution to measuring the value of geographic information, such as is most evident in the public sector, there exists a risk of under-investing in and under-valuing geographic data. Citing the earlier classic work of Arrow (1971) and Marschak (1972), Didier discusses approaches to evaluating non-market goods. He discusses information as a reducer of uncertainty in the environment and the necessity of an alternative approach to the measurement of the value of public goods. Didier notes the particular problem that arises with geographic information wherein the marginal disposition to pay is weak but the number of users is very large. In reference to this problem he cites the need to evaluate the impacts of geographic information on the economy as a whole, in other words, in terms of net social benefits, and introduces the concept of consumer surplus into the discussion. He ends with the admonition that it is necessary to bring the problem back to observable variables in real markets and warns that direct methods such as questionnaires may lead to biased results.

The second significant study on the economics of geographic information systems by an economist is a little known 1992 study completed for the Federal Geographic Data Committee (FGDC) by U.S. Geological Survey (USGS) National Mapping Division economist Stephen Gillespie (Gillespie, 1992). Digital Benefits Test is a study that, while more narrow in focus than the Didier study of 1990, made contributions to the discussion of the valuation of benefits that were perhaps just as significant.

Its primary purpose was to collect information about a variety of successful federal GIS applications. Sixty-two applications from 21 federal agencies were selected and reviewed. Of those 62 applications, it was reported that 15 generated efficiency benefits, 32 generated effectiveness benefits, and 15 generated both efficiency and effectiveness benefits. In all cases the author felt that efficiency benefits were successfully measured but that in seven of the cases claiming effectiveness benefits they were unable to produce reliable estimates of the value of those benefits.

The contributions of this study fall into three categories: the classification of benefits, the measurement of benefits, and the use of specific measurement techniques. Benefits are classified as efficiency, effectiveness, and unanticipated (that being possible in that all cases were studied *ex post facto*). Benefit classes are clearly defined and a series of tests devised to classify them as either the efficiency or effectiveness type. Efficiency benefits are seen to arise when GIS is used to produce the same quantity and quality of output as was previously produced using other methods. Effectiveness benefits allow for the increased quality, quantity, and type of output for the same investment of resources. Unanticipated benefits can be of either type and occur unexpectedly and often as positive externalities.

The study concluded that in 24 percent of the applications reviewed GIS had not changed what the federal agency did but that in another 24 percent of cases GIS led to an extension of what the agency did. Even more remarkable was the finding that in 52 percent of all case studies that the agency did something significantly different than it did in the past, prior to the use of GIS. This is one of several reasons why the strict reliance on cost avoidance methods for calculating benefits cannot be successful in measuring the full range of benefits.

In the 30 applications involving efficiency benefits, efforts to perform the same results with GIS showed savings ranging from 25 percent to 99.5 percent. The study gives a detailed description of the process by which these benefits were measured. The description of the process for measuring the value of effectiveness benefits is, of necessity, even more detailed and involves determining how the GIS output is different, what effect it has on the user of the output, and a series of examples by which the value of those effects can be derived. The study then shows a variety of direct and indirect methods for valuation of effectiveness benefits. Thirty-one of 54 effectiveness value measurements were made using information from market based transactions while the other 23 were arrived at through indirect observations of what consumers were willing to pay for similar results obtained through other methods.

"Comments on the Economics of Geographic Information and Data Access" in the 1995 National Center for Geographic Information and Analysis (NCGIA) publication *Sharing Geographic Information* (Taupier, 1995) suggested that earlier attempts at the use of benefit-cost analysis to estimate the benefits of GIS applications had been only partially successful and that without the use of measures of willingness-to-pay that "we are being successful primarily in establishing the lower bounds of value." The paper argued that early applications of benefit-cost looked only at efficiency benefits and failed to estimate the effectiveness benefits that often exist from the use of GIS in public agencies. It further observed that to set aside the type of social benefits that may accrue from the use of GIS may serve to set aside valuable gains in public welfare, precisely the kind of gains that should be expected from government institutions. The recommended approach to the estimation of effectiveness benefits was consistent with that of Didier (1990), Gillespie (1992), and Smith and Tomlinson (1992).

8.3.3 Research Issues Surrounding Benefits from the Use of GIS

The GIS literature surrounding benefits from the use of GIS addresses two primary issues, the classification of the benefits of GIS and the methods that are acceptable for the measurement and valuation of those benefits. A third primary consideration should be that of geographic information as a public good. Consideration of the issues associated with measuring the demand for public goods offers some insights into the difficulties associated with valuing the benefits of GIS. The distinction between the benefits of applications of GIS and the public benefits of geographic information in general is an important one. While the two issues have much in common, the latter is a significantly broader and more difficult issue to address. The problem of measuring the benefits of specific applications or even multiple applications of GIS is of lesser scope and more readily solved. Its resolution may well offer useful insights into the second more difficult problem. Means for assessment of the public benefits of geographic information in general and assessment of geographic data as a public good remain as significant economic research challenges.

8.4 INSTITUTIONAL ARRANGEMENTS

8.4.1 Federal Geographic Data Committee

OMB Circular A-16 describes spatial data as a national capital asset and establishes the FGDC as the interagency coordinating body for NSDI-related activities at the federal level. It is the authority by which FGDC exists. The circular describes the responsibilities of federal agencies in:

- Acquisition, maintenance, distribution, use, and archiving of spatial data by the Federal Government;

- Development of key national reference data (framework data) and other nationally significant data;

- Development of the NSDI and of related partnerships with Federal, State, Tribal, and local governments, academia, and the private sector;

- Development and implementation of spatial data-related standards including standards for documentation (metadata) of data and geospatial services, publication of all NSDI metadata in a NSDI Clearinghouse; and

- Elimination of duplication in spatial data acquisition, maintenance, archiving, distribution, and applications among all government organizations and between the government and private industry.

The FGDC in 2002 is composed of representatives from seventeen federal agencies and nine member organizations (including UCGIS). It plays an important role in promoting ideals such as working cooperatively across federal agencies and between federal, state, and local levels, sharing data through electronic means, following mutually accepted standards, and developing common base themes of data.

The general goals and objectives of the NSDI as published by the FGDC are highly laudable. The broad goals of the FGDC in advancing the NSDI include (1) increasing awareness and understanding of the vision, concepts, and benefits of the NSDI through outreach and education, (2) developing common solutions for discovery, access, and use of geospatial data in response to the needs of diverse communities, (3) using community-based approaches to develop and maintain common collections of geospatial data for sound decision-making, and (4) building relationships among organizations to support continuing development of the NSDI (FGDC, 1997a).

To encourage participation in the NSDI, the FGDC initiated the Cooperative Agreements Program (CAP) in 1994. These small grants were designed to provide incentives to GIS to use metadata following the FGDC Geospatial Metadata Standard and to create NSDI clearinghouses. Training material to assist users in learning how to create metadata was developed. The focus on metadata and clearinghouse development was intended to support a recognition of the value of organized data, encourage the beginning of an electronic library, force data management onto the table in local jurisdictions as an issue and a budget item, and to encourage the notion that information about data should be created and maintained by those who use and know the most about the data. The FGDC recognized that those who create and know geographic data are doing so for a purpose and have the greatest incentive to create and maintain metadata. FGDC also recognized that it is relatively easy to "publish" metadata in an electronic environment but that the self-discipline to learn how to produce and continually create structured metadata is much more difficult to achieve. It is harder yet to convince others to invest in its production.

Through 1998, the CAP had funded 130 projects involving hundreds of organizations. The funding was focused on cooperative projects that provided matching funds. The grants were intended as seed money for initiating clearinghouse nodes and related initiation activities and were awarded with sensitivity to their geographic distribution. By mid-1998, 80 clearinghouse nodes and 30 different metadata tools were created. In 1999 an emphasis was shifted to seeding metadata. This focus has continued along with support of framework demonstration projects and projects integrating clearinghouse nodes with the Open GIS Consortium's Web Mapping Testbed. By early 2002, there were over 250 clearinghouse node servers in 26 different countries and several thousand data sets in the system.

Contemporaneous with the metadata focus, FGDC initiated the Framework Demonstration Projects in 1996. Framework data themes provide a foundation for integrating other data themes. The following constitute the seven framework themes of NSDI:

- geodetic control,
- orthoimagery,
- elevation,
- transportation,
- hydrography,
- the definition of boundaries and names of government units, and
- cadastral information (FGDC, 1995).

The framework demonstration program of FGDC focused on multi-sector, multi-organization partnerships to develop framework layers and to develop preliminary technical, operational, and business contexts for distributed data collection and maintenance. FGDC also initiated programs for Community Demonstration Projects and Community-Federal Information Partnerships. Because framework data and thematic data are being accumulated by numerous federal, state, and local government agencies as well as by the commercial sector at widely varying scales of resolution it is difficult to assess the success of the framework concept in practice. Methods for tracking the progress of creation and access to framework data, metadata, and thematic data over time present substantial challenges.

NSDI was further identified as an important national priority in a 1998 report produced by NAPA (1998), *Geographic Information for the 21st Century: Building a Strategy for the Nation*. The NAPA report recommended establishing a national spatial data council to give all major players a role in NSDI policy and strategy, and to perform standards and clearinghouse functions. The report outlines the charter and potential membership, but much was left unresolved in the report in defining the make-up of the proposed council (or similar body), deliberation and voting procedures, financing, and membership selection. The Geodata Alliance initiative may be viewed as one attempt to incorporate a broader constituency than authorized and incorporated in OMB Circular A-16. However, that attempt did not track closely the NAPA report recommendations and its likelihood of success is yet in doubt.

The authors of the NAPA Report recognized that many of the laws authorizing federal geographic information programs were enacted long before the advent of computers, satellite imaging, global positioning system (GPS), and GIS and sensed that the wording of many of the laws has constrained agency attitudes. The panel recommended a new law for the NSDI and amendments to existing agency charters for geographic information but the panel did not spell out in detail just what changes were needed. Legislative and alternative authority options for promoting the goals of the NSDI may be a fruitful area for scholarly investigation.

8.4.2 Additional Institutional Initiatives

FGDC is next exploring the concept of a *Geospatial One-Stop* with the potential to revolutionize electronic government services through the addition of a geographic component. The USGS is proposing the creation of *The National Map*, a database of continually maintained, nationally consistent set of basic spatial data to provide a starting point and organized means for integrating, sharing, and using spatial data easily, consistently, and quickly (USGS 2001). While the details may be debated, both the Geospatial One-Stop and National Map initiatives appear to be rational next steps with potential high payoffs for the functioning of government. Because of U.S. policies of liberal access to the information developed by government, these and similar progressive steps by other agencies promise continued high economic and social benefits for citizens, the commercial sector, and the economy generally.

The Geography Network *(http://www.geographynetwork.com)* is an example of a noteworthy commercial service that extends from, utilizes, and complements the FGDC Clearinghouse node concept. The Geography Network is described as "a collaborative and multi-participant system for publishing, sharing and using digital geographic information on the Internet" (GISLounge, 2000). The Environmental Systems Research Institute's hope is that its service in combination with the network of clearinghouse nodes will result in increased use of previously isolated public data sets.

The Global Spatial Data Infrastructure *(http://www.gsdi.org)* serves as a point of contact for those in the global community involved in developing, implementing and advancing spatial information infrastructure concepts. This effort involves participants from a large number of nations and is dedicated to international cooperation and collaboration in support of local, national, and international spatial information infrastructure developments that will allow nations to better address social, economic, and environmental issues of pressing importance. This initiative holds substantial potential for promulgating consistency among nations in regard to some spatial information infrastructure developments while enhancing understanding among nations on why consistency is difficult to achieve in regard to other developments.

The concept of "geolibraries" was introduced earlier in the chapter. Research and development efforts to expand the ability to search and retrieve information from digital libraries based on location parameters are continuing (*e.g. http://www.alexandria.ucsb.edu*). While many technical challenges remain and most efforts are still focused within the research community, operational geolibrary capabilities may ultimately provide society with powerful new tools for searching, compiling, and understanding information through location correlation. The institutional and information policy challenges of sustaining geolibraries are just beginning to be explored.

Progress in establishing spatial data creation and sharing institutional arrangements in the government, commercial, and academic sectors through exploration of what works and what does not is worthy of tracking. Further, exploration should be pursued of prospective models incorporating technical, economic, and legal theory considerations for better achieving the goals of government, science, and industry.

8.5 STANDARDS

The development of standards raises issues of timeliness, social costs and benefits, and whether or not the standards are making a positive contribution to accessibility and the costs of geographic information systems technology. This section describes the structure and activities of U.S. and international GIS standards organizations and reviews the current status of proposed international standards.

8.5.1 GIS Standards Organizations

In the U.S., the American National Standards Institute (ANSI) is the national organization with primary responsibility for the development of information technology standards. ANSI standards are voluntary, developed by consensus with input from public, academic, and commercial sectors.

ANSI accredits certain organizations to develop standards on its behalf. In the domain of information technology, ANSI has designated the International Committee for Information Technology Standards (INCITS) as the accredited standards body. The INCITS Secretariat is part of the Information Technology Industry Council (ITI), which is composed of major computer manufacturers. Examples of projects developed by the INCITS committee and its predecessor include standards for computer languages such as FORTRAN, C, and SQL, standards for encoding such as ASCII, and more recently, standards for multimedia and geographic information systems.

INCITS organizes technical committees consisting of volunteers with expertise in a particular domain. Typically, the technical committees include members from public and private agencies and organizations, professional societies, and academic organizations that have an interest in that domain. One of these technical committees is L1, Geographic Information Systems. Current members of the L1 committee include:

- American Congress on Surveying and Mapping,
- Association of American Geographers,
- Environmental Systems Research Institute,
- Federal Geographic Data Committee,
- Geographic Data Technology,
- GeoResearch, Inc.,

- George Mason University,

- Getty Information Institute,

- Lockheed–Martin Corporation,

- NASA/Goddard Space Flight Center,

- National Imagery and Mapping Agency,

- Oracle Corporation,

- U.S. Army Corps of Engineers,

- U.S. Census Bureau,

- U.S. Department of Commerce, and

- U.S. Geological Survey.

The Open GIS Consortium has participated actively in L1, but has opted for observer status because it is an international organization and has a separate liaison status with TC 211 (see below).

The duties of the L1 committee include both promulgation of national standards and representation of the U.S. in the international arena. ANSI is the official member representing the U.S. in the International Organization for Standardization (ISO). One of the standing committees of ISO is the Technical Committee for Geographic Information/Geomatics (TC 211). As the accredited ANSI committee for GIS standards, INCITS has designated L1 to serve as the U.S. Technical Advisory Group (TAG) to TC 211. The members of L1 prepare U.S. positions for TC 211 meetings, respond to ballots on TC 211 issues, serve as technical experts on TC 211 working groups and project teams, and appoint members to serve on the U.S. delegation to plenary meetings of TC 211. Because members of L1 have been very active in TC 211, several of the international standards being prepared by TC 211 are closely linked to emerging U.S. national standards and vice versa.

L1 has recently been re-organized into three task groups: Maintenance and Liaison, Adoption and Implementation, and Emerging Standards. The Maintenance and Liaison Task Group is responsible for updating and revising existing INCITS/L1 standards and liaison with other national standards bodies. The Adoption and Implementation Task Group is responsible for review and possible adoption of standards proposed by industry (particularly Open Geographic Consortium–OGC), international standards organizations, and others. The Emerging Standards Task Group is responsible for initiating and coordinating new standards work items and finding U.S. experts to participate in their development.

The projects assigned to the Maintenance and Liaison Task Group currently include: liaison with the INCITS technical committee for Structured Query Language (SQL)—multimedia; maintenance of standards for U.S. county, city and place codes; the ANSI Spatial Data Transfer Standard; and standards related to spatial data quality and feature catalogs. Projects being reviewed by the Adoption and Maintenance Task Group include the TC 211 standards that have reached the stage of Draft International Standard (DIS). The Emerging Standards Task Group is concerned with an overview of the current family of proposed TC 211 standards and with newer TC 211 work items in areas such as remote sensing and web-based mapping. Further information about L1 activities can be found on the web at *http://www.incits.org/tc_home/l1.htm.*

Many of the organizations represented in L1 also have standards committees. For example, the Federal Geographic Data Committee has a Standards Working Group and the Association of American Geographers has its Standards for Geographic Data Committee. Among other tasks, these committees provide input to the representatives serving on L1 and, in turn, receive feedback from L1 through their representatives. Relationships between L1 and various other standards bodies are illustrated in Figure 8.1.

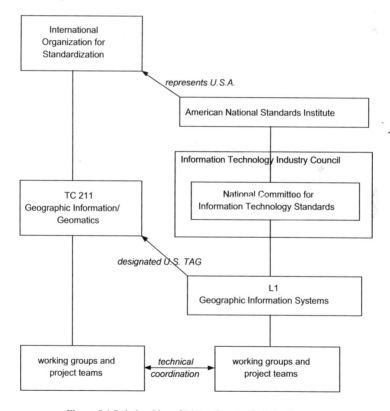

Figure 8.1 Relationships of L1 to other standards bodies

8.5.2 Current Status of International GIS Standards

TC 211 is responsible for developing a suite of standards "concerning objects of phenomena that are directly or indirectly associated with a location relative to the Earth".[1] There are currently more than 30 projects in the work plan for TC 211.[2] These mostly include projects to develop standards, and a few to develop technical reports. The standards include:[3]

- Reference model,
- Terminology,
- Conformance and testing,
- Profiles,
- Spatial schema,
- Temporal schema,
- Rules for application schema,
- Feature cataloguing methodology,
- Spatial referencing by coordinates,
- Spatial referencing by geographic identifiers,
- Quality principles,
- Quality evaluation procedures,
- Metadata,
- Positioning services,
- Portrayal,
- Encoding,
- Services,
- Schema for coverage geometry and functions,
- Simple feature access,
- Feature and attribute coding catalog data dictionary,
- Web map service interface,
- Sensor and data models for imagery and gridded data,
- Data product specification,
- Location based services; and
- Procedure for registration of geographic information items.

---

[1] (ISO) International Organization for Standardization, Technical Committee for Geographic Information/Geomatics. *Geographic Information – Part 2: Overview*, ISO/CD 15046-2, 1999-04-10, p.1

[2] ISO/TC 211 Programme of Work, updated 2002-02-16. Available on the web at *http://www.isotc211.org/pow.htm*

[3] ISO (*op. cit.*), p. 3–8

The technical reports include:[4]

- Conceptual schema language,
- Functional standards,
- Imagery and gridded data,
- Qualifications and certification of personnel,
- Imagery and gridded data components,
- Geodetic codes and parameters, and
- Imagery, gridded, and coverage data framework.

Most of the work items to develop standards have been underway since 1995. Many of the initial work items have reached the stage of DIS within the past 2 years. A DIS is approved by at least two thirds of the voting members of TC 211. Member countries may make comments when voting on a DIS. Such comments may be taken into account when preparing the "final DIS" or FDIS, which normally differs only slightly from the DIS. The FDIS is voted upon by all the member countries of ISO, whereupon it becomes an International Standard. To date, only "Conformance and Testing" has been adopted by ISO as an International Standard. This is a standard for how to test conformance to other TC 211 standards.

The current status of each of these projects can be tracked on the TC 211 web site, managed by the Secretariat in Oslo, Norway. The web address for TC 211 is *http://www.isotc211.org/*.

8.5.3 Implications of Standards and Suggested Research

Formal GIS standardization activities have been underway in the U.S. since 1980, and internationally since 1994. There is a lack of research on a number of issues related to these activities. To what extent does the emergence of informal standards meet the same needs as formal standards? As a result of the process by which formal standards are developed, to what extent have the proposed standards been driven by vendor requirements as distinct from user needs? Do a given set of standards have the effect of stifling competition and driving up prices, or have the opposite effect? Do they facilitate public access to geographic knowledge, or make access more difficult for naive users? To what extent have the proposed standards been affected, directly or indirectly, by standards and practices in related domains such as computer graphics and database technology? To what extent should they be developed in isolation from these other domains? It has also been suggested that "doing standards" has an educational value for participants and the user community that goes beyond the immediate application of the standards themselves. All of

---

[4] ISO/TC 211 (*op. cit.*), p. 5–6

these issues of indirect—and to some extent intangible—costs and benefits are potentially fruitful topics for further investigation.

One important fact stands out clearly—the length of time required between the initiation of a proposal for formal standardization and its final approval. For example, work on the Spatial Data Transfer Standard (SDTS) began in 1982. SDTS was finally approved as an ANSI standard early in 1999. New rules are supposed to speed up the process for developing and approving ISO standards, but it will have taken most of a decade, from an initial exploratory meeting in Paris in 1993, to final adoption as International Standards of the set of work items envisaged at that time.

Many individuals and public and private organizations have devoted considerable time, effort, and financial resources in support of the goal of formal GIS standardization, both within the U.S. and internationally. The amount of time required to achieve an end product, and the strong likelihood that the resulting standard may be technically out of date if not obsolete by the time it is finally approved, calls into question whether or not the investment is worthwhile. Is GIS a domain in which formal standardization is perhaps inappropriate? Skeptics will be quick to answer yes, but without further research into indirect costs and benefits, no one can be sure that the social benefits of formal GIS standardization do not significantly outweigh the costs.

8.6 PRIORITY RESEARCH AREAS

We propose four broad areas in which research will help strengthen the future of the nation's spatial information infrastructure:

- *Information policy*—The factors that shape the development of spatial information policy and law reflect traditional and contemporary culture and technology. Research is needed to identify the most advantageous government information policies and practices for promoting a robust spatial information infrastructure. Basic policy issues include intellectual property rights, information privacy, and liability as they pertain to spatial data. Perspectives from local to global vantage points and perspectives from public, commercial, scientist, citizen, user, and data subject vantage points will need to be considered. Emerging models for expanding the public commons of geographic data and services would appear to complement traditional means of production and are worthy of exploration.

- *Access to government spatial information*—Research is needed to examine how government information policies affect access to and use of spatial data for a broad spectrum of public and private sector stakeholders for a variety of public and private purposes. Approaches for balancing the competing interests of freedom to access government spatial information, community security, and personal information privacy are particularly worthy of attention. The full range of technological, legal, economic, and institutional issues should be considered in the exploration of alternative models for providing access. Public and private roles in information creation through cooperative arrangements and partnerships should be addressed.

- *Economics of information*—Geographic information is an unusual commodity of great value. Issues of cost recovery, pricing, and markets for geographic data and their relationship to intellectual property rights are of central importance. We need to achieve a better understanding of the economic characteristics of information, especially government information, through such concepts as public goods theory, network externalities, and value-adding processes.

- *Integration and local generation of spatial information*—The power of geographic information systems arises from their ability to integrate spatially referenced information from multiple sources. Locally generated information is an increasingly important component of this mix because new developments in technology make it possible for local people to more readily gather local geographic data germane to their own needs, draw data from library depositories, develop information products they need for themselves, use such data for decision-making, and contribute their locally gathered data to networked depositories. Contributions of data can be systematic or *ad hoc*, coming from civic groups, schools, local institutions, and informed individuals. Local users are often best positioned to provide updates of spatial data, identify gaps in existing data resources, and identify errors. Developing the technical and institutional means as to support creation of local knowledge and providing incentives to make such knowledge accessible to those beyond the local community present novel challenges to technologists and decision makers alike. Neither technological nor policy solutions alone will suffice. The scientific community, both technological and social and in collaboration with each other, can aid in advancing the knowledge base allowing more efficient and effective integration and sharing of data.

8.6.1 Example Research Topics

The following are examples of specific projects that might be undertaken within the context of a spatial information infrastructure research program.

- Conduct real-time case studies designed to measure the effects of different legal, economic, and information policy choices on the development of spatial information infrastructures.

- Assess geodata production and maintenance projects funded by federal, state, and local governments evaluating costs, benefits, effectiveness, and efficiencies and identifying aspects of current government information policies that need to be revised or improved.

- Explore and develop a range of institutional and legal arrangements for geolibraries that meet the needs and desires of all interested parties in accessing geographic resources, ranging from public domain to private holdings and from public to private provision of services for digital library patrons.

- Develop alternative strategies for increasing public access to government information, using digital and other emerging dissemination and retrieval technologies as a basis.

- Examine the role that pricing and cost recovery practices play in public access and commercial uses of data.

- Compare local, state, and national government dissemination policies as a means for analyzing alternative approaches for allocating public and private funds to sustain government investments in a spatial information infrastructure.

- Develop guidelines for increasing public participation in the identification, creation, use, and exchange of relevant spatial information resources to inform community decision making.

- Experiment with collaborative projects that are based on local knowledge and incorporate various types of information to support public awareness and enhance decision-making processes.

- Model the components and dimensions of an expanded view of the spatial information infrastructure focusing on technology and institutional developments and how they are embedded in other processes and media.

8.7 CONCLUSION

Research on spatial information infrastructures and geolibraries spans a broad range of technical, social, and institutional issues. The purpose of this chapter has been to focus on priority social and institutional knowledge gaps that are likely to impede the growth of the nation's spatial information infrastructure. Results achieved from the recommended research should help policy-makers, scientists, business leaders, and community groups better understand the relationships between information policies and spatial information resources, products, and services—and by so doing to facilitate the accelerated growth and utilization of geographic information resources in meeting society's future needs.

8.8 ACKNOWLEDGEMENTS

The authors wish to thank Xavier Lopez, Anne Hale Miglarese, and David Tulloch who contributed insights or made comments that helped direct the writing of this chapter. We also wish to thank the editors and anonymous reviewers.

8.9 REFERENCES

Agre, P.E. and M. Rotenburg, 1998, *Technology and Privacy: The New Landscape*, Cambridge: MIT Press.

Arrow, K., 1971, The Value and Demand for Information. *Essays in the Theory of Rick-Bearing*, Chicago, IL: Markham.

Brown, K., 1990, *Local Government Benefits from GIS*, Lexington, KY: PlanGraphics.

Budic, Z.D., 1994, Effectiveness of Geographic Information Systems in Local Planning. *Journal of the American Planning Association*, Vol. 60, pp. 244–263.

Cornes, Richard and Todd Sandler, 1986, *The Theory of Externalities, Public Goods, and Club Goods*, New York: Cambridge University Press.

DeLone, W.H. and McLean, E.R., 1992, Information Systems Success: The Quest for the Dependent Variable. *Information Systems Research*, Vol. 3, pp. 60–95.

Didier, M., 1990, *l'Utilite et Valeur de l'Information Geographique*, Paris: Editions Economic.

Federal Geographic Data Committee (FGDC), 1995, *Development of a National Digital Geospatial Data Framework*, Washington, D.C.

FGDC, 1997a, *A Strategy for the NSDI, http://www.fgdc.gov/NSDI/strategy.html*

FGDC, 1997b, *Framework: Introduction and Guide.*

General Accounting Office, 1990b, *Geographic Information System: Forest Service Not Ready to Acquire Nationwide System*, Washington, D.C.: GAO.

General Accounting Office, 1992, *Geographic Information System: Forest Service Has Resolved Concerns about Its Proposed Nationwide System*, Washington, D.C.: GAO.

Gillespie, S., 1992, *Digital Benefits Test*, Reston, VA: USGS National Mapping Division.

GISLounge, 2000, *The Geography Network*, July 5, *http://gislounge.com/features/aa070500.shtml*

Janelle, D. and Hodge, D., 1998, *Measuring and Representing Accessibility in the Information Age: Research Conference Report*, *http://www.ncgia.ucsb.edu/varenius/access/access-rpt.htm*

Marschak, J. and Radner, R., 1972, *Economic Theory of Teams*, New Haven, CT: Yale University Press.

Maurer, S.M., Hugenholtz, P.B. and Onsrud, H.J., 2001, Europe's Database Experiment. *Science*, Vol 294, 26 October 2001.

National Academy of Public Administration (NAPA) 1998, *Geographic Information for the 21st Century : Building a Strategy for the Nation*, Washington, D.C.

National Research Council (NRC) Mapping Science Committee, 1993, *Toward a Coordinated Spatial Data Infrastructure for the Nation*, Washington, D.C.: National Academy Press, *http://www.nap.edu/readingroom/records/0309048990.html*

NRC Mapping Science Committee, 1994, *Promoting the National Spatial Data Infrastructure through Partnerships*, Washington, D.C. National Academy Press, *http://www.nap.edu/readingroom/records/030905141X.html*

NRC Mapping Science Committee, 1995, *A Data Foundation for the National Spatial Data Infrastructure*, Washington, D.C.: National Academy Press, *http://www.nap.edu/readingroom/records/NX005078.html*

NRC Mapping Science Committee, 1997, *The Future of Spatial Data and Society: Summary of a Workshop*, Washington, D.C.: National Academy Press, *http://www2.nas.edu/besr/22d6.html*

NRC Steering Committee on Research Opportunities Relating to Economic and Social Impacts of Computing and Communications. Computer Science and Telecommunications Board, 1998, *Fostering Research on the Economic and Social Impacts of Information Technology: Report of a Workshop*, Washington, D.C.: National Academy Press.

NRC Committee for a Study on Promoting Access to Scientific and Technical Data for the Public Interest, Commission on Physical Sciences, Mathematics and Applications, 1999, *A Question of Balance: Private Rights and the Public Interest in Scientific and Technical Databases,* Washington, D.C.: National Academy Press.

NRC Panel on Distributed Geolibraries (PDG), Mapping Science Committee, 1999, *Distributed Geolibraries: Spatial Information Resources,* Washington, D.C.: National Academy Press.

NRC Committee on Intellectual Property Rights and the Emerging Information Infrastructure, Computer Science and Telecommunications Board, 2000, *The Digital Dilemma: Intellectual Property in the Information Age,* Washington, D.C.: National Academy Press.

NRC Committee on Geophysical and Environmental Data, Board on Earth Sciences and Resources, 2001, *Resolving Conflicts Arising from the Privatization of Environmental Data,* Washington, D.C.: National Academy Press.

NRC Mapping Science Committee, 2001, *National Spatial Data Infrastructure Partnership Programs: Rethinking the Focus,* Washington, D.C.: National Academy Press.

Office of Management and Budget (OMB), 1994, *Referenced in NAPA report or Congressional breakfast materials.*

OMB, Revised OMB Circular A-16, Final Draft, July 3, 2001.

Office of the President, 1994, *Coordinating Geographic Data Acquisition and Access: The National Spatial Data Infrastructure*, Presidential Executive Order 12906 (April 11, 1994), Federal Register (April 13, 1994), 59 (71), pp. 17,671–17,674, Washington, D.C.

Onsrud, H.J., 1998, The Tragedy of the Information Commons. *Policy Issues in Modern Cartography*, Elsevier Science, pp. 141–158.

Pira International, 2000, *Commercial exploitation of Europe's public sector information,* (Luxembourg: Office of Official Publications of European Communities).

Smith, D.A. and Tomlinson, R.F., 1992, Assessing costs and benefits of geographical information systems: methodological and implementation issues. *International Journal of Geographic Information Systems*, Vol. 6, pp. 247–256.

Taupier, R.P., 1995, Comments on the Economics of Geographic Information and Data Access in the Commonwealth of Massachusetts. *Sharing Geographic Information*, New Brunswick NJ: Center for Urban Policy Research.

Torkzadeh, G. and Doll, W.J., 1988, Test-Retest Reliability of the End-User Computing Satisfaction Instrument. *Decision Science*, Vol. 22.

Tveitdal, S. and Hesjedal, O., 1989, GIS in the Nordic Countries—Market and Technology, Strategy for Implementation—A Nordic Approach. *Proceedings of GIS 89 Symposium*, Vancouver, B.C.

U.S. Congress, 1986, *Intellectual Property Rights in an Age of Electronics and Information,* Washington, D.C.: Office of Technology Assessment.

U.S. Geological Survey (USGS), 2001, *The National Map*: Draft for Public Comment April 26, 2001, *http://nationalmap.usgs.gov/*

CHAPTER NINE

Distributed and Mobile Computing

Michael F. Goodchild, University of California, Santa Barbara,
and Environmental Systems Research Institute
Douglas M. Johnston, University of Illinois
David J. Maguire, Environmental System Research Institute
Valerian T. Noronha, Geographic Research Corporation

9.1 INTRODUCTION

Over the past few years a large number of advances in computing and communications technology have made it possible for computing to occur virtually anywhere. Battery-powered laptops were one of the first of these, beginning in the mid 1980s, and further advances in miniaturization and battery technology have reduced the size of a full-powered but portable personal computer dramatically—in essence, it is the keyboard, the battery, and the screen that now limit further miniaturization and weight reduction in the laptop. More recently, the evolution of palmtop computers and other portable devices (Portable Digital Assistants or PDAs), as well as enhanced telecommunication devices have further stimulated the trend to mobile computing. The evolution of new operating systems (MS Windows CE and JavaSoft Java OS) and software components (MS COM and JavaSoft Java Beans are the main standards) has also had major impacts. The Internet has vastly improved inter-computer connectivity, making it possible for people to share data, visualizations, and methods while separated by great distances. Wireless communication technologies, both ground- and satellite-based, now make it possible to connect from places that have no conventional connectivity, in the form of copper or fiber. In short, we are rapidly approaching a time when computing will be:

- *itinerant*, maintaining full connectivity and computational power while moving with a person, in a vehicle, or on an aircraft or ship;
- *distributed*, integrating functions that are performed at different places in a way that is transparent to the user; and
- *ubiquitous*, providing the same functionality independent of the user's location.

0-8493-2728-8/05/$0.00+$1.50

Of course we are still some way from achieving this ideal, and it is still easier and cheaper to compute in the conventional manner at a desktop using only the data, software, and hardware present in the workstation. Wireless communication is still more expensive, and more constrained by bandwidth, at least in those areas of the developed world that are well served by copper and fiber. But the pace of change is rapid, and in some countries we are already approaching the point where the majority of telephone communication will be wireless.

New devices and methods of communication suggest the need for a fundamental rethinking of the principles of geographic information systems design. Computers are likely to be used in new kinds of places that are very different from the conventional office; to communicate information using new modes of interaction, such as the pen; and to require much better use of restricted displays and computing power. On the other hand, devices are already available that offer the power of a desktop machine in a wearable package. All of these developments challenge the research community to investigate entirely new applications, and to address technical problems.

This discussion is motivated largely by technological change, for three major reasons. First, technological change induces change in society, its practices and arrangements, and in the conduct of science, and it is important to anticipate such changes. By moving quickly in response to anticipated change, society and science will be better able to take advantage of the benefits of new technology, following the pattern that has typified the history of geographic information systems over the past three decades (for histories of geographic information systems [GIS] and the role of technology in their development see Coppock and Rhind, 1991; Foresman, 1998). Second, it is to be expected that generic technologies such as those of itinerant, distributed, and ubiquitous computing (IDU for short) will require specialized adaptation to exploit their potential fully in different domains. The geographic domain is sufficiently distinct and complex that substantial research and development activity will be needed, and substantial efforts will have to be made to introduce these new technologies into the domain's many application areas. Finally, research will be needed to explore the impacts these new technologies will have, and the issues they will raise, in areas such as personal privacy, community-based decision-making, and the accumulation of power.

IDU computing is by its nature generic, so it is not immediately clear that it merits special attention by the geographic information science community, or that substantial research problems exist that if solved will have value to the GIS application domain. But two arguments seem especially compelling in this regard. First, all three characteristics of IDU computing—the ability to be itinerant, distributed, and ubiquitous—are inherently geographic. Mobility specifically means with respect to spatial location; there are clear advantages to being able to integrate computing functions across many locations; and being able to compute anywhere in space is clearly an advantage. It is implicit in this chapter, therefore, that IDU refers to capabilities in space, rather than to capabilities in time or in any other framing dimensions. IDU computing is a geographic problem, and part of this chapter is devoted to the associated question: if computing can occur anywhere, *but must occur somewhere*, where should it occur? Answers to this question are likely to have profound influence on the geographic organization of society and its activities as we enter the information age.

Second, GIS analysis is focused on geographic location, and the ability to distribute computing across geographic locations is clearly valuable. It helps to make better decisions if the associated computing can occur where it is most helpful. Effective emergency management, for example, is likely to require that decisions be made at or close to the emergency, rather than in places more traditionally associated with computing, such as the desktop or the computer center. Collection of data on geographic phenomena is often best conducted in the presence of the phenomena, where more information is available to be sensed, than remotely, though satellite remote sensing is often a cost-effective option. Scientific research is often better done through controlled experiments in the field than through simulation in the laboratory. Thus applications of geographic information technologies stand to benefit disproportionately from IDU computing.

In summary, geographic information science and the UCGIS in particular recognize distributed and mobile computing as a significant area of research because the problems that geographic information technologies are designed to address are better solved in some places than others, and because in a distributed world it is possible to distribute the software, data, communications, and hardware of computing in ways that can convey substantial benefits.

This topic of distributed and mobile computing clearly has much in common with interoperability, another UCGIS research agenda topic, and there have been discussions over the past few years over the advantages and disadvantages of merging the two. Interoperability is clearly an important requirement for IDU computing. As will become clear from reading both chapters, the technical issues of standards and problems of semantics are discussed in the chapter on interoperability, while the emphasis in this chapter is on the geographic dimensions of computing, and on how new technologies are allowing computing to occur in different places, and in different environments, with significant implications for geographic information science and GIS.

The chapter is structured as follows. The next section addresses the nature of computing, proposing that it is best understood today as a form of communication. The following section briefly reviews the history of computing from a locational perspective, and the changes that have led to substantial IDU capabilities. The fourth section discusses the technologies of IDU computing, including hardware, software, and communications. The fifth section places IDU computing within an economic framework, to address questions of costs and benefits. This is followed by a section on distributed computing from the perspectives of libraries and central facilities location theory, in an effort to cast IDU computing within the framework of traditional arrangements for the production and dissemination of information. The chapter ends with a summary of research issues, and the benefits to be anticipated if they are solved.

9.2 THE NATURE OF COMPUTING

9.2.1 What is Computing?

Computers began as calculating machines, designed to process very large numbers of arithmetic operations at speeds far beyond the capabilities of humans using tables of logarithms, slide rules, or mechanical calculators (Wilkes, 1957). Massive calculations had become necessary in many human activities, including surveying, nuclear weapons research, physical chemistry, and statistics, and the computers that began to appear on university campuses and in government laboratories in the late 1940s and 1950s permitted very rapid advances in many areas of science and engineering. Cryptography provided another well-funded and highly motivated application, in which computers were used not for arithmetic calculations but for very large numbers of simple operations on codes. In essence, the development of computing was the result of a convergence of interests between the military and intelligence communities that flourished in WWII and the early days of the Cold War, and the more general needs of science. Languages like FORTRAN reflected these priorities in their emphasis on calculation, and the representation of mathematical formulae in computer languages.

Until the 1980s the community of computer users was small, and the vast majority of the population had little need to interact directly with computing machinery. Even on campuses, the community of users formed a small elite, with its own largely inaccessible language, unusual hours of work, and seemingly endless demands for funds to purchase computing time. But the advent of the personal computer changed this dramatically. Academics with no interest in calculation suddenly found the need for a computer; and computers entered the classroom, in areas far removed from science and engineering. Computers entered the family household, and became a significant source of entertainment. Today, of course, numerical processing occupies very little of the capacity of society's computers, especially if weighted by number of users, and very few users are concerned with the code transformations of cryptography. Instead, computing is about accessing information, presenting it in visual form for easy comprehension, searching through databases, and sending information to others. Early computing was dominated by processing and transformation in the service of the user; today it is dominated by *communication*, either between users, or between the user and the machine as information storehouse.

Today, one *computes* if one uses a computing system to:

- perform a series of arithmetic operations (*e.g.*, run a population forecasting model);
- display or print information (*e.g.*, print a map or driving directions);
- transmit information to another computer (*e.g.*, share a database with another user);

- search a network of computers for information (*e.g.,* browse the World Wide Web [WWW]);

- request information from a local or remote database (*e.g.,* download an image from an archive);

- transform information using a set of rules (*e.g.,* convert data from one map projection to another);

and many other possibilities. All of these in some way involve communication of information, possibly combined with transformation, between computers and users. Users may communicate remotely with computers, and computers may communicate remotely with each other.

9.2.2 The Location of Computing

In the early days of computing there were no communication networks, and there were very strict limitations on the possible distance between the input and output peripherals, typically card readers and printers, and the central processing unit. While a user could travel away from the location of the input and output peripherals, there was a distinct cost to doing so: input and output had to be communicated physically, by shipping cards, tapes, or printed output, and thus incurred a substantial time penalty. Nevertheless, this was often considered worthwhile, since computing capacity was so expensive relative to the costs associated with delay. Thus, a user might willingly suffer a one week delay in order to obtain the processing services of a computing system at a remote campus.

Today, of course, such delays are no longer normal or acceptable. Computing has become so ubiquitous and cheap that delays are rarely experienced, and one expects to be able to connect to significant computing resources from almost anywhere. The locational pattern of computing has changed substantially in 30 years.

Nevertheless, every bit of information and every operation in a central processing unit have a well-defined location, at the scales of geographic experience (Heisenberg's uncertainty principle applies only at scales well below these). It is clear where the user is located, where bits are located on hard drives, where communications networks run, and where transformations of data occur as a result of the operation of software. From the perspective of communication, the important locations include:

- the location of the user, where information finally resides as a result of computing;

- the location of the user's workstation, which may travel with the user or may be fixed at some defined location such as an office desk;

- the locations of the network used to transmit information to and from the user's workstation;

- the locations where information is transformed or processed into the form requested by the user;

- the locations where the necessary data are stored;

- the locations where the data are input to the network or storage locations;

- the locations where data are interpreted, processed, compiled, or otherwise prepared for storage;

- the locations where the data are defined or measured; and

- the locations that are represented, in the special case of geographic data.

The last two bullets are of course particularly relevant to geographic data, since such locations always exist.

In the early days of computing all of these locations except the last must have been the same, or penalties would have been incurred through delays in shipping data on physical media. Today they can be widely separated, since data can be communicated effectively instantaneously and at little cost between locations connected by copper or fiber to the Internet, and at low cost between any other locations. A database could be in one place, fully distributed (spread over several locations), or federated (partitioned into several separate, but linked databases). In other words, computing has evolved from an activity associated with a single location to one of multiple locations, raising the interesting question: what factors determine the locations of computing?

The value of computing accrues only when information is provided to the user. As computing and communication costs fall (Moore's Law, propounded by Intel Corporation co-founder Gordon Moore, predicts that the power of a central processor chip will double every 18 months at constant cost, and similar observations apply to storage devices), the locations of the human actors in the computing task become more and more important. At 1960s prices, it was cost-effective to move the user to the computer; but at 1990s prices, it is far more cost-effective to move the computer to the user. Today, a high-end workstation costs little more than a plane ticket, and the value of the user's time is far higher than the cost of computer rental. Of the nine locations listed above, those involving human intelligence (1, 7, and possibly 8 and 9) now dominate the locational equation to a degree that would have been inconceivable three decades ago.

In a sense, then, the locations of the remaining tasks in the list do not matter. It makes little difference to the costs of computing whether data are stored in Chicago or Paris, given the existence of a high-speed network between them and the minimal costs of its use. However, there are still significant *latencies* or delays on the Internet, and they are strongly correlated with distance, at least at global scales (Murnion and Healey, 1998). Often the time-delay costs associated with long-distance communication on the Internet are sufficient to justify *mirroring* storage, or duplication of information at closer sites. No site on the Internet is

100 percent reliable, and there are therefore costs associated with expected site failure. In such cases, the costs of providing duplicate sites are perceived to be less than the costs associated with latency and down-time. Mirroring is rare within the U.S. (the Federal Geographic Data Committee's National Geographic Data Clearinghouse is a notable exception, *http://www.fgdc.gov*), but mirroring is more common internationally, reflecting the fact that much apparently distance-based latency is actually attributable to crossing international borders.

But while economic criteria may have little significance, there are nevertheless strong locational criteria associated with computing. Goodchild (1997) has argued that the following factors are important in determining where geographic data are stored and served to the Internet:

- *jurisdiction*: or the association between information about an area and the governmental responsibility for that area: information about a state, for example, is likely to be provided by a state-sponsored server;

- *funding*: since a server requires funding and creates a certain amount of employment, servers are likely to be located within the jurisdiction of the agency that funds them;

- *interest*: since geographic data are most likely to be of interest to users within the spatial extent or *footprint* of the data, they are most likely to be served from a location within that spatial extent; and

- *legacy*: since data servers often replace traditional services, such as map libraries or stores, they often inherit the locations of those services, and the institutions that sponsored them.

Nevertheless, with the trend to out-sourcing and facility management contracts it is sometimes the case that computing occurs at a third-party location unaffected by any of these factors.

Each of the other tasks listed above also has its associated locational criteria. Compilation of geographic data often occurs in public sector agencies, such as the U.S. Geological Survey, at its national headquarters or at regional facilities. Increasingly, it is possible and cost-effective to compile geographic data locally, in county offices of the U.S. Department of Agriculture, or in City Halls, or even in the farm kitchen. Ultimately, such locations are constrained by the locations of the human intelligence that is an indispensable part of the compilation process. But with wireless communication and portable hardware, that intelligence may be best located in the field, where phenomena are best observed and where observations are most comprehensive and uninhibited.

In a world of IDU computing, therefore, the locational patterns of computing are likely to adapt to those of the human institutions and intelligence that define the need for computing and use its products. This is in sharp contrast to the historic pattern, when the scarcity and cost of computing were the defining elements. Although computing in an IDU world *can* occur anywhere, its various stages and tasks must occur somewhere, and there are both economic and less tangible criteria to provide the basis for locational choice. Exactly how, and with what long-term consequences, is a research issue that geographic information science should solve, so as to anticipate the long-term effects of IDU computing on the distribution of human activities. IDU computing may alter the locational patterns that evolved prior to the use of computers to communicate information; or further alter patterns that adapted to earlier and less flexible forms of computing, such as the mainframe.

9.3 THE LOCATIONAL HISTORY OF COMPUTING

IDU is the latest of a series of forms of computing technology that have followed each other in rapid succession since the early days of computing in the 1940s. In this section, four phases of development are identified, each with a distinct set of locational imperatives.

9.3.1 Phase I: The Single-User Mainframe

From the 1940s through the mid 1960s computing technology was limited to mainframes, each costing upwards of $1 million, and financed by heavy charges on users based on the number of cycles consumed. Each user would define a number of instructions, to be executed in a batch during an interval while the user had been granted control of the machine.

High-speed communication was expensive and limited to very short distances. In essence, then, the locations of computers were determined by the distributions of their users in patterns that are readily understood within the theoretical framework of *central facilities location theory*. In classical central place theory (Berry, 1967; Christaller, 1966; Lösch, 1954)) and its more recent extensions, a central facility exists to serve a dispersed population if the demand for the service within its service area is sufficient to support the operation of the service. The minimum level of demand needed for operation is termed the *threshold*, measured in terms of sales for commercial services or size of population served for public services. The distance consumers are willing to travel to obtain the service or good is termed its *range*.

In principle, mainframes were spread over the geographic landscape in response to the distribution of demand. In practice, this demand was clustered in a few locations, such as university campuses. A few users not located in such clusters were willing to travel to obtain computing service, or to wait for instructions and results to be sent and returned by mail, because no alternative was available. The pattern of mainframes that emerged was thus very simple: one mainframe was located wherever a cluster was sufficiently large or influential to be able to afford one. In time as demand grew and the costs of mainframes fell, it became possible to locate multiple mainframes where clusters were sufficiently large. In summary, the characteristics of Phase I were:

- very high fixed costs of computers;
- a highly clustered pattern of users; and
- costs of travel for those users not located in large clusters that were low in relation to the costs of computing.

These conditions became increasingly invalid beginning in the mid 1960s, and today it is difficult to identify any legacy of this period, with the possible exception of certain large scientific data centers which still occupy locations that were determined initially by the presence of mainframes, and which still benefit from substantial economies of scale through co-location of staff and servers.

9.3.2 Phase II: The Timesharing Era

By the mid 1960s, developments in operating systems had made it possible for computers to handle many users simultaneously, through the process known as time-sharing. Although very large numbers of instructions could be executed per second, only a fraction of these would be the instructions issued by one user. As a result, it became possible for users to issue instructions and receive results *interactively* over an extended period of time, without being constrained to batch operation. This mode of operation required only relatively slow communication speeds between users and computers, speeds that could be supported by existing teletype technology over standard telephone lines. *Terminals*, consisting initially of simple teletype machines and evolving into combinations of keyboards and cathode ray tube displays, provided for local input and output. More sophisticated displays appeared in the 1970s that could display simple graphics. Only batch and advanced graphics applications required high-speed communication and thus were restricted to very short separation between user and computer.

Time-sharing changed the locational criteria of computing substantially. Computers were still massive mainframes, representing very large investments with high thresholds. But the range of their services increased dramatically. Users were no longer required to pay the costs of travel, but could obtain computing service through a terminal and a simple telephone line connection. Because a dedicated connection was required it was difficult to justify toll charges for interactive computing, but connections were free within local calling areas.

Nevertheless, batch interaction remained attractive well into the 1980s. Large data sets still had to be stored on site at the mainframe computer, since it was not possible to communicate large amounts of data in reasonable time using slow teletype technology. Remote use was feasible only for programming, the input of parameters, and relatively small amounts of output. As a result, there was little incentive to change the locational patterns that had evolved in Phase I, except in a few specialized cases.

9.3.3 Phase III: The Workstation Era

The early computers of the 1940s used vacuum-tube technology, required massive cooling, and were highly unreliable because of limited tube life. Very high costs were incurred because every component required manual assembly. Reliability improved substantially with the introduction of solid-state devices in the 1950s, but costs remained high until the invention of integrated components, and their widespread adoption beginning in the 1970s. Today, of course, millions of individual components are packaged on a single chip, and chips are manufactured at costs comparable to those of a single component of the 1950s.

Very-large-scale integration of components led to rapid falls in the costs and sizes of computers through the 1970s, until by 1980 it had become possible to package an entire computer in a device not much larger than a terminal, and to market it at a cost of a few thousand dollars. The *threshold* of computing fell by three or four orders of magnitude, until a single user could easily justify the acquisition of a computer, and the *personal computer* was born. The *range* also fell close to zero, because computing resources had become so common that it was almost never necessary for a user to travel to use a computer. Portable and laptop computers, which appeared quickly on the heels of the desktop workstation, removed the need to travel to a computer altogether. Vast numbers of new users were recruited to computing, the range of applications expanded to include such everyday tasks as word processing, and computers appeared as standard equipment in the office. Because of the economic advantages it took very little time for the necessary changes to be made in work habits: the stenography positions of the 1960s quickly disappeared, and keyboard skills became an essential part of most desk-job descriptions.

Nevertheless the mainframe computer survived well into this era. Early workstations had much less power than their mainframe contemporaries, could store and process relatively small amounts of data, and had limited software.

9.3.4 Phase IV: The Networked Era

Although the idea of connecting computers had its origins in the 1950s, and although many of the technical problems had been solved by the 1960s, high-speed communication networks capable of carrying bits at rates far above those of conventional telephone networks finally became widely available at reasonable costs only in the 1980s, with the widespread installation of fiber, microwave, and satellite links. Fiber networks were installed on university campuses, through industrial sites, and between major cities, although connections to the individual household were still largely restricted to telephone networks operating at a few thousand characters per second. Internet technology permitted bits to flow over these hybrid networks in ways that were essentially transparent to the user.

Computer applications evolved quickly to take advantage of the development of networking. *Client-server* architectures emerged to divide computing tasks between simple client systems owned by users, and more powerful servers owned by a range of providers. For many applications, software and data could reside with the server, while instructions were provided by the client and results returned. The World Wide Web represents the current state of evolution of client-server technology, with powerful servers providing services that range from sales of airline tickets and information about movies to geocoding and mapping.

Today's communication networks present a very complex geography (Hayes, 1997). A computer that is located at a node on the network *backbone* may be able to communicate at speeds of billions of characters per second, while another location may require telephone connection to a high-speed node, restricting communication to a few thousand characters per second. Other locations may lack fixed telephone connection, so communication will have to be carried over wireless telephone links. The most remote locations will be outside the coverage of wireless telephone systems, and will have to use relatively expensive and unreliable satellite communication. Economies of scale are highest for the backbone, resulting in very low costs per user or per bit; while the costs per user or bit of the so-called *last mile* or most peripheral connection may be far higher. The location theory of networks (*e.g.*, Current, 1981) provides a comprehensive framework for optimization of communication network design.

In summary, four phases of evolution of computing have completed a transition from a location pattern based on provision of service from point-like central facilities of high fixed cost, to a pattern of almost ubiquitous, low-cost facilities located with respect to a fixed communications network. In Phase I, computers established themselves wherever a sufficient number of users existed; in Phase IV it is connectivity, rather than the existence of users, that provides the most important economic determinant of location, along with a large number of less tangible factors. Over the 40-year interval the costs of computing have fallen steadily; in the context of GIS, the cost of hardware and software to support a single user has fallen from around $100,000 to $100 in today's dollars.

The next section discusses the nature of communication technologies in more detail, and also introduces new technologies of computing that extend its significance in IDU applications.

9.4 IDU TECHNOLOGIES

9.4.1 Communications

As noted earlier, the first widely available form of communication with computers was provided by a simple adoption of existing teletype technology, which allowed characters to be sent over standard telephone lines using ASCII code, at rates of a few hundred bits per second (essentially determined by the rates of typing and printing achievable in mechanical teletypes). The coded signals of the teletype were made compatible with the analog technology of telephone networks by use of *modems* (modulator/demodulator), which converted streams of bits into acoustic signals. Today, rates as high as several thousand characters per second have become routine through incremental improvements in telephone systems.

Local area networks (LANs) are communication systems that connect computers over limited geographic domains, such as a single office complex or a campus, using combinations of copper wire and fiber, and achieving rates of millions of bits per second. They are largely transparent to users, to whom the network appears as an extension of the desktop workstation, with additional storage devices and processors.

Over the past 10 years the Internet has become the most widely known instance of a *wide area network* (WAN), with services that approach those of the LAN in connectivity and transparency. Indeed it has become common to compare the Internet and the computers connected to it to a single, vast computer. Internet services, such as the WWW, provide additional layers of functionality.

At the periphery, connection is provided by a series of technologies that are much less reliable, and less and less transparent to the user. When computers move out of the office and into the vehicle or field, data communication can take place only over wireless media. This currently presents implementation hurdles, increased cost, and constraints on data communication rates. But the field of wireless communication is developing rapidly, both in the technologies available, and in the number of subscribers. In developing countries where the telephone cable infrastructure is not extensively developed, wireless voice telephony is attractive, and increasingly cost-competitive with traditional wire line. It is reasonable to expect that within a decade wireless technologies will be far more advanced and more readily available, facilitating an explosion of IDU computing.

9.4.2 Technology

Wireless communication relies on radio waves, which are part of the spectrum of electromagnetic radiation (microwaves, visible light, and x-rays are examples of waves in other ranges of the spectrum). The frequencies currently employed for data communication are about 300 kHz to 6 GHz, corresponding to wavelengths of 5cm–1,000m. Weak electrical signals—music broadcasts, taxi dispatcher instructions, GPS transmissions from satellites, or cell phone conversations—are loaded onto stronger radio waves by a process called *modulation*. The wave is transmitted at a certain frequency; a receiver tunes in to this frequency, demodulates the wave, and retrieves the electrical signal. The signal fades while traveling through the air, weakened by interference with other radio waves, and confused by bouncing off physical obstacles. Intermediate *repeater* stations may therefore amplify and relay the transmission as required. Low frequencies travel further; higher frequencies require more repeaters, which translates to more physical infrastructure. The history of wireless communication reflects a progression from lower to higher frequencies.

In early 20th century equipment, a *channel* width was about ± 60 kHz; current equipment is more precise, with widths in the range of ± 10 kHz, with ± 60 kHz widths reserved for data-rich transmissions such as video. Transmissions on adjacent frequencies interfere with each other, hence even a two-way communications channel employs two well separated frequencies, one for each direction. Clearly there is a limit on the number of simultaneous transmissions that can be accommodated within the confines of the radio spectrum. A government body (the Federal Communications Commission or FCC in the U.S.) governs the use of frequencies. In the early days of radio, frequencies were assigned only to emergency services and public agencies. It is only over the last 30–40 years that the public have been allowed to transmit in wireless media, and this has led to unprecedented demand for finite spectrum space. Two broad approaches are used to conserve the radio spectrum: (a) a frequency is assigned to a defined geographic area, typically 10–50 kilometers in radius, and the same frequency can then be re-used at several other distant locations—this is the basis of cellular telephony; and (b) messages are multiplexed, that is, divided into time or other slices, and combined into a single stream with other messages on the same frequency.

An important distinction is that between analog and digital media. In *analog* transmission, the electrical signal most closely resembles the original acoustic profile as spoken into the microphone—picture this as a smooth sine wave. Interference degrades the signal as it travels, and while repeaters can amplify the signal, they do not improve the signal-to-noise ratio. In *digital* transmission, the signal is quantized into discrete values by a process called Pulse Amplitude Modulation—picture the result as a wave stair-cased into a rectilinear path. About 8,000 samples are taken each second, so that acoustic degradation, while perceptible, is not overwhelming (by comparison, music on a CD is sampled 44,000 times per second). The next step is Pulse Code Modulation, whereby the sampled wave amplitude values are converted into binary strings. The strings are organized into frames or packets, with origin and destination tags, and auxiliary data to enable error detection and correction. Due to the error correction ability, fidelity of the signal can be maintained. Repeaters receive a corrupted signal, correct it digitally, and re-transmit the clarified signal. Digital signals are easily encrypted and stored, and fraudulent use can be limited.

Note that there are few boundaries between voice and data in terms of mode of carriage. Voice can be transmitted digitally, and conversely data can be transmitted by analog technology. There are several operational technologies for wireless communication: radio beacons, AM/FM, shortwave and TV broadcasts, two-way radios (as used in taxi dispatch), walkie-talkies and citizen's band radios, pagers, cordless and cellular telephones. They are distinguished by the portion of the radio spectrum they occupy, the power of transmission, and the effective speed of data transfer. Below we discuss three of the most likely technologies for digital data exchange: cellular telephones, spread spectrum, and FM subcarrier.

9.4.3 Cellular

Cellular telephony is currently the most popular medium for private two-way voice and data communication. An area is organized into honeycomb-like cells, and a base station in each cell operates a transceiver (transmitter and receiver) with an operating radius of 10–50 kilometers. Micro-cells and pico-cells can be set up in specific zones such as tunnels and buildings. Mobile phones communicate with the transceiver, which in turn is connected to the wired network. As a mobile unit nears the cell boundary, the base station senses a reduced signal strength, and hands off control to the neighboring cell. Within a fraction of a second, the frequency of communication changes, and the call resumes, the switch being transparent to the user.

The first wireless telephones were offered in the 1940s, but the concepts of cells and frequency re-use were developed later, and it was only in the late 1970s that automatic switching was sufficiently developed, and licensing authorities permitted widespread public use. North American service began in 1983 with the ~800 MHz Advanced Mobile Phone System (AMPS), which remains the most popular service. It is primarily an analog system, but recently there have been digital outgrowths.

9.4.4 Analog Cellular

To transmit digital data over an analog network, the data are modulated into audio signals (the "chirps" heard when a modem or fax machine establishes a connection); the audio is then transmitted in exactly the same way as a voice. Transmission rates are low, in the 2,400–9,600 bits per second (bps) range, and analog is subject to fading and other problems described above.

An alternative was introduced in 1993: Cellular Digital Packet Data (CDPD) or Wireless IP. CDPD is carried mostly over the analog cellular network. Data are organized into Internet Protocol (IP) *packets*, which are transmitted in short bursts over analog lines during lulls in voice traffic. Transmissions hop between channels in search of vacant slots. CDPD is currently one of the most effective forms of wireless data exchange, particularly for intermittent transmission, up to

1 kb at a time (circuit switching is more appropriate for continuous data communication). CDPD operates at 19.2 kbps; actual throughput rates are 8–11 kbps. Encryption, combined with channel hopping, makes CDPD extremely secure. Service is billed by the number of packets transmitted, rather than air time, making it particularly appropriate for multiple mobile units (*e.g.*, vehicle fleets). CDPD also operates on digital networks, on dedicated frequencies.

9.4.5 Digital Cellular

Digital service is new in North America, largely because the continent was relatively well served by analog AMPS in the 1980s; by contrast, in Europe, multiple analog protocols were implemented simultaneously, and a lack of standards inhibited interoperation between countries. Europe therefore took an early lead in the switch to digital technology. The Groupe Speciale Mobile (GSM) established a standard that has now been adopted in many countries. Unfortunately GSM is only marginally compatible with AMPS. Hence North America went a different route with digital cellular in the 1990s, employing voice digitization and multiplexing to create Digital-AMPS (D-AMPS), which was backward-compatible with AMPS. There is also some GSM in America, under the label PCS-1900. GSM is currently the only digital cellular technology in America that supports data transmission (D-AMPS does not), albeit at a relatively slow 9.6 kbps. In this context it is worth noting that the term PCS is used with liberty, and that some services sold under the title PCS are based on D-AMPS technology. Figure 9.1 summarizes the cellular options in North America today.

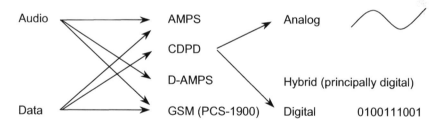

Figure 9.1 Cellular options in North America. AMPS is the most widely available

9.4.6 Spread Spectrum

The principle of spread spectrum (SS) is to spread a signal very rapidly over a wide range of frequencies (similar to frequency hopping in CDPD), according to a specific pattern. If a receiver knows this pattern, it can de-spread the signal and recover the data. Current technology allows frequency hopping at a rate of about

40 per second. The method is resistant to jamming and interference, and is inherently secure because the receiving party must know the frequency hopping schedule. For this reason, although SS was originally developed about 50 years ago, it was largely restricted to the military until about 1985. It is now growing rapidly because of the high data rates it enables, and because it allows multiple users to share the same frequency space. Wireless wide and local area networks are being developed based on spread spectrum, and it is being proposed as a generic means for users to access Information Service Providers (ISPs). The drawback is cost, which is currently high. Three radio bands are now reserved for spread spectrum, at approximately 900 MHz, 2,400 MHz, and 5.8 GHz, and licensing requirements are minimal.

9.4.7 FM Sub-Carrier

FM sub-carrier technology is the basis of the Radio Data System (RDS) in Europe, or Radio Data Broadcast System (RDBS) in the United States. Just as CDPD makes use of unused voice cellular bandwidth, RDS exploits an unused portion of commercial FM broadcast waves. Unlike cellular and spread spectrum, this is a one-way broadcast technology, with data targeted to subscribers to an information service, rather than to the individual user. RDS is used for broadcasting digital traffic reports from Traffic Management Centers (TMCs), hence the acronym RDS-TMC. A decoder in the vehicle can be programmed to filter out traffic messages that do not pertain to the current route, and to accept text data transmissions only in a given language. RDS also has a feature to override an audio station, and to wake up a radio from sleep mode. It is therefore well suited to disaster warning. Another popular RDS application is real-time differential GPS correction, where 100 MHz RDS competes against 300 kHz radio beacons.

RDS is not yet widely established in the U.S. There are message standards issues to be resolved, and location referencing in particular is an area in which standards are still under development (*e.g.*, Noronha *et al.*, 1999). Currently there are two RDS protocols vying for adoption in the U.S.: Sub-Carrier Traffic Information Channel (STIC), developed specifically for use in Intelligent Transportation Systems, and Data Audio Radio Channel (DARC). STIC receivers are expensive and require more power, but enable slightly higher data rates, 18 kbps versus DARC's 16 kbps.

9.4.8 Implementation Issues

It is clear from the discussion above that there are numerous considerations in selecting a technology for wireless communication: (a) is communication one-way or two-way; (b) is it directed to an individual, or a community of subscribers, or is it broadcast to the general public; (c) is encryption required, and are there security concerns; (d) are voice and data both required; (e) how portable is the transceiver; and what are its power requirements; (f) what data speeds does the technology support; and (g) what are the costs and benefits compared with local data storage (as opposed to real-time data transmission), wireline, and other wireless technologies.

Consider a police fleet of say 100 vehicles, communicating with a base station in an Automated Vehicle Location (AVL) application. With some technologies, each mobile unit requires say 10 seconds to connect, transmit data, and sign off. If all 100 vehicles use the same communications channel, it would be 15 minutes between location reports for a given vehicle—this does not qualify as real-time tracking. On the other hand, technologies such as CDPD do not require sign-in for each transmission, and are clearly more appropriate to the application.

9.4.9 Field Computing Technologies

There are many reasons for wanting to access powerful computing facilities in the field. The field is where many geographic data are collected, by direct measurement, interview, observation, photography, or other suitable means. These data must be made available to their eventual users, either by processing in the field and transmitting interpreted information, or by transmitting raw data to some other site for processing. Data collection may require other supporting information, such as base maps or results of earlier observation to support change detection, and these data may be in digital form, allowing them to be downlinked and used in the field if suitable technology is available. Decisions of a geographic nature are often best made in the field, and in cases such as emergency management may require extensive and rapid processing of appropriate data.

Technologies suitable for field computing have advanced rapidly in recent years. The first truly portable computers appeared in the mid 1980s, following important developments in battery technology, lowered power consumption, and greater computational power and storage. Laptops are now commonly used in field settings, by scientists and decision makers who require access to information technology. More recently, improvements in wireless communications have made it possible to operate virtually anywhere, though at relatively low communications speeds, and reliability of communications remains an issue.

Field computing technologies now include many devices besides the laptop. To allow further size reductions, many systems have dispensed with the keyboard, replacing it with other interaction modalities such as the pen and sensitive screen, and making use of software for limited recognition of handwriting. Egenhofer (1997) has explored the potential for communication of geographic data by sketch, and voice recognition and speech synthesis are also promising technologies for human-computer interaction in the field. Screen size remains an issue for geographic data, though, given the need for visual communication and high visual resolution.

Clarke (1998) reviews recent advances in various types of wearable field devices. Entire computing systems have been constructed to be worn under clothing, and have become notorious for their use in increasing the odds of winning in various forms of gambling. It is possible, for example, to input data by pressing sensitive pads with the toes, and to receive output through miniature devices concealed in eyeglasses. Several gigabytes of storage can be concealed in clothing, along with powerful processing units. Of particular relevance to field GIS are devices that make it possible to see visual data displayed *heads up* in a unit that resembles a heavy pair of eyeglasses, and combines actual and virtual views. These devices are now used routinely in assembly plants, since they allow their users to see blueprints while working on assemblies. They could be used in field GIS to provide visual access to base maps, images, maps showing the landscape as it existed at some earlier time, or simulations of future landscapes. Systems such as these that *augment* reality are similar in some respects to *immersive* technologies, which replace reality with virtual renderings (*e.g.*, Earnshaw *et al.*, 1995).

Goodchild (1998) argues that GIS should be seen as an interaction not only between human and computer, but between human, computer, and geographic reality (HCRI rather than HCI). A GIS database cannot be a perfect rendition of reality, since reality is infinitely complex but the database must always be finite, and thus some degree of approximation, generalization, or abstraction will always be necessary. Given this, effective GIS will always require interaction with the reality that the database imperfectly represents. Field GIS is an instance of HCRI, in which human and computer interact in the presence of the phenomenon, allowing the human to interact with reality while interacting with the computer. Field GIS allows direct access to ground truth, as well as to any digital representations of existing and prior conditions. As such, it promises to vastly improve the effectiveness of field research, and to open much more efficient channels of communication between field workers and the eventual users of their data.

9.4.10 Distributed Computing

The networking functions provided by the Internet and its services support a wide range of modes of interaction between computers (for a review of GIS applications of the Internet see Plewe, 1997). But these are largely limited today to binary interactions between two computers. In a client-server environment, operations are divided between the client, which is commonly a machine of limited processing, storage, and software capabilities, and a server, which may have access to large resources of data, and much more powerful software and processing. Suppose, however, that a GIS user wishes to obtain two sets of data for the same area, and these are located on different servers. It is possible to download one set of data from one server, disconnect, and reconnect to the second server to download its data. It is not possible, however, to access both servers simultaneously, or to make use of services that avoid download of data. If a map is needed, it must be computed at the client from two downloaded data sets. This is in contrast to a map based on a single data set, which can be computed at the data set's server, avoiding the need to download data or to have mapping software at the client.

Thus while the Internet and WWW offer many powerful functions, they fail at this time to support many important but advanced capabilities that are part of the vision of truly distributed computing:

- simultaneous access to multiple servers from a single client;

- support for true distributed databases, such that the user sees a single database, but tables or even parts of tables are resident at different server sites; and

- support for truly distributed software, such that the user sees a single software environment but modules remain resident at different server sites.

In all three of these areas there are active research projects and prototypes, but no comprehensive solution yet exists.

The WWW Mapping Special Interest Group of the Open GIS Consortium (*http://www.opengis.org*) seeks to develop prototype demonstrations of solutions to the first bullet above, for GIS applications. Its aim is to make it possible for a client to access layers of data resident on multiple servers, while allowing the user to work with them as if they were all resident locally. For example, a user should be able to display and analyze a layer of soils and a layer of roads as if they were in the user's own GIS, without downloading them from their respective servers. Many problems will have to be overcome, including specification and adoption of standards, before this vision becomes a practical reality, despite the fact that it is already reality in the case of data on a single server.

In the case of the second bullet, there is much interest in the GIS community in making it possible for different agencies to own and maintain different parts of a single, unified database. For example, responsibility for the fields in a street's database might be divided between a mapping agency, with responsibility for defining and maintaining the basic geometric and topological framework of streets; and a transportation agency responsible for adding dynamic information on levels of congestion and construction projects. Since the latter will be represented as attributes in tables defined by the former, it is clearly preferable that parts of the database exist on different servers. But there are numerous problems in developing the technology of distributed databases (Burleson, 1994). Maintenance of integrity is particularly difficult, if different users are to be allowed to access and modify parts of the same tables simultaneously. An open question is the extent to which distributed GIS databases will be based on generic distributed database technology, or supported using conventional technology through administrative arrangements and protocols.

In the case of the third bullet, much effort is currently under way in the GIS software industry to exploit modern capabilities for distributed components. The underlying standards and protocols are being developed by the Open GIS Consortium (*http://www.opengis.org*), a group of companies, agencies, and universities dedicated to greater interoperability in GIS. Already it is possible to integrate GIS and other software environments, by allowing:

- a user in a GIS environment to access directly the services of some other environment, provided both are resident locally;

- the reverse, for a user in a non-GIS environment, such as a spreadsheet, to access GIS functions without leaving the current environment; and

- a user to combine the services of a remote host by integrating them with the services of a local client.

Much more research needs to be done, however (see, for example, Goodchild *et al.*, 1999), before true distributed processing will be possible in GIS.

The benefits of distributed processing are clear, however. GIS applications are often complex, in areas such as environmental modeling or vehicle routing. A developer of capabilities in such areas may find it much more attractive to make software available in modules than as part of monolithic software, and may even insist on retaining complete control, requiring users to send data to the owner's host. The possibility of "sending the data, not the software" raises numerous issues of ownership, institutional arrangements, and protection of intellectual property.

9.4.11 Distributed Production

Traditionally, production of high-quality, reliable geographic data has been the almost exclusive domain of central governments. Every country has invested in a national mapping agency, sometimes under military command and sometimes in the civilian sector (Rhind, 1997). Today, several trends suggest that we are moving away from that model into a more complex, distributed set of arrangements with an emphasis on local production and use. The simple model of a flow outwards from the center to a dispersed user community is being replaced by a much more complex model involving various forms of data sharing (Onsrud and Rushton, 1995). This is occurring for several reasons:

The economics of geographic data production have changed dramatically. New sensors, new instruments, and new software make it possible for geographic data production to take on the characteristics of a cottage industry, with farmers, for example, able to afford sophisticated technology for mapping and monitoring their own fields and crops (Wilson, 1999). The fixed costs of mapping in particular have fallen as the set of applications has expanded, leaving little economic incentive or advantage in centralized production.

Contemporary political trends are against large central government, and for privatization of what previously were largely government functions. In the United

Kingdom, for example, the Ordnance Survey has been forced to operate on commercial lines, and similar patterns can be observed in most countries. In the United States, commercial production of data is now considered profitable even though much government data is available free, because the private sector is seen as more responsive to the needs of users.

Modern technology allows much more flexible approaches to geographic data production. The "wall-to-wall" coverage mandated for government mapping agencies is no longer appropriate when most users access data through technologies that can accommodate data of varying accuracy and resolution. Future coverage is likely to be on a *patchwork* basis, rather than the uniform coverage that is the expressed objective of national mapping agencies.

If much geographic information is produced locally, and much of it is also consumed locally, there is little reason to integrate data at a broader scale. The interests of national governments are more likely to be satisfied by coarse-scale generalizations, leaving detailed data to be produced, distributed, and consumed locally.

9.5 AN ECONOMIC FRAMEWORK

The previous section on locational history has already hinted at how the location of computing might be placed within an economic framework. This section expands on that theme, and presents a basis for research on the costs, benefits, and economic value of distributed and mobile computing.

From the communications perspective established earlier, computing is seen as a process of information transfer from one person, group, or agency to another. Locations are associated with both sender and receiver, since the human intelligence associated with both must be located somewhere. Locations are also associated with storage, processing, and all of the other stages identified in Section 9.2.2.

9.5.1 Transport Costs

Various costs are associated with the need to overcome the separation between the locations of sender and receiver, and with other separations such as that between the location of geographic ground truth and that of the sender. In classical location theory these costs are tangible, being determined by the costs of moving people or goods from place to place. In the case of information, however, there are both tangible costs, related to renting communication links and the fixed costs of establishing them, and intangible costs related to delay and unreliability.

Consider the costs associated with sending information between locations i and j. Several distinct conditions are relevant in today's environment.

- There exists a fixed communications link of high capacity, whose fixed costs have been absorbed by some agency and for which rental is free. Tangible costs are zero, as are intangible costs. This is the normal case when the Internet is accessible and both sender and receiver are in the U.S.

- There exists a fixed high-capacity communications link, but it has significant latency; while tangible costs are zero, there are substantial intangible costs to using the link.

- No fixed high-capacity communications link exists. Communications must rely on low-capacity links such as telephone lines, which may incur rental charges; or on wireless links with rental charges and substantial unreliability. There are both tangible and intangible costs associated with communication of information.

- When costs exist, they begin to influence locational decisions, and either the sender or receiver may choose to relocate, other locations needed for the communication of information may be affected, or communication links may be chosen, to minimize transport costs.

9.5.2 Facility Costs

Various facilities are needed for communication to occur between sender and receiver. Computer processing may be needed to modify or transform data, or to provide analysis or modeling. Processing will also be needed at the locations of servers, and at other nodes in the communications network. As in the previous section, many of these processing resources are available free because they are part of the Internet, or provided through other arrangements to the sender or receiver. Others must be covered by the users through rental or fixed cost charges. Locational decisions may be involved if there is the possibility of selection among alternative processing locations, or alternative computing resources. Costs may be tangible, when processing power must be rented or purchased, but they may also be intangible, as when choices exist between the computing resources of different agencies, and issues such as security, privacy, or intellectual property are important.

9.5.3 Human Intelligence as a Locational Factor

Finally, the locational decision will be influenced by the need to consider the locations of human actors in the system. If GIS is a communication problem, as suggested here, then the locations of sender and receiver are both important. Other human actors may also be involved, as interpreters, custodians of ancillary data, or developers of software. In Section 9.3 it was argued that in earlier phases of the history of computing it was common for human intelligence to move in response to the availability of computing resources. The decision to compute in the office rather than the field may also be an instance of moving human intelligence to the location of computing resources, rather than the reverse. As computing becomes cheaper and the costs of communication lower, it will be increasingly common to move computing to human intelligence, rather than the opposite; and arguably that process is already almost complete with respect to locations provided with power supplies and Internet connections. Changing economics of computing and emerging field technologies will have substantial influence on the locational decisions made by GIS users.

Locational decisions such as those discussed in this chapter will clearly impact where computing is done, and where users choose to locate, both for scientific and for decision-making applications of GIS. An important area of research is emerging, in the development of models and frameworks that allow such decisions and options to be explored in a rigorous, well-informed framework that can make use of our increasing understanding of the costs of computing and communications.

9.6 LIBRARIES AND ARCHIVES

Like any other institution, libraries are feeling the influence of the shift to digital communication, and the concept of a *digital library* has received much attention in recent years. In principle a digital library is usable entirely through the Internet, and thus achieves universal access. Its catalog is digital, and all of its books and other information resources are also in digital form. Novel forms of searching are possible; information can be sorted and delivered independently of the media on which it was originally stored; and information can be processed and analyzed either by the library or by its users. In short, the digital library holds enormous promise for increasing humanity's access to its information resources. This section discusses several aspects of digital libraries of relevance to geographic information, in the context of distributed and mobile computing.

9.6.1 The Search Problem

Much of the information used in GIS analysis is *framework* data—largely static data produced and disseminated for general purposes by government agencies and corporations (MSC, 1995). Framework data sets include digital imagery and digital topographic maps; more specifically, they include representations of terrain, hydrography, major cultural features, and other information used to provide a geographic frame of reference.

Such data sets are ideally suited to dissemination through libraries and similar institutions, where they can be stored, maintained, and lent in a reliable manner. Many libraries maintain extensive *collections* of maps and their digital equivalents, and increasingly the services of libraries are available online via the Internet. Many data archives, data centers, and clearinghouses have emerged in recent years, largely following the library model, for the purpose of serving specialized communities of users, and the National Geospatial Data Clearinghouse (NGDC; *http://www.fgdc.gov*) is an excellent example.

Today the WWW includes some 100 million servers, any one of which might contain and be willing to provide a collection of geographic data sets. Any given data set may be available from several sites. In effect, the WWW serves as a distributed library of data resources, and is expected to grow massively in this role in the coming years. At the same time the WWW presents a growing problem for its users: how to find relevant information among the myriad sites of the network? Users must work through a two-stage process of search, first finding a site within the set of Internet sites, and then finding data within the site.

The second step in this process is often relatively straightforward, because owners of sites are able to set up effective cataloging and searching mechanisms within the sites under their control. But the first step is much more problematic, since few standards and protocols exist for the WWW as a whole. How, for example, is a GIS user in need of a digital base map for a small area of Utah to find one among the vast array of servers and services?

At this point in time, and in the absence of effective search mechanisms for geographic data, the user is forced to rely on one or another of the following heuristics:

- Assume that all servers possess the needed data. This is a reasonable strategy for major libraries, since every major library attempts to include all of the most important books in its collection, but is clearly absurd for the WWW.

- Use one of the WWW search services. The current generation of services provides powerful means for searching across the WWW, but relies almost entirely on key words. Since it is difficult to express location in words, and since these services are not effective at detecting the existence of geographic data, this strategy is generally ineffective.

- Go to a site that serves as a clearinghouse. The NGDC provides a mechanism for searching across a large number of collections, using common protocols for data description. If all geographic data could be registered with it, this would be an effective method of search. But it is very unlikely that all owners of geographic data will be willing to invest the necessary time and effort; and it is likely that more than one clearinghouse will emerge, and that protocols will vary among them.

- Rely on personal knowledge. In practice most searches for data rely on some form of personal knowledge, personal network, or other form of informal communication.

Effective mechanisms for searching distributed archives of geographic data would be very useful, in helping users with little personal knowledge to exploit the massive resources of the WWW. The National Research Council (MSC, 1999) has defined the concept of a *distributed geolibrary*, a network of data resources that is searchable for information about a given geographic location, or *place-based search*. Search based on location is already possible within certain sites, including the Alexandria Digital Library (*alexandria.ucsb.edu*), but not across the WWW itself. Development of such methods should be a major research priority of the geographic information research community.

9.6.2 Libraries as Central Places

Although the previous section has laid out the principles underlying search in the era of the WWW, the distribution of geographic data sets among WWW sites reflects far more the legacy of previous technologies and approaches. For example, the Alexandria Digital Library has as one of its objectives the provision of access to the rich holdings of the University of California, Santa Barbara's Map and Imagery Laboratory, a large collection of paper and digital maps and imagery. This collection has been built up over the years to serve a population of users largely confined to the UCSB campus and its immediate surroundings. It reflects the fact that it is the only accessible store of geographic data for that user community, and thus the collection has attempted to prioritize acquisitions on that basis. But this strategy is the exact opposite of what is needed in the WWW era, where distance is relatively unimportant, and where the user's problem is to find the collection most likely to contain a given data set. In the earlier era, it made sense for every major collection to try to include all important items; today, it makes more sense for collections to specialize, so that an item is present in a very small number of collections, and so that there are clear guidelines regarding where to search.

This transition is readily understood within the context of central place theory, as discussed earlier in Section 9.3. The WWW has increased the range of archival service, and has reduced its threshold. Rather than every site offering the same good, it is now possible for each site to specialize, and specialization also helps the search process by providing ready definitions of, and limits to, the contents of each collection. In the language of central place theory (Berry, 1967), a dense pattern of offerings of a low-order good is being replaced by a sparse pattern of offerings of a large number of high-order goods.

The literature of central place theory provides few clues as to how this transition will occur. The *adoptive hypothesis* (Bell, 1970) argues that it will occur by a form of Darwinian selection, in which those sites that are most pro-gressive will ultimately force out the others. The speed with which this process operates depends very much on the far-sightedness of individuals, because the economic and political forces underlying it are comparatively weak, allowing inefficient and inappropriate sites to survive well into the new era.

9.7 CONCLUSIONS

9.7.1 Research Issues

Many research issues have been identified in the previous discussion. This section summarizes them, as a series of priority areas for research.

- Examine the status and compatibility of standards across the full domain of distributed computing architectures and geographic information at national and international levels; identify important gaps and duplications; examine the adaptability of standards to rapid technological change; evaluate the degree to which geographic information standards and architectures are compliant with and embedded in such emerging frameworks as Java, CORBA, and COM/OLE; recommend appropriate actions.

- Build models of GIS activities as collections of special services in distributed object environments to support their integration into much broader modeling frameworks. This will help promote the longer-term objective of making GIS services readily accessible within the general computing environments of the future.

- Develop an economic model of the distributed processing of geographic information; include various assumptions about the distribution of costs, and use the economic model to develop a model of distributed GIS computing.

- Modify commonly used teaching materials in GIS to incorporate new material about distributed computing architectures.

- Develop methods for the efficient use of bandwidth in transmitting large volumes of geographic data, including progressive transmission and compression; investigate the current status of such methods for raster data; research the use of parallel methods for vector data.

- Develop improved models (*i.e.*, structure and format) of geographic metadata to facilitate sharing of GIS data, to increase search and browse capabilities, and to allow users to evaluate appropriateness of use or allow compilers to judge fitness of data for inclusion in GIS.

- Develop theory that addresses the optimal location of computing activity, building on existing theories of the location of economic and other activities and on the economic model described earlier.

- Study the nature of human–computer interaction in the field, and the effects of different interaction modalities, including speech, sketch, and gesture.

- Develop new adaptive methods of field sampling that are directed by real-time analysis in the field.

- Study the role of contextual information gathered in the field by new technologies and used to inform subsequent analysis of primary data.

- Examine the implications of IDU computing with respect to intellectual property rights to geographic information and within the context of broader developments in this area.

- Examine the social implications of IDU computing and its impacts on existing institutions and institutional arrangements.

- Conduct case studies examining the application of IDU computing in GISs, including horizontal applications (with data distributed across different locations), and vertical applications (with data distributed at different levels in the administrative hierarchy).

- Monitor the progress of research addressing the technical problems that IDU computing architectures pose with regard to geographic information, including maintenance of data integrity, fusion and integration of data, and automated generalization.

- Examine the various architectures for distributed computing and their implications for GIS. This will include consideration of distributed database design, client-server processing, database replication and versioning, and efficient data caching.

9.7.2 Anticipated Benefits from Research

- *Access.* By decentralizing control, distributed computing offers the potential for significant increases in the accessibility of information technology, and associated benefits. There have been many examples in recent years of the power of the Internet, wireless communication, and other information technologies to bypass the control of central governments, linking citizens in one country with those with common interests around the world. Wireless communication avoids the restrictions central governments impose through control over the installation of copper and fiber; digital communication avoids many of the restrictions imposed over the use of mail.

- *Cost reductions.* Modern software architectures, with their emphasis on modularity and interoperability, work to reduce the cost of GIS by increased competition and sharing, and by making modules more affordable than monolithic packages.

- *Improved decision-making.* Current technologies virtually require decisions that rely on computing support to be made at the desktop, where powerful hardware and connectivity are concentrated. IDU computing offers the prospect of computing anywhere, resulting in more timely and more accurate data and decisions.

- *Distributed custodianship.* The National Spatial Data Infrastructure (NSDI) calls for a system of partnerships to produce a future national framework for data as a patchwork quilt collected at different scales and produced and maintained by different governments and agencies. NSDI will require novel arrangements for framework management, area integration, and data distribution. Research on distributed and mobile computing will examine the basic feasibility and likely effects of such distributed custodianship in the context of distributed computing architectures, and will determine the institutional structures that must evolve to support such custodianship.

- *Data integration.* This research will help to integrate geographic information into the mainstream of future information technologies.

- *Missed opportunities.* By anticipating the impact that rapidly advancing technology will have on GIS, this research will allow the GIS community to take better advantage of the opportunities that the technology offers, in timely fashion.

9.8 ACKNOWLEDGMENT

The authors thank Dawn Wright of Oregon State University for her help in synthesizing the discussions that occurred on this topic in the UCGIS Virtual Seminar in Fall 1998.

9.9 REFERENCES

Bell, T.L., 1970, A test of the adoptive hypothesis of spatial-economic pattern development: The case of the retail firm. *Proceedings of the Association of American Geographers,* 2: 8–12.

Berry, B.J.L., 1967, *Geography of Market Centers and Retail Distribution,* Englewood Cliffs, NJ: Prentice Hall.

Burleson, D.K., 1994, *Managing Distributed Databases: Building Bridges between Database Islands,* New York: Wiley.

Christaller, W., 1966, *Central Places in Southern Germany* (Translated by Baskin, C.W.), Englewood Cliffs, NJ: Prentice Hall.

Clarke, K.C., 1998, Visualising different geofutures. In Longley, P.A., Brooks, S.M., McDonnell, R. and Macmillan, W. (eds.), *Geocomputation: A Primer*, London: Wiley, pp. 119–138.

Coppock, J.T. and Rhind, D.W., 1991, The history of GIS. In Maguire, D.J., Goodchild, M.F. and Rhind, D.W. (eds.), *Geographical Information Systems: Principles and Applications*, Harlow, UK: Longman Scientific and Technical, Vol. 1, pp. 21–43.

Current, J.R., 1981, *Multiobjective design of transportation networks*. Unpublished PhD Dissertation, Johns Hopkins University.

Earnshaw, R.A., Vince, J.A. and Jones, H. (eds.), 1995, *Virtual Reality Applications*, San Diego: Academic Press.

Egenhofer, M.J., 1997, Query processing in spatial-query-by-sketch. *Journal of Visual Languages and Computing,* 8(4): 403–424.

Foresman, T.W., (ed.), 1998, *The History of Geographic Information Systems: Perspectives from the Pioneers*, Upper Saddle River, NJ: Prentice Hall PTR.

Goodchild, M.F., 1997, Towards a geography of geographic information in a digital world. *Computers, Environment and Urban Systems,* 21(6): 377–391.

Goodchild, M.F., 1998, Rediscovering the world through GIS: Prospects for a second age of geographical discovery. *Proceedings, GISPlaNET 98,* Lisbon, CD–ROM.

Goodchild, M.F., Egenhofer, M.J., Fegeas, R. and Kottman, C.A. (eds.), 1999, *Interoperating Geographic Information Systems*, Norwell, MA: Kluwer Academic Publishers.

Hayes, B., 1997, The infrastructure of the information infrastructure. *American Scientist,* 85 (May–June): 214–218.

Lösch, A., 1954, *The Economics of Location* (Translated by Woglom, W.H.), New Haven, CT: Yale University Press.

Mapping Science Committee, National Research Council, 1995, *A Data Foundation for the National Spatial Data Infrastructure*, Washington, D.C: National Academies Press.

Mapping Science Committee, National Research Council, 1999, *Distributed Geo-libraries: Spatial Data Resources*, Washington, D.C.: National Academies Press.

Murnion, S. and Healey, R.G., 1998, Modeling distance decay effects in Web server information flows. *Geographical Analysis,* 30(4): 285–303.

Noronha, V.T., Goodchild, M.F., Church, R.L. and Fohl, P., 1999, Location expression standards for ITS applications: Testing the Cross Streets Profile. *Annals of Regional Science.*

Onsrud, H.J. and Rushton, G. (eds), 1995, *Sharing Geographic Information*, New Brunswick, NJ: Center for Urban Policy Research, Rutgers University.

Plewe, B., 1997, *GIS Online: Information, Retrieval, Mapping, and the Internet*, Santa Fe: OnWord Press.

Rhind, D.W. (ed.) 1997, *Framework for the World*, New York: Wiley.

Wilkes, M.V., 1957, *Automatic Digital Computers*, London: Methuen.

Wilson, J.P., 1999, Local, national, and global applications of GIS in agriculture. In Longley, P.A., Goodchild, M.F., Maguire, D.J. and Rhind, D.W. (eds.), *Geographical Information Systems: Principles, Techniques, Management and Applications*, Second Edition. New York: Wiley, Vol. 2, pp. 981–998.

CHAPTER TEN

GIS and Society: Interrelation, Integration, and Transformation

Gregory A. Elmes, West Virginia University
Earl F. Epstein, The Ohio State University
Robert B. McMaster, University of Minnesota
Bernard J. Niemann, University of Wisconsin–Madison
Barbara Poore, U.S. Geological Survey
Eric Sheppard, University of Minnesota
David L. Tulloch, Rutgers University

10.1 INTRODUCTION

No agenda could have anticipated the influences on priorities for research into geographic information systems (GIS) and society brought about by the terrorist attacks on the United States of America on September 11, 2001 (9/11). The strategic importance of geographic information was demonstrated by the "Emergency Response and Homeland Security: World Trade Center Experiences in the Use of GIS" at the UCGIS Winter Meeting in Washington D.C. in February 2002. A National Science Foundation (NSF) funded study has resulted in the publication of "The Geographical Dimensions of Terrorism," which considers the role of geospatial data, geoinformation technology infrastructure, and hazards research in emergency preparedness (Cutter *et al.*, 2003.) In the immediate aftermath of the attack, emphasis was placed on the practical assistance afforded by GIS during rescue and recovery. There are significant lessons for medium- and long-term research into the roles of geographic information and security, threat detection and risk analysis, disaster preparedness, mitigation, and sustainability. Issues of surveillance, privacy and confidentiality are raised simultaneously. The interrelationships and integration of geographic information and society will be resolved in many different ways. It is the authors' contention that for the GIS and society research agenda these ways will differ in degree and relative priority rather than in kind; priorities will have been altered. The reader should therefore assess the following contribution to the debate on research priorities in GIS and society as being the results of discussions which preceded 9/11, but should be cognizant that the underpinning themes of interrelation, integration, and transformation have assumed new importance, and indeed urgency.

GIS is inextricably interwoven into society. The widespread and increasingly routine use of GIS enables the efficient storage, display, analytical spatial modeling, and exchange of diverse types of map, image, and other spatial data. Advances in computing technology provide efficient GIS interfaces to Internet map servers, mobile telephones, and to embedded computer applications in machinery and household appliances. GIS is applied to contexts and scales spanning from the global information economy, through local business uses, to the individual. Used extensively for land-cover mapping, land use and land parcel applications, and urban and regional planning, GIS found wide early acceptance in public sector agencies. Natural resource applications grew rapidly for land and water inventory assessment, monitoring, taxation, sustainable harvest management, and resource conservation practices. Increasingly, private businesses apply geodemographic and other forms of spatial analysis to problems of facilities location, production, and marketing. Transportation networks are represented in GIS for traffic engineering, transportation management, scheduling, and routing for deliveries and emergency response. Utility companies map digital networks for communications, gas, water, and electricity facility maintenance and management. Since 1995, GIS has found a regular role in areas as diverse as public health surveillance, product vending, pilot training, vehicle routing and navigation and precision agriculture. Today there is a growing demand for location-based services (Goodchild, 2003). Location-based services capitalize on the value of real-time, location-specific data for customized service, productivity and response. GIS users include private firms and individuals, national, regional, and local governments, nonprofit organizations, grassroots and community groups, students, schools, colleges, universities, and research institutes. Yet, in spite of its apparent ubiquity and diversity of use, GIS adoption remains sporadic, the quality and degree of its application vary, and the term "Geographic Information System" remains unknown to the great majority of the American public even though their daily lives are increasingly affected by its use.

Following a background review of the principal research perspectives on GIS and Society, this chapter seeks to frame a series of nationally important research challenges based on discussions held by the UCGIS community of scholars. We then explore the various positions taken by UCGIS participants in assessing the importance of GIS and Society research to national needs. Special emphasis is paid to issues arising in the private sector, as academic researchers have by and large neglected this topic. Next, the chapter visits the four themes identified in the background review and proposes a selection of related research questions initially composed during the 1998 UCGIS Summer Assembly and subsequently revised. Based on these questions, the penultimate section provides examples of specific research topics and methods. The paper ends with an evaluation of relative priorities of the research agenda in GIS and Society and an assessment of the principal research challenges in this field.

10.2 BACKGROUND

GIS technologies co-evolve with the societies in which they are embedded. In understanding the societal use of GIS, the overarching concern is how the science and technology of geographic information will influence—and be influenced by—societal circumstances and processes. Knowledge of the differing levels of GIS adoption, development, and use within various societal contexts is important to assure efficient, equable, and ethical use. The ways GIS shapes society and, in turn, is shaped by society—and by which segments of society—is crucial information for the enlightened development of this technology and for its prudent application. Many people view access to information technology as a key to improving the quality of life, contributing to the efficiency of the economy, and the effectiveness of government and education (Gore, 1992; Gringrich, 1995). This is no less true for geographic information (Tulloch *et al.*, 1998). Geographic information's more recent penetration into the business sector promises far-reaching economies, new commodities and services, and increased productivity in traditional and high technology industries. Conversely emerging problems stimulate calls for a thorough examination of the highly disparate access of people to GIS technology and its products. Equally important is the study of the relationship of spatial information to power and control and the increasing difficulty in preserving individual privacy and autonomy in an electronically connected and spatially enabled world (Onsrud, Johnson, and Lopez, 1994; Curry, 1998; Dobson, 2002). Some have argued that geography itself is destroyed by information technology. Although Curry and Eagles (1999) emphatically deny that electronic communication annihilates geography, they seriously question the form that privacy and individual identity will take under these new technological conditions, a question sharpened by the events of 9/11. Niles and Hanson (2003) also conclude that the identity of place and the importance of context are undiminished by cyberspace. Indeed they suggest that the patterns of access of pre-Internet geographies are likely to be accentuated post-Internet. As a result of these multiple uncertainties, the GIS and Society theme calls for an extensive examination of the direct impact of its applications and a rigorous evaluation of the construction of new geographies engendered by information technology.

Since the writing of the UCGIS White papers (UCGIS, 1996, 1998) the study of GIS and Society has become essential to the GIS research agenda. The theme gained considerable momentum through the NCGIA-sponsored Varenius Project when "Geographies of the Information Society" was selected as one of three research areas. The Varenius Project sought to encourage scholarly inquiry into the actual, virtual, and conceptual geographies of information and resulted in three lines of inquiry (Sheppard *et al.*, 1999). These agenda setting activities included "Place and Identity in an Age of Technologically Regulated Movement," "Empowerment, Marginalization, and Public Participation GIS," and "Measuring and Representing Accessibility in the Information Age." See Curry and Eagles (1999) for examples of problems of place and identity; Craig, Harris, and Weiner (1999) for examples of empowerment, marginalization, and participation; and Janelle and Hodge (1999) for examples of the meaning of accessibility in the information age and its consequences.

Continuing this impetus has been an issue of *Environment and Planning B* focused on public participation using web-based GIS (Carver, 2001), and URISA has sponsored two international conferences on public participation GIS in 2002 and 2003. A direct outcome of UCGIS–AGILE cooperation was the European Science Foundation/National Science Foundation jointly sponsored workshop on access and participatory approaches (Craglia and Masser, 2001; Weiner *et al.*, 2001). The stated aims of the workshop were to assess the current state of research on geographic access theory and access to geographic information; to evaluate the impact of evolving policy and legal trends in the United States and Europe on access to scientific and technical data generally and geographic data specifically; to assess the current state of research on participatory approaches surrounding the use of geographic information; to explore commonalities and differences in United States and European directions of research within these arenas; and to develop a joint United States/European research agenda on geographic information access and participatory issues. A special issue of the *URISA Journal* (Volume 15, 1, 2003) begins this process.

Research addressing the many interrelationships between GIS and society is often a component of other GIS research (programmatically and at the project level) and of more general research into the relationships between information technologies in general and society. Notwithstanding the overlap with other UCGIS research challenges, a specific focus on GIS and Society is intrinsically important. Research questions are being formulated from greatly differing perspectives. These vary from how GIS and its uses alter the construction, perception, and experience of time and space, through how social processes affect the form taken by the technology itself, to how the spread of GIS technology affects the geographic political, economic, legal, and institutional structures of society. Of particular relevance to and overlap with the GIS and Society theme are the six emerging UCGIS challenges relating to research in mobile and distributed computing, interoperability, cognition of geographic information, extensions to geographic representation, uncertainty in spatial data, and the future of the spatial information infrastructure (UCGIS, 2000).

Certain theoretical and methodological approaches to GIS and Society have received considerable visibility in the literature while other, equally deserving, perspectives have received less attention. Critical social theory rose to prominence with the publication of "Ground Truth" (Pickles, 1995). A number of articles assailed GIS as the successor to logical positivism, the means to preserve a defunct quantitative revolution, and made outright attacks on the intellectual credibility of GIS (for some of the more sober assessments, see Sheppard, 1995; Harris and Weiner, 1996; Pickles, 1997, 1999). Investigations into the status of GIS in society and the mutual influences between GIS and society have been enriched by insights coming from many philosophical and epistemological approaches addressing a variety of topics. One substantial philosophical debate subtends the types of questions and problems raised by the question of whether GIS is a "science" or a "tool" (Pickles, 1997; Wright *et al.*, 1997). Often represented as an over-simplified dualism between practitioners of the scientific method and those who choose alternative approaches, this debate has nevertheless been the source of many original insights and new, frequently relativist, research directions (Pickles, 1999). The creative tensions generated by this discourse promise to stimulate further progress through critical reflection on the technology, challenges to unquestioned concepts embedded within the theory and practice of GIS and, most productively, from the recognition of the necessity of multiple intellectual perspectives. Research programs are needed that embody intensive and extensive, and quantitative and qualitative methodologies. Methods include, but are not limited to, hypothesis testing, survey research, case studies, participant observation, and ethnography. Calls have been made within the academic community to explicate the philosophical and theoretical underpinnings of GIS more rigorously and to extend the conceptualization and dimensionality of georepresentation (Martin, 1999; Raper, 2000). These calls should not go unheeded as the discipline matures, particularly in the social, political, economic, and cultural domains.

10.3 MAJOR RESEARCH PERSPECTIVES

The diversity of research themes reflects the multiplicity of approaches emerging from the debate. The following five perspectives serve to illustrate the major research trends identified during the UCGIS research agenda process.

10.3.1 The Institutional Perspective

This perspective focuses on the magnitude and rate of adoption and influence of GIS within government, education, and business. One set of themes and questions arising from this perspective addresses the status and magnitude of GIS implementation by public and private institutions measured as expenditures and benefits, and their rates of adoption of, and subsequent participation in, GIS. Costs and benefits associated with GIS implementation are assessed along with accessibility and the pricing of data, services, and products. The perspective calls for a comprehensive determination of the costs and benefits as distributed among individuals and social groups, and an assessment of the degree of equity in that distribution. Theories, tools, and techniques continue to be developed to determine the impact of GIS on policy decisions. The perspective assesses the influence of GIS on interactions between public agencies, and GIS's effects on citizens' relationships with government. These issues are reflected in people's beliefs and actions regarding the management of land and resources (Kishor *et al.*, 1990; Ventura, 1995; Epstein *et al.*, 1996; Tulloch and Neimann, 1996; Moyer and Neimann, 1998; Tulloch *et al.*, 1998).

10.3.2 The Legal and Ethical Perspective

This perspective is concerned with the changing legitimacy and ethical setting of GIS, the various formal and informal mechanisms governing access to spatial data, and the consequences of the proliferation of, and access to, proprietary spatial databases. In particular, this perspective pays attention to whether and how the georeferencing characteristics of geographic information invites questions about the norms of privacy and increases in the surveillant powers of different social institutions. It is also concerned with investigating how these changes are rooted in governmental and legal regulation, the ethical and moral implications of these changes, and possible legal remedies (see, for example, the work of Chrisman, 1987; Onsrud, 1995; Onsrud and Rushton, 1995; Curry, 1997, 1998).

10.3.3 An Intellectual History Perspective

An intellectual history perspective addresses the evolution of geographic information science and the dynamics through which dominant technologies are selected from a set of options at critical points in time. Of considerable interest are the technological choices **not made** at a given time or by given groups. The intellectual history perspective is engaged in revealing the societal, institutional, and personal influences governing these selection processes and in questioning whether and why productive alternative technologies have been overlooked. Recent advances draw on the work of the French philosopher Bruno Latour and the examination of the everyday practice of science. These new directions draw on Actor–Network theory stressing the influence of societal structures on the evolution of geographic information technologies without entirely abandoning structure as a concept (see for example work by Latour, 1987, 1993; Chrisman, 1988; Mark, 1997; Foresman, 1998; Harvey and Chrisman, 1998).

10.3.4 A Critical Social Theory Perspective

A critical social theory perspective examines the limitations inherent in the ways that populations, location conflicts, and natural resources are represented within GIS and the extent to which these limits can be overcome by extending the possibilities of geographic information technologies. The critique first under-scores the problems associated with a search for a single truth or meaning, and second emphasizes ways in which the nature of, and access to, GIS simulta-neously marginalize and empower different social groups, which may have over-lapping and/or opposing interests. Furthermore, the perspective studies the evolu-tion of geographic information technologies as reflecting societal structures and priorities as well as the practices of those who develop and utilize the technol-ogies (see for example Chrisman, 1987; Pickles, 1995; Poiker and Sheppard, 1995; Harris and Weiner, 1996; Harris and Weiner, 1998).

10.3.5 A Public Participation GIS Perspective

A public participation GIS (PPGIS) perspective investigates how a broader effective use of GIS by the general public and by community and grassroots groups can be attained. Public participation has implications for broadening the base of empowerment within groups using GIS and broadening the ability of social movements seeking to use GIS. The PPGIS approach accentuates develop-ment of alternative ways of conceptualizing geographic information and the adap-tation of new geographic technologies to address problems arising from the use of current GIS technologies (Schroeder, 1997; Craig *et al.*, 1999, 2002).

A brief overview of five current research perspectives serves to introduce a selection of the broad variety of conceptual and methodological approaches used to assess propositions about GIS and Society. This range of approaches and under-lying theoretical frameworks reflects the complexity of the questions being raised and highlights the fact that many of them are not fully tractable using conven-tional modes of analysis. As a consequence of the shear magnitude of the task, many questions have yet to receive adequate attention and many of those quest-ions that have received attention are still in the early stages of research. Even fundamental questions, such as the assessment of the economic contribution of GIS to a firm, an agency, or the domestic economy, remain at best educated guesses, rather than meticulously researched answers.

10.4 THE UCGIS APPROACH

The University Consortium for Geographic Information Science (UCGIS) seeks to facilitate a broad interest and involvement in GIS and Society research and education, which requires insightful contributions from economists, political scientists, ethicists, lawyers, and psychologists and other social scientists, in addi-tion to the main contributory disciplines such as geography, planning, policy analysis, geomatics, and computer science. A cross-disciplinary discussion is essential to interpret the breadth and depth of this research field. Those who understand human cognition and perception; those who understand the means by which cultural and natural spaces can or should be represented; and those who use this information for social, political, legal, and economic purposes, and in the resolution of disputes, should work together to understand the complex relationship between societal processes and GIS. UCGIS exists in part to facilitate interest and involvement in the topic of GIS and Society research by diverse disciplines through encouragement of greater diversity in its membership and acting to persuade researchers from scientific and technical disciplines to participate in this undertaking. Without a complete, multifaceted understanding of the societal context within which GIS is used, much money and effort may be wasted on this technology and good intentions for its use may result in extremely limited benefits or actual loss or harm.

In its 1996 White Paper on GIS and Society, UCGIS proposed the following research priorities developed, in significant part, from issues proposed at a 1995 specialist meeting sponsored by the National Center for Geographic Information and Analysis (Harris and Weiner, 1996). The priorities were as follows:

- In what ways have particular logics and visualization techniques, value systems, forms of reasoning, and ways of understanding the world been incorporated into existing GIS techniques, and in what ways do alternative forms of representation remain to be explored and incorporated?

- How has the proliferation and dissemination of databases associated with GIS, as well as differential access to these databases, influenced the ability of different social groups to utilize this information for their own empowerment?

- How can the knowledge, needs, desires, and hopes of non-involved social groups adequately be represented as input in a decision-making process, and what are the possibilities and limitations of GIS technology as a way of encoding and using such representations?

- What possibilities and limitations are associated with using GIS as a participatory tool for democratic resolution of social and environmental conflicts? What implications does research on the relationship between GIS and society reveal with regard to the types of ethical and legal restrictions that should be placed on access to and use of GIS?

This first set of UCGIS priorities does not exhaust the set of possible approaches to GIS and Society and the rapid progress in the field calls for their continual revision. Indeed, an argument can be made that many of the questions posed above about GIS and Society cannot be answered definitively until fundamental baseline research into the status of the technology has been completed (Tuloch *et al.*, 1998; Geographic Information Science and Technology Research Definition Workshop, 2002). Economic efficiency and effectiveness are primary concerns in both contemporary business and government, and comprehensive methods for estimating the full range of costs and benefits of GIS should be considered in spite of the difficulties inherent in measuring the monetary value of digital information. Furthermore, commercial GIS developers and users outside government and academia are poorly represented in the literature. Even the manner and language in which the 1996 UCGIS research issues were expressed speaks of a lack of engagement between GIS and society research and the main goals and uses of GIS in the private sector.

In order to redress the relatively sparse attention paid by the academic research community to for-profit activities, the following section assesses some principle themes and trends arising primarily, but not exclusively, from the private sector. It is apparent that firms have been adopting GIS technology at relatively high rates over the last 8 years, at the expense of a very large amount of time and money. There is a trade magazine (Business Geographics), an association (The GeoBusiness Association), an annual conference, and, to some extent at least, a shared paradigm. It is highly likely that the effects on society of the private sector use of GIS technology could exceed the effects of the public sector's use. Nevertheless, academic assessment of GIS and Society barely recognizes the existence of the private sector, let alone discussing the implications of this sector's widespread adoption of GIS. For a variety of reasons, academic geographers tend to know little about the state of GIS from the commercial perspective, so have not spent much time trying to understand them. The firms that have adopted this technology have changed many things and often significantly. Firms have changed the way they conceptualize problems, the way (departmentally, regionally, *etc.*) they allocate their resources, the geographical configurations of their infrastructure and facilities, their target markets (or social groups), and their business and marketing strategies.

The firms that have adopted the technology at the highest rates are the larger firms that already have a penchant for new technology. Usually these firms are located at or near the core (bigger cities in more developed regions) rather than the periphery. Although there have been varying uptake rates over the past couple of years in various sectors, almost all industrial groups ("verticals") within the private sector are involved. Industries are not using the technology for the same purposes. While some firms are using it largely for direct marketing, others apply it to new site selection, still others for logistics and delivery, and a few for long range planning. A classic example is provided by the use of GIS for appliance delivery logistics by Sears, Roebuck and Co., realizing a 15 percent reduction of truck mileage, improved customer satisfaction, and upwards of $100 million through GIS-guided warehouse operations (Miller, 1999). All users expect that they will increase their revenues, decrease their costs, or both as a result of using GIS technology.

Until recently, the technology has not been as successful with middle and smaller sized firms as most industry observers had predicted several years ago. There is considerable resistance in many sectors of smaller firms to adopt the technology. Small retailers and financial services firms are good examples of slow adopters, despite the fact that some of the most aggressive users are big retailers and big banks and insurance companies. What are the implications of differential adoption? To the extent that the technology is helpful in improving the competitive position and "bottom line", the technology tends to promote fewer, larger firms (industrial concentration). The implications of this trend as it has been encouraged by the adoption of technology in general have been studied, but little has been published on the distinctive implications for industrial concentration (if any) which result from the adoption of GIS. There are some situations in which business users of GIS have such a significant competitive advantage that it is clear (at least to informed observers) that it is only a matter of time before the competitors "leave the field." The rise of location-based services may well accelerate this competitive effect. There has been much anecdotal research about the benefits, problems, and implications of adopting GIS conducted by GIS software and data vendors, and also by firms considering adoption. Some of the former is entirely self-serving as one would predict; some of the latter is surprisingly thorough. Unfortunately this research is considered proprietary and is very unlikely to be published.

The use of this business GIS paradigm has a significant impact on the questions that are asked and the priority and resources devoted to each question. Firms that are full adopters and casual adopters of GIS technology are different, and both are different from the firms that are non-adopters, which still form a very significant part of the economy (smaller firms). In general, the adoption of this technology by private firms brings with it:

- A greater emphasis on quantitative or quantifiable type questions and issues (for example, geographically oriented models are now very common in business and are especially common in direct marketing, database marketing, retail site evaluation, sales territory optimization, logistical systems models);

- A strongly increased predisposition to focus on the types of questions which are at least partly addressed by colorful maps and dynamic displays;

- An increased interest and analysis of geographical differentiation in the internal operations of the firm (for example, differences on costs, revenues, marketing effort and response, productivity, competition, *etc.*);

- An emphasis on questions that epitomize the KISS ("keep it simple, stupid") principle widely followed by business persons;

- A greater appreciation for the importance geography in general; despite this appreciation, a rather simplistic view of spatial distributions and processes exist; and

- A remarkably simplistic view of neighborhood socioeconomic structure and diversity, which may serve to reconstruct social space, and thereby directly influence future development.

Based on an assessment of research reported during the last 3 years, the 1996 UCGIS priorities required revision to ensure that:

- Sufficient studies of the magnitude and rate of use of geographic information are available to demonstrate the levels of effectiveness and efficiency of its applications in local, national, and international contexts for both public and private sectors;

- Attention given to the influence of geographic information systems on the private sector is counterbalanced by attention to its influence on the evolution of geographic information technologies;

- Attention to the determinants and consequences of the broadening use of geographic information science in public agencies and institutions is balanced by the study of their use by private firms, individuals, and communities; and

- Attention to empirical questions regarding the societal determinants and consequences of the increasing use of digital geographic information is counterbalanced by attention to ethical and legal implications.

Besides an increased engagement in the comprehensive appraisal of economic aspects of information technology, an essential task is to ascertain if and how geographic information is fundamentally different from other types of information in each of these areas of inquiry.

10.5 IMPORTANCE OF NATIONAL RESEARCH NEEDS

For a multitude of reasons, basic research into the relationship between GIS and Society is of significance to the national research agenda. GIS technology is now found in nearly all federal and state government agencies, educational institutions, and large private firms, and is rapidly being adopted by local governments, environmental organizations, neighborhood organizations, and small firms. It is often repeated that 80 percent of all business data have a geospatial component, and, recently, that geographic information may contribute directly and indirectly to as much as 50 percent of the gross domestic product (Lane, 1999). Increasingly, spatial data and analytical methods are being shared among these organizations and since President Clinton's Executive Order 12906 of April 1994, the National Spatial Data Infrastructure (NSDI) has been formally under development to meet the critical need for finding and sharing geographical data. The NSDI is being created to "support public and private sector applications of geospatial data in such areas as transportation, community development, agriculture, emergency response, environmental management, and information technology." (FGDC, 1997) Currently the spatial data infrastructure is expanding its role by adding new portals for data and services sharing, including Geospatial One Stop and the ESRI-sponsored Geography Network. It is essential to evaluate the contribution of digital spatial technologies to the working and transformation of the national economy. The first steps in this direction were in a survey of the national distribution of the production, storage, and dissemination of geospatial data, undertaken by the National States Geographic Information Council (NSGIC) on behalf of the Federal Geographic Data Committee in 1998 (FGDC/NSGIC, 1998). Further surveys will contribute to the national ability to increase the benefits of, and equity of access to, geospatial information. These should be aimed at understanding the magnitude and rates of GIS adoption, the value of geographic data in a national context, evaluating the impacts of components of the NSDI (Clearinghouse, Framework, Standards, Metadata), and potential regulatory barriers to the spread of GIS technology and its use.

Access to geographic information over the World Wide Web and new modes of distributed computing has significant societal implications. Federal announcements of FY 2001 interagency research and development priorities, initiatives to explore the implications of "Digital Earth" and "Digital Government," exemplify the variety of scales and applications (see *http://digitalearth.gsfc.nasa.gov/* for the NASA perspective on Digital Earth, *http://www.nsf.gov/od/lpa/news/ press/images/diggov.pdf* for the NSF initiative on Digital Government, and *http://www.fgdc.gov/fgdc/coorwg/1999/lane_lew.html* for the Executive Office of the President's point of view of the role of geospatial information within National science policy). GIS is fully expected to contribute to education and to the reorganization of federal, state, and local government. While there are many ways in which human activities can be carried out more effectively and democratically through the application of GIS, it is equally clear that the introduction of GIS can lead to unintended consequences. There are particular dimensions associated with the visual power and locational precision of GIS, which may result in intrusion into private lives. Inequity of access or benefits may reinforce existing social and spatial inequalities. The study of GIS and society is essential to maximize benefits across all segments of society, to identify and limit the undesirable conesquences, and to direct the development of new geographic information technologies that are relevant and useful to all members of society.

The paper now revisits the themes identified in the background review, the UCGIS approach and National needs and suggests a series of related questions that appear to be significant for short-term and medium-term research. Setting long-term research goals in such a rapidly advancing technical field would have little practical value, especially given the complexity of the relationships within GIS and within society, and their cross product.

10.6 THE MAGNITUDE AND STATE OF USE OF
GEOGRAPHIC INFORMATION SYSTEMS

Several basic economic questions should be answered regarding the overall cost of societal investment in order to assess benefits to society at large as well as by economic sector, and to evaluate the economic effectiveness of GIS at the institutional, sectional, and national levels. From a baseline of the magnitude and distribution of expenditure on GIS it becomes possible in turn to evaluate rates of expenditure on GIS infrastructure and to describe rates of adoption by institutions, and by types of usage. The GIS industry extends far beyond hardware, software, and data sales. In addition to data capture, conversion and maintenance, staffing and training costs, it also includes support industries such as GPS. The position on baseline economic research and the nature and effects of the GIS technology workforce is especially important, as emphasized by Tulloch *et al.* (1998), and the findings of a workshop "Geographic Information Science and Technology in a Changing Society" (2002).

The following types of questions are important with respect to the status and value of GIS in society:

- Determine the total monetary value of investment made in GIS/LIS technology (hardware, software, data, and people (salaries and training)), over some all-inclusive time period;

- Determine the annual dollar investment made in GIS/LIS to establish the rate of investment for any geographic area (*i.e.*, neighborhood, city, county, region, state, *etc.*) including investment for the following sectors. local, state and federal government, and the private sector;

- Determine the value of an individual GIS or collective set of systems in a community, state, or nation;

- Assess the value placed by the user on spatial data (*e.g.*, Dickinson and Calkins, 1988; Dickinson, 1989; Epstein and Duchesneau, 1990; Moyer and Niemann, 1991; Steger, 1991; Poe, Bishop, and Cochrane, 1992). What is the rate of adoption of geographic information technologies locally, statewide, nationally, and globally; and

- Determine what factors are affecting the levels or rates of investment in GIS/LIS technology in the private sector and at the local, state, and federal government levels (Moyer, 1990; Cullis, 1995).

Various forces and factors affect societal adoption of a technology and its diffusion in time and space. Adoption theory, historically, has been studied in depth across many disciplines (Brown, 1981). A number of scholars within the GIS community have also devoted research into the topic (Azad, 1993; Onsrud and Pinto, 1993; Budic, 1994; Onsrud, 1995; Anderson, 1996). Additional research is needed to continue the process of validating the various adoption models that have been proposed or documented (Tulloch *et al.*, 1997, 1998). Long-term monitoring of adoption rates requires longitudinal research methods. Assessing types and extent of overall use also requires longitudinal methods. These long-term methodologies are essential for providing a sound basis from which to assess and extrapolate societal impacts.

What are the factors and forces that accelerate and/or inhibit the adoption of geospatial technology?

To what degree can these factors be manipulated to control the rate of system adoption?

Is there a relationship between the rate of system adoption, factors influencing adoption, and the quality of system development?

Are there policy, investment, or technical steps that can or should be taken to improve the adoption and system development process?

10.7 THE RECIPROCAL RELATIONSHIP BETWEEN GIS AND SOCIETY

With respect to the reciprocal relationship between GIS and society, the types of questions fall into two main groups:

1) Questions about the capabilities and limitations of prevalent GIS software, and how these capabilities have evolved:

- How can social phenomena and processes be better represented in GIS? What are the appropriate spatial forms of phenomena such as unemployment, health, migration, *etc.*?

- What can and cannot be done easily with current software, and who is most comfortable using it?

- Who has access to spatial data, and how does accessibility affect the influence of users over social processes?

- How can various non-Euclidean geometries, complex and ambiguous spatial concepts, and representations of social and physical space be embedded within a GIS? Is the currently dominant GIS software more appropriate for some cultures and social groups than others?

- Can the dominance of certain types of GIS practice be explained on efficiency grounds? By societal priorities? By historical contingencies? By the needs of large public and private institutions? By the ways in which complex networks of GIS developers and users have created a standardized set of ideas about what makes GIS important to society?

- What alternative GISs are possible, and who would be best served by their development?

2) Questions about the implications of these capabilities for different social groups and society in general:

- How is GIS affecting the relationships among and between different types of users and non-users, the ability of individuals to achieve their goals, and the relative influence of different groups over society?

- In which ways is GIS empowering social groups and individuals, making them aware of their rights (for example, to land) and increasing their participation in and influence over democratic processes?

- In which ways is GIS marginalizing social groups and individuals, by preventing equal access to information, by establishing or normalizing values, by downplaying particular views of the world, by creating unequal capacities for surveillance, and by creating inequalities in access to appropriate and effective tools for geographical analysis?

- How does the use of GIS affect users' social practices and their views of society and nature?

- How is the use of GIS changing the geographical organization and the ecological and social sustainability of human societies, in different parts of the world?

- How are geographic information and geographic information technologies altering the nature of space and place as social constructs?

10.8 BROADENING THE SCOPE OF THE USERS OF GIS

A core set of questions needs to be asked about each of the different groups of potential user organizations identified in section two (government agencies, research and educational institutions, private corporations and firms, community and grassroots social organizations, and individuals). Representative types of question are:

- Who is and who is not adopting GIS, and what are they doing with it?

- How is the adoption of GIS affecting the tasks an organization undertakes, its ways of thinking and learning about the environment within which it operates, the ways in which it goes about its tasks, and the effectiveness and efficiency with which these tasks are completed?

- How does the adoption of GIS affect the relative influence of different participants within an organization?

- What are the implications of within-group inequalities in GIS adoption for the overall organizational structure of institutions (*e.g.*, is it reinforcing the concentration of economic power within the private sector, the power of local government in the public sector, or the educational outcomes of schools)?

- How do the networks of users that develop (both within but also across the different groups) affect the views and norms held about the use and utility of GIS and influence the direction of development of GIS technologies and databases?

- Beyond the group of organizations affected by its use of GIS, who else is affected by this use and which of them benefit from or bear the costs of it?

- What are the social and economic factors that accelerate or inhibit access and use of GIS technology to enhance or protect their specific needs and values (Harris and Weiner, 1998)?

- What are the concepts, technical factors, and representations that accelerate or inhibit the adoption of GIS technology by different types of users (Onsrud and Pinto, 1993; Cullis, 1995; Tulloch *et al.*, 1997)?

10.9 ETHICAL AND LEGAL IMPLICATIONS

With respect to ethical and legal implications, the following questions are important. Information systems develop in a legal and institutional context including legislatures, government agencies, judiciary, proprietary and commercial, professional standards and practice, and behavior as customary law. The processes encompassed by the elements define the rights and interests people have in data, technology, and expertise. Distinct questions are raised by spatial data and technology including individual rights to examine or acquire publicly held data, the transition of public spatial data, and technological investment into a privatized environment.

- Interaction at the individual level underpins all other human relationships. What, if any, are the interpersonal implications of GIS?

- In which ways does GIS enhance surveillance capabilities, and which regulatory mechanisms are necessary or possible to limit surveillance?

- What additional intrusions into privacy result from the capacity to map geographic information, and what cartographic techniques can be used to maintain confidentiality of individuals?

- What are the ethical implications of geographic information technologies? Should software design and GIS use be governed by ethical considerations? How might these be implemented?

- How accessible will spatial data and related GIS analysis tools be to all types of social actors and institutions (Curry, 1994; Onsrud and Rushton, 1995)?

- What is the status of legal regimes that determine who and under what conditions has access to public data, considering both the letter of the law and actual practice?

- How does the commercial use of public information by private groups and individuals influence access to public data?

- Can GIS provide citizens with an increased ability to monitor and hold government accountable for proposals and actions?

- Will GIS provide citizens with a better understanding of their rights and interests in land or other resources?

- What is the impact on other parts of the world of the diffusion of GIS and associated regulatory and legal norms developed elsewhere?

10.10 EXAMPLE PROJECTS

Based on these general research questions the following examples of more detailed projects may be used to articulate the importance of basic research about the relationship between GIS and society.

10.10.1 Case Study Research

Since the evolving relationships between society and GIS can take many directions depending on their context, and given also that we know less about actual consequences than we do about potential consequences, initial progress is perhaps best pursued through a series of carefully selected case studies. These case studies should be focused on individuals, organizations, and geographic contexts, chosen either to be representative or to illustrate specific anomalies. Since less is known about GIS in the private sector and in community organizations, case studies of these contexts, such as that of Sears, Roebuck and Co. mentioned previously, will be particularly useful to develop and challenge and improve our understanding of theoretical scenarios. Possible case studies could address:

- The use of geodemographic marketing by firms: Its effect on the success of those firms, and its effect on the attitudes, purchasing behavior, and social make-up of the neighborhoods and social groups targeted by such GIS software.

- The use of GIS by neighborhood organizations in low income and minority communities, seeking to improve the social and physical environment available to community residents. Its effect on the ability of these organizations to make or negotiate improvements; and its effect on the internal coherence of these organizations and their ability to represent the diversity of views of local residents.

- The ways in which norms about where and how GIS should be used, about how GIS is thought and talked about, and about the putative benefits of GIS result from the practices of GIS and the networks of GIS users and GIS organizations.

- The influence of GIS, and especially the NSDI, on the actions of government agencies, and on the capacity of the general public to assert democratic influence over those agencies.

- Analysis of testbeds for sharing geospatial data and for establishing exchange standards, including the digital earth reference model (DERM), and for identifying benefits of data sharing for government reform for initiatives such as "livable communities".

- In-depth, longitudinal studies of controversial applications of GIS, such as those where conflict resolution or NIMBY issues are to the fore, paying attention to what can be learned about appropriate principles for GIS use in negotiation, ethics, and legal regulatory mechanisms.

10.11 COMPARATIVE ANALYSIS OF CASE STUDY RESULTS

In concert with, and drawing on individual case studies, comparative analysis across case studies will be important to tease out which kinds of contextual conditions affect which kinds of outcomes. Comparative analysis will be as important for the study of how social practices influence the evolution of GIS technologies as it is for the study of the social implications of GIS. Such analysis should compare case studies of similar organizational contexts in different places and case studies of different organizational contexts in similar places, for example:

- Compare and understand the role of "culture" in GIS adoption, as defined in the broadest sense to include the concept of corporate identity and affiliation, as well as more conventional geographic and anthropological interpretations.

- International comparison of spatial development infrastructures and the development of GIS analysis is a significant means to cast new perspectives on successful methods of data sharing and use.

- A successful outcome of such comparisons would be the development of mild-range generalizations about the relationships between GIS and society, and about ethical and legal principles, which may be capable of further examination through a combination of extensive empirical analyses and new, targeted case studies. Progress on these questions will depend crucially on fostering collaborative research networks.

10.11.1 Alternative Geographic Information Systems

This set of projects spans both the highly technical and the completely socio-logical. It requires an unusual combination of computational knowledge and skills, and an appreciation of the subtleties and nuance of the human condition. As such is it emblematic of the need for multidisciplinary expertise and teamwork in the study of GIS and Society issues and points up the difficulty of preparing scholars for these types of task. It will be important to develop parallel areas of research into new types of GIS technologies reflecting the flexibility, interaction, and communicative logic of Java and the World Wide Web rather than the logic of expert programs over which users have little influence. To be effective in designing GIS that are appropriate for all groups in society, such developments should combine the practical experiences of new users struggling with currently domi-nant GIS, the expertise of programmers, graphic artists and communications specialists, and that of individuals skilled in the study of GIS and Society. Focused research in this area will increase the possibility of lateral development of new approaches to GIS, which can qualitatively enhance their relevance for an equitable and democratic society. Clearly animation, multi-media presentation, and advances in human-computer-reality interfaces are essential components of this work.

There is a very strong tie-in of this topic of study with other UCGIS research challenges, including cognition of geographical and spatial information, and exten-sions to the representation of geographic information. Advances in the nature of language, ontology, and semantics will be necessary to underpin advances in alternative GIS.

10.11.2 Priority Areas for Research

The breadth suggested by the proposed projects in the previous section provides no excuse for lack of depth and rigor of inquiry into GIS and Society. The inter-disciplinary, multi-faceted approach required for successful inquiry into such diverse subject matter has been stressed, recognizing the inevitability of different, sometimes competing, theoretical and methodological frameworks. GIS and Society research will inevitably produce ambiguous, confronting, and disputed results. UCGIS calls for a careful focus on the significant differences engendered by geographic and spatial aspects of the many relationships between information technology and society, placing some questions firmly in the mainstream of current concerns for assessment of economic effectiveness and efficiency, and others as only of peripheral academic interest. At the same time the role of GIS in the reinvention, restructuring, and renovation of government, business, and edu-cation provides numerous opportunities for evaluation of interactions other than purely financial and commercial measures of impact and effects.

UCGIS has identified the following areas of study as representative of the principal issues and as priorities facing the community of scholars interested in GIS and Society research.

1) Assessment of some basic contextual questions—Assessment, evaluation and interpretation of the status and trends of GIS adoption and utilization, especially as this relates to economic benefits and expenditures, at the national, local, and enterprise scales. The mutual influences between GIS and Society should be assessed, evaluated, and interpreted for status, nature, magnitude, and trends in different local, regional, and national contexts. Assess the magnitude and impact of differential access with respect to spatial data, analytical power, and information, including economic, and cultural inequity of pricing practices and quality.

2) Development of different practices and a range of GIS technologies—Influences of society on GIS technology and practice, especially georepresentation of human and social concepts; alternative GIS development and practice among low income, minority, and indigenous peoples; and implications of using GIS education and training.

3) Ethical and moral implications, especially those relating to individual privacy—Impacts on social organizations, groups, and places affected by uses and outcomes of GIS and issues of the structure and acquisition of power raised by control over spatial data and information systems by business monopolies or as a result of political dominance.

The overriding concern in the UCGIS GIS and Society research theme is how developments in this science and technology will influence—and be influenced by—the events, processes, mechanisms, and structures of society. Consequently research in this field addresses extensive, interrelated, and complex areas of the diffusion of GIS technologies and practices, how its spread influences political, economic, legal, and institutional structures of society, and how these social processes affect the form taken by the technology itself. On the other hand, it is equally important to investigate intensive, local, contextualized, and individual relations to the technology. At the same time, while there is a requirement for precise and rigorous quantitative work, there is an equal need for the very best theory, concepts, and practice in qualitative analyses. These challenges are as ambitious as any faced by the GIS research community.

10.12 REFERENCES

Anderson, C.S. 1996, GIS Development Process: A Framework for Considering the Initiation, Acquisition, and Incorporation of GIS Technology. *URISA Journal*, 8, 1: 10–26.

Azad, B., 1993, Theory and Measurement in GIS Implementation Research: Critique and Proposals. *3rd International Conference of Computers in Urban Planning and Management*, July 23–25, Atlanta, GA.

Brown, L.A., 1981, *Innovation Diffusion: A New Perspective*, London: Methuen.

Budic, Z.D., 1994, Effectiveness of Geographic Information Systems in Local Government Planning. *Journal of the American Planning Association*, 60 (Spring): 244–263.

Carver, S., 2001, Guest editorial. *Environment and Planning B: Planning and Design*, 28, 6, 803–804.

Carver S., 2001b, *Participation and Geographical Information.* Position paper for the ESF–NSF Workshop on Access to Geographic Information and Participatory Approaches Using Geographic Information, Spoleto, 6–8 December 2001.

Chrisman, N.R., 1987, Design of Geographic Information Systems Based on Social and Cultural Goals. *Photogrammetric Engineering and Remote Sensing*, 53 (10): 1,367–1,370.

Chrisman, N.R., 1988, The Risks of Software Innovation· A Case Study of the Harvard Lab. *The American Cartographer*, 15 (3): 291–300.

Craglia, M. and Masser, I., 2001, *Access to Geographic Information: A European Perspective*, Position paper for the ESF-NSF Workshop on Access to Geographic Information and Participatory Approaches Using Geographic Information, Spoleto, 6–8 December 2001.

Craig, W.J., 1994, The Rising Tide of GIS in Minnesota. *URISA Journal*, 6 (1994) 1, 75–80.

Craig, W.J., Harris, T. and Weiner, D., 1999, *Empowerment, Marginalization and Public Participation GIS: Report of a Specialist meeting held under the auspices of the Varenius Project* (October 15–17, 1998), National Center for Geographic Information and Analysis: Santa Barbara, CA.

Craig, W.J., Harris, T.M. and Weiner, D. (eds.), 2002, *Community Participation and Geographic Information Systems*, New York: Taylor & Francis.

Craig, W. and Johnson, D., 1997, Maximizing GIS Benefits To Society. *Geo Info Systems*, 7 (March) 3: 14–18.

Cullis, B.J., 1995, *Modeling Innovation Adoption Responses: An Exploratory Analysis of Geographic Information System Implementation at Defense Installations.* Ph.D. Dissertation. Columbia, SC: University of South Carolina.

Curry, M., 1994, Image, Practice and the Hidden Impacts of Geographic Information Systems. *Progress in Human Geography,* 18: 44, 441–459.

Curry, M., 1997, The Digital Individual and the Private Realm. *Annals of the Association of American Geographers*, 87 (4): 681–699.

Curry, M., 1998, *Digital Places, Living with Geographical Information Technologies*, London: Routledge.

Curry, M. and Eagles, M., 1999, *Place and Identity in an Age of technologically Regulated Movement: Report of a Specialist meeting held under the auspices of the Varenius Project* (October 8–10, 1998), National Center for Geographic Information and Analysis: Santa Barbara, CA.

Cutter, S., Richardson D.B. and Wilbanks, T.J. (eds.), 2003, *The Geographical Dimensions of Terrorism*, Routledge: New York and London.

Dickinson, H.J., 1989, *Selective Bibliography: Value of Information*, NCGIA Technical Paper, National Center for Geographic Information and Analysis, Santa Barbara, CA.

Dickinson, H.J. and Calkins, H., 1988, The Economic Evaluation of Implementing a GIS. *International Journal of Geographic Information Systems,* 4, 4: 307–327.

Dobson, J.E., 2002, *Geoslavery*, presented at Association of American Geographers Annual Meeting, Los Angeles, CA, March 19–22, 2002.

Epstein, E.F. and Duchesneau, T.D., 1990, Use and Value of a Geodetic Reference System. *Journal of the Urban and Regional Information Systems Association,* 2, 1: 11–25.

Epstein, E.F., Tulloch, D.L., Niemann, B.J., Ventura, S.J. and Limp, F.W., 1996, Comparative study of land records modernization in multiple states. *Proceedings GIS/LIS '96*, ACSM/ASPRS Bethesda, MD.

Federal Geographic Data Committee, 1997, *A Strategy for the National Spatial Data Infrastructure,* Federal Geographic Data Committee, U.S. Geological Survey, Reston, VA.

Federal Geographic Data Committee, 1999, *Preliminary Results of NSGIC/ FGDC framework survey, January 15 1999, http://www.fgdc.gov/whatsnew/ whatsnew.html#survey*

Foresman, T.W. (ed.), 1998, *The History of Geographic Information Systems: Perspectives from the Pioneers,* Upper Saddle River, NJ: Prentice Hall PTR.

Geographic Information Science and Technology in a Changing Society: A Research Definition Workshop, 2002, Center for Mapping and The School of Natural Resources, The Ohio State University, October, Columbus, OH.

Gingrich, N., 1995, *To Renew America,* New York: Harper Paperback.

Goodchild, M.F., 2003, *Augmenting Geographic Reality,* Distinguished Lecture, Hong Kong Polytechnic University, March 2003.

Goodchild, M.F., 1995, Geographic information systems and geographic research. In Pickles, J. (ed.), *Ground Truth: The Social Implications of Geographic Information Systems*, New York: The Guilford Press.

Gore, A., 1992, *Earth in the Balance*, New York: NAL/Dutton.

Harris, T.M. and Weiner, D., 1996, GIS and Society: The Social Implications of How People, Space, and Environment are Represented in GIS. *Technical Report 96-7*, National Center for Geographic Information and Analysis: Santa Barbara, CA.

Harris, T. and Weiner, D., 1998,. Empowerment, Marginalization and "Community-integrated" GIS. *Cartography and Geographic Information Systems*, 25 (2), 67–76.

Harvey, F. and Chrisman, N.R., 1998, Boundary Objects and the Social Construction of GIS Technology. *Environment and Planning A*, 30, 1,683–1,694.

Janelle, D. and Hodge, D., 1999, *Measuring and Representing Accessibility in the Information Age: Report of a Specialist Meeting held under the auspices of the Varenius Project* (November 19–22, 1998), National Center for Geographic Information and Analysis: Santa Barbara, CA.

Kishor, P., Niemann, Jr., B.J., Moyer, D.D., Ventura, S.J., Martin, R.W. and Thum, P.G., 1990, Lessons from CONSOIL: Evaluating GIS/LIS. *Wisconsin Land Information Newsletter,* 6:11–13.

Lane, N., 1999, *Making Livable Communities a Reality*, unpublished luncheon speech at National Geodata Forum, June 8 1999, Washington, D.C.

Latour, B. 1987, *Science in Action: How to follow scientists and engineers through society*, Cambridge: Harvard University Press.

Latour, B., 1993, *We Have Never Been Modern*, Cambridge: Harvard University Press.

Mark, D.M., 1997, The History of Geographic Information Systems: Invention and Re-Invention of Triangulated Irregular Networks (TINS). *Proceedings, GIS/LIS'97*, ACSM/ASPRS: Bethesda, MD.

Martin, D.J., 1999, Spatial representation: The social scientist's perspective. In Longley, P., Goodchild, M.F., Maguire, D. and Rhind, D. (eds.), *Geographical Information Systems*, 2nd edition, New York: John Wiley & Sons, pp. 71–80.

Miller, J., Senior Vice President and Chief Information Officer, Sears, Roebuck and Co., 1999, *Hearing on Geographical Information Systems Policies and Programs*, Committee on Government Reform, Wednesday, June 9, 1999, 1 p.m., Room 2154 Rayburn House Office Building, Washington, D.C.

Moyer, D.D., 1993, Economics of MPLIS: Concepts and Tools. In Brown, P.M. and Moyer D.D. (eds.), *Multipurpose Land Information Systems: The Guidebook,* Washington, D.C: Federal Geodetic Control Committee, Chapter 15.

Moyer, D.D. and Niemann, B.J., 1991, Economic Impacts of LIS Technology Upon Sustainable Natural Resource and Agricultural Management. *Surveying and Land Information Systems,* 51, 1: 17–21.

Moyer, D.D. and Niemann, B.J., 1998, Land Information Systems: Development of Multipurpose Parcal-based Systems. In Foresman, T.W. (ed.), *The History of Geographic Information Systems: Perspectives from the Pioneers,* Upper Saddle River, NJ: Prentice Hall PTR.

Niles, S. and Hanson, S., 2003, *A New Era of Accessibility? Journal of the Urban and Regional Information Systems Association,* 15, APA 1, 35–42.

Onsrud, H.J., 1995, The Role of Law in Impeding and Facilitating the Sharing of Geographic Information. In Onsrud, H.J. and Rushton, G. (eds.), *Sharing Geographic Information,* New Brunswick, NJ: Center for Urban Policy Research.

Onsrud, H.J., Azad, B., Brown, M., Budic, Z., Calkins, H., Godschalk, D., French, S., Johnson, J., Niemann, B., Pinto, J., Sandberg, B., Ventura, S., Wetherbee, R. and Wiggins, L., 1995, Experiences in Acquisition, Implementation, and Use of GIS in U. S. Local Governments: A Sampler of Academic Studies and Findings. *Proceedings from the Annual Conference of the Urban and Regional Information Systems Association,* pp. 626–636.

Onsrud, H.J. and Craglia, M. (eds.), 2003, Introduction to the Special Issues on Access and Participatory Approaches in Using Geographic Information. *Journal of the Urban and Regional Information Systems Association.* Volume 15, pp 5–8.

Onsrud, H.J., Johnson, J.P. and Lopez, X.R., 1994, Protecting Personal Privacy in Using Geographic Information Systems. *Photogrammetric Engineering and Remote Sensing,* 60(9), 1,083–1,095.

Onsrud, H.J. and Pinto, J.K., 1993, Evaluating Correlates of GIS Adoption Success and the Decision Process of GIS Acquisition. *URISA Journal,* 5, 1: 18–39.

Onsrud, H.J. and Rushton, G. (eds.), 1995, *Sharing Geographic Information Systems. Sharing Geographic Information Systems,* New Brunswick, NI: Center for Urban Policy Research Center for Urban Policy Research.

Pickles, J. (ed.), 1995, *Ground Truth: The Social Implications of Geographic Information Systems,* New York: The Guilford Press.

Pickles, J., 1997, Tool or Science? GIS, Technoscience and the Theoretical Turn. *Annals Association of American Geographers*, 87 (2): 363–372.

Pickles J., 1999, Arguments, debates and dialogues: The GIS-social theory debate and the concern for alternatives. In Longley, P., Goodchild, M. F., Maguire, D. and D. Rhind (eds.), 1999, *Geographical Information Systems, 2nd edition*, New York: John Wiley & Sons, pp. 49–60.

Poe, G.L., Bishop, R.C. and Cochrane, J.A., 1992, Benefit-Cost Principles for Land Information Systems. *URISA Journal*, 4, no. 2, 20–31.

Poiker, T. and Sheppard, E. (eds.), 1995, GIS and Society. *Cartography and Geographic Information Systems*, Special Issue 22 (1).

Raper, J., 2000, *Multidimensional Geographic Information Science*, London: Taylor & Francis.

Sayer, A., 1992, *Method in Social Science: A Realist Approach*, London: Hutchinson.

Schroeder, P.C., 1997, *GIS in Public Participation Settings*, Paper presented at University Consortium for Geographic Information Science (UCGIS) 1997 Annual Assembly and Summer Retreat, Bar Harbor, Maine, June 15–21, 1997.

Sheppard, E., 1995, GIS and Society: Toward a Research Agenda. *Cartography and Geographic Information Systems*, 22(1): 5–16.

Sheppard, E., Couclelis, H., Graham, S., Harrington, J.W. and Onsrud, H.J., 1999, Geographies of the Information Society. *International Journal of Geographic Information Science*, 13 (8).

Sicber, R., 1997, *Computers in the Grass Roots: Environmentalists, Geographic Information Systems, and Public Policy*, Ph.D. Dissertation. Rutgers University, New Brunswick, NJ.

Steger, W.A., 1991, A Framework to Value Information Systems. *Journal of the Urban and Regional Information Systems Association*, 3, no. 1, 33–42.

Tulloch, D.L. and Epstein, E., 2002. Benefits of Community MPLIS: Efficiency, Effectiveness, And Equity. *Transactions in Geographic Information Systems*, 6 (2): 195-212.

Tulloch, D.L. and Niemann, B.J., 1996, Evaluating Innovation: The Wisconsin Land Information Program. *Geo Info Systems*, 6 (October) 10, 40–44.

Tulloch, D.L., Epstein, E.F., Niemann, B.J. and Ventura, S.J., 1997, *Multipurpose Land Information Systems Development Bibliography: A Community-wide Commitment to the Technology and its Ultimate Applications*, National Center for Geographic Information Analysis Technical Report 97-1. Santa Barbara, CA.

Tulloch, D.L., Epstein, E.F., Moyer, D.D., Niemann, B.J., Ventura, S.J. and Chenowyth, R., 1998, *GIS and Society: A Working Paper*, Land Information and Computer Graphics Facility, College of Agricultural and Life Sciences, University of Wisconsin-Madison, Madison, WI.

UCGIS, 1996, Research Priorities for Geographic Information Science. *Cartography and Geographic Information Systems*, 23(3).

UCGIS, 1998, *White Paper on Research in GIS and Society, http://www.ucgis.org*

UCGIS, 2000, *Emerging Themes in GIScience Research. http://www.ucgis.org/ emerging/*

URISA, 2003, *2nd Annual Public Participation GIS Conference*, July 20–22, 2003 Portland State University, Portland, Oregon.

Ventura, S.J., 1995, The Use of Geographic Information Systems in Local Government. *Public Administrative Review,* 55 (5): 461–467.

Weiner, D., Warner, T.A., Harris, T.M. and Levin, R.M., 1995, Apartheid Representations in a Digital Landscape: GIS, Remote Sensing and Local Knowledge in Kiepersol, South Africa. *Cartography and GIS,* 22(1): 30–44.

Weiner, D., Harris, T.M. and Craig, W.J., 2001, *Community Participation and GIS*, Position paper for the ESF-NSF Workshop on Access to Geographic Information and Participatory Approaches Using Geographic Information, Spoleto, December 6–8 , 2001.

Wright, D., Goodchild, M.F. and Proctor J.D., 1997, GIS: Tool or Science? Demystifying the Persistent Ambiguity of GIS as "Tool" versus "Science". *Annals Association of American Geographers*, 87 (2): 346–362.

CHAPTER ELEVEN

Geographic Visualization

Aileen R. Buckley, Environmental Systems Research Institute
Mark Gahegan, The Pennsylvania State University
Keith Clarke, University of California, Santa Barbara

11.1 INTRODUCTION

The human visual system is the most powerful processing system known. By combining technologies such as image processing, computer graphics, animation, simulation, multimedia, and virtual reality, computers can help present information in new ways so that patterns can be found, greater understanding can be gained, and problems can be solved. "Geographic visualization", also referred to as "geovisualization" (GVis), focuses on visualization as it relates to spatial data. With vast increases in the availability of digital geospatial data, coupled with a need to address ever more complex questions, it becomes increasingly difficult to explore, understand, analyze, and communicate information. Visualization offers new methods to tackle these difficulties in a manner that keeps the expert (geographer) within the problem-solving loop (Gahegan, 2000, 2000a). GVis can be applied to all stages of the problem-solving process in geographical research, from development of hypotheses, through knowledge discovery, data analysis, presentation of results, and evaluation.

While computational scientists have for a number of years proposed an annually updated set of research priorities for visualization, it is only recently that geographers, specifically cartographers, have defined a set of research priorities for geographic visualization. Although the previous lack of a stated research direction may have been acceptable or at least understandable as GVis matured and developed over the past decade, the time has now come to review the advancements of GVis and assess its research needs for the future, as well as identify links with other research priorities in GIScience. This need is due in part to current and expected demand for GVis capabilities. Not only has the volume of data available increased, the technological capabilities are more advanced. As a result, there are more data to visualize, more ways to visualize them, and more need to understand how visualization works. A thoughtful and directed approach to structuring our understanding of the efforts to, need for, and issues relating to geographic visualization will uniquely position spatial scientists to contribute to the development of visualization in general.

In this chapter, we suggest a number of research priorities that should be recognized in a geographic visualization research agenda. Although there are a large number and a wide variety of issues relevant to the advancement of research in geographic visualization, the ones presented here reflect those that are closest to the core of GVis rather than those that might be more easily categorized under one of the other UCGIS research themes. For example, while scale is integral to map-based visualization and there are clear research needs that lie at the intersection of these two themes (such as the support of scale dependent displays and the use of visualization for understanding how patterns and processes relate between scales), the UCGIS research theme on scale captures many of the issues that are also relevant in GVis. On the other hand, there are some research priorities of such great importance to GVis that they could not be omitted from this theme in anticipation of possible but not certain inclusion in other themes. For example, cognitive issues that underlie the design and use of geographic visualization tools are central to the unique approach that geographers offer to visualization research and therefore deserve special attention despite the existence of a "Cognition of geographic information" research theme in the UCGIS agenda.

Geographic visualization is worthy of our attention because of the need for useful guidance in the improvement of curricula and courses, as well as the development of course materials and software for use in GVis. Additionally, there is a need to identify specialists in this area, especially given the potential for collaboration with others in a variety of disciplines, including computer science, statistics, and the arts, as well as non-academic enterprises such as the entertainment industry. A clearly identified research agenda for GVis will help to illuminate potential pitfalls for the advancement of studies in this area, such as inadequate resource allocation for projects in GVis which may involve significant time and cost in order to progress substantially.

This chapter describes how geographers and other spatial scientists are uniquely positioned to contribute to the advancement of GVis (and visualization in general), and the importance of accepting the challenge to make significant contributions in these areas. We describe current developments of the theory and methods of geographic visualization as a precursor to outlining the research challenges in GVis. Finally, we describe a number of projects that demonstrate GVis developments to date and illustrate the promise of future developments in geo-visualization research

11.2 BACKGROUND

In its 1987 commissioned report to the United States National Science Foundation, the Panel on Graphics, Image Processing, and Workstations defined *visualization* as "a method of computing...a tool both for interpreting image data fed into a computer, and for generating images from complex multi-dimensional data sets..." the goal of which is "...to leverage existing scientific methods by providing new insight through visual methods" (McCormick *et al.*, 1987: 3). MacEachren *et al.* (1992: 101) expanded that view by arguing that "visualization...is definitely not restricted to a method of computing...it is first and foremost an act of cognition, a human ability to develop-mental representations that allow us to identify patterns and create or impose order." Geographic visualization is now considered to encompass not only the development of theory, tools, and methods for the visualization of spatial data, it also involves understanding how the tools and methods are used for hypothesis formulation, pattern identification, knowledge construction, and the facilitation of decision making.

Information visualization is generally considered to involve the use of computers to generate interactive, often animated, representations of multiple variables in multiple linked formats, the goal of which is to develop greater understanding of interactions of components of a system or distribution (Buckley, 1997). Often this understanding comes from the exploration of data rather than through problem solving, and the process generally involves one or a few "expert, highly motivated viewers who are often engaged in ill-defined tasks such as hypothesis formulation" (DiBiase *et al.*, 1992: 213). For this chapter, our concept of geographic visualization is broader than others definitions of visualization (*i.e.*, computer-dependent methods of data display for small groups of highly trained individuals who are primarily interested in exploration of very large spatial data sets). For example, we also consider other media (*e.g.*, paper, film, projected displays, holography), other data sets (non-spatial), user groups ranging in size from individuals to crowds, the cognitive process of visualization, and other computer configurations (*e.g.*, mobile computing, wearable computers).

11.3 RELATED EFFORTS

Drafting a research agenda for the UCGIS in geographic visualization parallels extensive efforts at the international level by the International Cartographic Association (ICA). In 1995, the ICA formed a Commission on Visualization: "The Commission's focus is on use of dynamic maps as prompts to thinking (dynamic maps are maps that change in response to user action or to changes in data to which they are linked)" (MacEachren and the ICA Commission on Visualization, 1998). Important activities of the group include, in addition to the elaboration of a research agenda, a number of commission meetings to advance work on the research agenda, development of comprehensive bibliographies in each of four major focus areas identified in the research agenda, and special journal issues.

The ICA approach appropriately reflects the cartographic character of the working group, focusing on geographic visualization as an area of research and application that is closely aligned with cartography (*e.g.*, directing attention to the opportunities for cognitive research in visualization, which is largely if not entirely absent from research approaches in other domains of science). A UCGIS research agenda will complement the ICA agenda, providing a slightly different perspective by directing attention to how visualization integrates with other research priorities in GIScience as a whole. The agenda in this chapter recognizes the important contribution of the ICA agenda, but it also builds on it to include additional research priorities, presents a number of links to example projects, and focuses on some specific needs that the UCGIS is uniquely positioned to encourage and support. For example, we specifically address the need for collaborative development of this area of research both within the discipline of geography and between geography and other domains of science, as well as non-academic sectors (*e.g.*, commercial industry, private organizations, and governmental agencies).

The ICA effort to this point has resulted in extremely comprehensive efforts and supporting literature in scientific journals and on-line owing to the focused work by the commission for a number of years. Duplicating this effort would not only be a waste of time, it is also unlikely that our results would be either vastly different or of superior quality. Therefore, in some sections of this chapter, we refer readers to the work of the ICA and remind the audience of the extensive efforts at the international level on a variety of fronts to advance a research program in geographic visualization. In this chapter, we neither attempt to synthesize that work nor to provide an explicit complement—rather this is an alternate compilation of research priorities that focus on distinct research opportunities in GVis with respect to the UCGIS.[1]

[1]The ICA Commission on Visualization's impressive effort in the compilation of bibliographies reference work in the four focus areas that they have developed for their research agenda: (1) representation, (2) interface design and interaction, (3) integration of visualization with databases, data mining, spatial analysis, and geocomputation, and (4) cognitive issues in visualization (MacEachren and the ICA Commission on Visualization 1998). Although different categories could be used to compile sources of information about relevant research and advancements in geographic visualization, it is likely that most of the works would be cited in one or more of the ICA working bibliographies. The bibliographies are available on-line at the following web addresses:

- Representation (Team leader: Marc Armstrong)
  http://www.geovista.psu.edu/icavis/biblios/representation.html
- Interface design & interaction (Team leader: Bill Cartwright)
  http://www.geovista.psu.edu/icavis/biblios/interface.html
- Integration of Visualization with databases, data mining, spatial analysis, and geocomputation (Team leader: Mark Gahegan)
  http://www.geovista.psu.edu/icavis/biblios/dataMining.html
- Cognitive issues in visualization (Team leader: Terry Slocum)
  http://www.geovista.psu.edu/icavis/biblios/cognitive.html

We enthusiastically refer our readers to these resources as they detail more historic and recent research in geographic visualization than we can present herein. Rather than duplicating the ICA effort, we instead focus on a number of ongoing research projects that we consider to be contributing significantly to the advancement of geographic visualization.

11.4 MAJOR RESEARCH PERSPECTIVES

Although there is a diverse array of research themes within geovisualization, the following perspectives illustrate the major research trends in the UCGIS research agenda for GVis.

11.4.1 A Spatial Analysis Perspective

Visualization is part of the scientific analysis process from exploration to confirmation, and indeed it can be incorporated in all phases of analysis from data collection to data analysis to presentation of results. In the early stages of analysis, GVis can be used for exploratory spatial data analysis. GVis can help facilitate discussions between scientists collaborating on multidisciplinary projects, especially in projects where exploratory data analysis is involved. Data that are collected can be explored and the uncertainty associated with the data can be visualized. Models of spatial-temporal phenomena, such as process models, can be represented visually. Visualization of spatial data can help researchers summarize complex spatial data arrays and/or the complex processes that underlie them. The format of visual communication can extend well beyond graphic images on paper of computer screens—immersive environments, mixed reality, virtual reality, and tele-immersion offer alternatives to traditional communication modes. In the presentation of results, attention should be directed at some level toward the potential misuse (or uninformed use) of visualization to convey deceptive results; that is, efforts to better understand GVis will allow us to identify visualizations of surrogate artifacts rather than appropriate representations of the underlying complexity of the data or process. Recognition of geovisualization as a central part of the entire research process highlights the practicality and importance of this research theme in the realm of GIScience.

11.4.2 A Knowledge Discovery and Data Mining (KDDM) Perspective

A number of research issues relate to visualization with particular emphasis on KDDM. These issues include the visualization of scientific information other than findings communicated as output, facilitation of the scientific process, testing the complex tools that can be used in this facilitation, exploring the role of interactivity (or lack thereof) in visualization for KDDM, and developing different tools for different tasks (can/should we limit options for users in a KD context?) A related issue is understanding what *a priori* knowledge is introduced in a KD context.

Many examples of geographic research questions demonstrate the need for GVis research relating to this perspective. Often we necessarily collapse dimensions on maps—how can we generate better explanation/understanding of the effect of these distortions on representations? Can visualization be useful in defining for more people what is "interesting" and what "things" we want to learn more about? How does visualization support the application of the "knowledge" that is gained at various levels, *e.g.*, public/private, low-, and high-levels? What new forms of analysis does visualization provide at every stage in the analytic process? What are the relevant display size (generalization/symbolization/ interaction) issues for various display modes (*e.g.*, Personal Digital Assistants [PDAs] vs. immersive virtual reality)? What are the related hardware issues and the computational needs for further advancement of GVis for KDDM? Evidence from outside our domain of science also provides evidence that GVis research relating to this perspective is needed; for example, information visualization researchers often employ geographic visualization although they are not well informed about research in GVis.

Needs relating to this research perspective include examining: differences in representation for KD in expert/novice users, the potential to develop expertise in users for understanding complex "patterns" or clusters of interest, the creation of visual tools for analysis and visualization "possibilities" or simulation, and the development of a multistage approach to KD (leading users through the selection of databases, variables, and models, as well as communication options).

11.4.3 A Spatial Data Representation Perspective

Spatial data representation and its integration with geovisualization is also a central research perspective. Despite the obvious connection between GVis and spatial data exploration and analysis, a disconnect currently exists between visualization and spatial databases, due primarily to the inadequacy of spatial databases to adequately represent the complex nature of geographic phenomena. For example, the complexity of representing a phenomenon as conceptually simple as water flowing through a weir impedes the utility and advancement of visualization of spatial and spatiotemporal phenomena. Currently, there are few adequate methods to represent the many attributes and the interactions between them that would be required to model and visualize this type of flow. Therefore, an important issue relating to this perspective is the ability to represent different time and space scales, both of which are also connected. Also related to this perspective is representative of complex data and representation of the uncertainty of spatial and spatial-temporal data. Research in GVis is also needed to better understand alternative representations of the multisensory nature of spatial data.

11.4.4 A Cognitive Perspective

Another research perspective relates to the relationships between visualization and cognition. This perspective is related to representation through the linkages to spatial databases and their ability to adequately represent spatial phenomena.

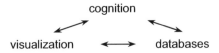

There is a need to better understand the cognitive process of GVis and how visualization leads to understanding. At its best, visualization provides creative insights not previously imagined, and at the least, visualization can be used to confirm what is suspected or hypothesized. The cognitive processes that are involved in these visual endeavors are not well understood. In order to advance the tools and techniques of visualization, we need to better understand the human visual processing system. Geographers and other spatial scientists can provide insight into the visual processing of spatial and spatiotemporal data in particular. With a greater insight into this issue, we will be better able to evaluate whether visualization techniques are powerful enough to portray the data and processes we want to represent, and perhaps more importantly, we will be able to identify the constraints of the techniques. Additionally, this perspective encompasses issues relating to communication of spatial and spatiotemporal data through both traditional and non-traditional modes, such as virtual environments, mixed reality, teleimmersion, and immersive environments.

11.4.5 A Collaborative Research Perspective

Large-extent, multivariate, spatial, or spatiotemporal data sets present scientists and researchers with special challenges for displaying the data in some form that is intelligible and insightful. Advancements in the past decade in the development of concepts, theory, and methods of geographic visualization have earned GVis recognition as a viable and valuable tool for the exploration of these large data sets. However, the development of GVis methods and techniques is often uncoupled from the practice of data exploration within a specific scientific domain, as computer scientists, statisticians, and/or GIS/cartographic scientists work in isolation from the domain experts. Although technologically advanced, the techniques developed may not address the specific requirements of the domain scientists, or the complexity or novelty of the methods may be confusing, distracting, or even frightening to the domain experts. Too often, the result is the slow (or lack of) adoption of the visualization methods by the targeted scientific community. A research agenda in GVis should, therefore, foster communication and collaboration between developers and the practitioners.

A central research need is for groups of experts, each from different perspectives, to work together. The pressing need here is to understand Earth's complex physical and social systems from an integrated perspective, necessarily involving a shared understanding across many traditional domains in geoscience/ geography. Collaborative visualization might be used to facilitate this need (Brodlie *et al.*, 1998). In addition to collaborating with other academics, researchers in GVis stand to benefit greatly from interaction with non-traditional collaborators. For example, some of the most impressive advancements in visualization have come from outside the academy. The entertainment industry has contributed to the advancement of computer animation for film, abstract and schematic graphic representations in cartoons, virtual reality and immersive environments for recreation, and game animation for video and other media. Additionally, commercial industries have developed methods for creative and innovative visualization in areas such as car navigation systems, animated and interactive graphics design for the Web, and voice-activated methods for hands-free wearable computation. Collaboration between GIScientists, domain scientists, computational scientists, and non-scientists offers potential promise for rapid innovative advancements in geographic visualization.

11.4.6 An Emerging Technologies Perspective

In addition to paying attention to who is working in the visualization arena, GIScientists should be aware of emerging technologies and their potential impact on the advancement of GVis. This research perspective recognizes emerging technologies, such as those below, that will influence geographic visualization to varying degrees.

- Computer hardware and software continues to increase in capability and decrease in cost; we are still obeying Moore's law—double in performance in 18 months. Advancements relating to the GVis include improvements to graphics hardware and the ability to maximize the advantages of fast bus architecture.

- Alternative display media are being explored to overcome the limitations of CRT displays. This is of critical concern in visualization since the low resolution and limited extent of these media exert unavoidable controls on the quality of graphic displays. Alternatives being explored include scalable display wall systems for large-format display, high-resolution wall displays for clearer projector-based displays, and multiprojector displays for multiple views (Funkhouser and Li, 2000).

- Internet capabilities and limitations will continue to drive visualization as many tools are now being developed for the Web to take advantage of the capabilities of networked communication.

- Personal Digital Assistants currently provide low resolution, affordable, mobile displays with limited data storage and data processing capabilities. These devices continue to become more advanced, offering more communication and display capabilities.

- Immersive environments, such as computer-assisted virtual environments (CAVEs), walls, and workbenches, will continue to be developed and have an impact on GVis. The realism of immersive environments continues to increase and the potentials for substitution for realism with abstraction are yet to be fully explored.

- Mobile computing will release users from desk-bound influence on not only when and where we can access displays, but also how we access them. Wireless area networks (WANs) support untethered computing over potentially limitless ranges.

- Real-time displays of spatial location are made possible through the integration of global positioning systems (GPS) and visualization and directly influence navigation capabilities.

- Advancements in navigation tools will create further demands for innovative displays that allow users to quickly and easily orient themselves in frequently unknown spaces.

- Hands-free computing that takes advantage of voice recognition and interactivity will enable additional navigation solutions.

- Wearable computers, such as head mounted displays, are being used to augment the real landscape with abstract information that is spatially aligned in superimposed real/virtual displays.

- New data coding schemes, such as XML (extensible Markup Language), GML (the geographic XML implementation), SMIL (Synchronized Multimedia Integrated Language), and SVG (Scalable Vector Graphics), will improve the functionality of New Media.

11.5 THE UCGIS APPROACH

"Geographic information science" is accepted by the UCGIS as an umbrella term to capture the fundamental problems surrounding the effective capture, interpretation, storage, analysis, and communication of geographic information. While various applications of geographic visualization can be incorporated into all of these facets of GIScience, GVis is primarily a form of both *spatial data analysis* and *communication*. It can also be applied to all stages of geographic problem solving and potentially to all types of geographic problems. Therefore, it is possible to find linkages between geographic visualization and all of the other UCGIS research priority areas. However, there is a distinct core to GVis and research issues that are related to it that are not engulfed in the other themes.

The key objectives of a UCGIS research initiative in visualization should be to:

- Channel efforts to develop inventive and effective methods of geographic visualization;

- Support the application of geovisualization methods to the significant problems in GIScience;

- Foster cooperation between researchers in geovisualization and other areas of GIScience;

- Promote geovisualization research in academia with respect to relevant efforts in private, industry, and government arenas;

- Support methods to disseminate advancements in geovisualization to the wider scientific community; and

- Investigate how visualization can be applied successfully to all the stages of problem solving in geographical analysis, from development of initial hypotheses, through knowledge discovery, analysis, presentation, and evaluation.

11.6 IMPORTANCE TO NATIONAL RESEARCH NEEDS

Although GIScience is a nascent science, it draws heavily upon the rich historical development of cartography, an art and science that evolved around the display and use of spatial data representations for thousands of years. Geographic visualization is closely linked to cartography, and much of the research in GVis is able to exploit the advancements in cartography. Robinson's seminal *The Look of Maps* laid the foundation for the development of cartography as a graphic science. Subsequent works by Bertin (1967) and Macinlay (1986) further defined the basic building blocks used to represent information visually through cartography. More recent contributions by Vasconcellos (1992), Krygier (1994), and Mac-Eachren (1995) extended that representation framework to include touch, sound, and advanced graphics. GVis, however, is not cartography, and we are currently at a point where *basic research* on the *theoretical* and *conceptual*, as well as *methodological* aspects of geographic visualization, is necessary and *critical*.

While information visualization is being developed in nearly all branches of science, geographers are uniquely positioned to contribute to its further development. A core has emerged, primarily within the discipline of geography, that begins to define the scope of GVis research and its research trends through a unifying set of theoretical concepts and increasing sophisticated methodological advancements. The important early theoretical and conceptual contributions of MacEachren (1994), Taylor (1991), DiBiase (1990), and Buttenfield (1996), among others, provided a framework for GVis research within the spatial sciences that may, to a large extent, be absent in other disciplines. This historical foundation and the recent advancements of geographic visualization in the past decade place geographers and other spatial scientists in a unique and important position to contribute to the national, and indeed international, research agenda in not only GVis, but also information visualization. With the contributions that have and remain to be made from more traditional approaches in cartography, especially in areas such as symbolization, representation, abstraction, and various cognitive aspects of visualization, GVis should maintain a stronghold in the development of scientific computing in the broader scientific and computing communities. We must take up the challenge of setting a GVis research agenda or risk losing the opportunity to make significant contributions from our unique perspective and at the same time help to set the larger research agenda. This paper outlines research directions that will allow us to approach that challenge in a thoughtful and strategic way.

11.7 PRIORITY AREAS FOR RESEARCH

The research priorities presented here roughly reflect the immediacy of the need for research in the short, medium, and long term. At the same time, we recognize that advancements, particularly those related to computation, could shift the relative importance of the research priorities over time.

11.7.1 Short-Term Visualization Research Challenges

1) Interactivity. Interactivity can be viewed as either internal (*e.g.*, the user alters the map display) or external in the sense that the interface provides linkages to resources outside the immediate display (*e.g.*, linked databases, hyperlinked web sites) (Andrienko and Andrienko, 1999). Interactivity is key in the ability to "mine a data set" effectively, and there is a serious need in the development of tools that allow users more interaction with both the data and the displays.

2) Cognitive issues. Research in cognition includes a number of pressing research needs, including greater understanding of the impact of visualization on knowledge discovery, extending representations for visualization to include qualitative, intangible, multisensory, and conceptual data, and the development of evaluation methods to test the effectiveness of visualization tools.

3) Computer interface design. It has been suggested that the interface is in fact the visualization since, at least for computer-based visualization, the user only interacts with the data through the interface, and the interface defines the ways that the users can interact with the data. Multimodal and natural interfaces attempt to mimic the way humans interact with one another and appear particularly promising for the visual-spatial data (Oviatt, 1997; Oviatt and Cohen, 2000). The development of new interface metaphors is also related to this research theme.

4) Media issues. The format in which the visual displays are developed or presented (*e.g.*, paper, screen, projective displays, virtual environments) can influence the effectiveness of visualization. Research is needed on the impact of visualization in and by new media formats, including geospatial virtual environments (GeoVE), immersive environments, CAVEs, augmented reality, and head-mounted displays (HMDs).

5) Multi-user/collaborative visualization. An emerging area of research in GVis is collaborative visualization for multiple users. This involves the development of "environments that facilitate the use of manipulable visual displays for exploration of ideas and/or decision making by two or more individuals (perhaps located at a distance)" (MacEachren and Kraak, 2000). Tools for collaborative visualization of georeferenced information have considerable potential for use in contexts such as urban planning, environmental management, and scientific interpretation of models of climate or other environmental processes.

6) Differences between users. Various characteristics of users, including expertise, sex, age, culture and language, and sensory limitations, have the potential to greatly affect the effectiveness of tools (Gilmartin and Patton, 1984; Nyerges 1993; McGuinness, 1994). The effects of individual and group differences such as these are not clearly understood and require further research (Nyerges *et al.*, 1998).

7) Visual design issues. Issues include the ability to incorporate a number of capabilities into visualization, including dimensionality (spatial, symbolic, temporal), dynamism, animation, and, as mentioned above, interaction. Further research is also needed in the representation of non-spatial data in spatial formats, such as spatialization, or the process of converting abstract non-numerical information into a viewable spatial framework (Skupin and Buttenfield, 1996, 1997; Fabrikant and Buttenfield, 1997; Kuhn, 1997; Couclelis, 1998; Fabrikant, 2000).

8) Non-conventional graphics. Rather than using the computer to replicate traditional methods of data display, the computational power of computers needs to be further exploited to develop heretofore previously unconceived of methods for displaying spatial information. Additionally, further development of non-conventional techniques, such as morphing (the transformation of the geometry of one image to that of another), superimposition (overlaying images with varying degrees of transparency), and multiple viewpoints, is required.

11.7.2 Medium-Term Visualization Research Challenges

1) Multisensory GeoVE. To further our understanding of the capabilities that can be developed for virtual and immersive environments, research is needed in how visualization can be adapted for the other senses to complement or enhance methods that have been designed for visual communication. Geographic virtual environments should be developed to make use of all the human senses, requiring related research in tactile (sensations on the skin surface), haptic (relating to pressure and resistance), kinematic (relating to motion), olfactory, and other sensations. Of these, mapping for the visually impaired is perhaps the most advanced at this point (see, for example, Olsen and Brewer, 1997; Jacobsen, 1998; Kitchin *et al.*, 1998).

2) Effectiveness of visualization. There is need for further research into perceptual and cognitive aspects of communication, and we need to understand more about what affects the effectiveness of communication through visualization in particular (Buckley, 1998). Visualization offers the means to combine and interactively change many of the visual properties of a display, and also new techniques to show ever-increasing amounts of data; yet research into the effectiveness of visualization has not kept pace.

3) Sensory limitations and variations. Related to the above theme is the effect that sensory limitations may have on the effectiveness of visualization tools and techniques. For example, limitations in or lack of the ability to see color can negate otherwise sound graphic design decisions in visual displays. As visualization begins to take advantage of other senses, the effects of limitations in those senses will also have to be considered.

4) Abstracting away from reality. The interplay between realism and abstraction needs to be further explored. Augmented reality, for example, couples these two to create new perspectives of previously unseen displays. The challenge is to automatically create a sketch to convey a desired action more effectively than can a real picture. Much research has been directed toward realism, and indeed, more is required. "At the same time, realism is in some sense easier, as we can draw on the quantitative tools of physics and mathematics. When we want to create an abstraction that somehow conveys key ideas while suppressing irrelevant detail, we need to draw on the less-quantified tools of perceptual psychology and cognitive psychology, and the vast knowledge of cartographers and animators" (Foley, 2000). This is likely to vary with the application. We need to be able to determine the right amount of realism for visualization to be maximally effective. For example, imagine an "abstraction" control, interactively varying the realism of the scene.

5) Displaying more data. There has already been some work by the visual data mining community to further the developments that allow massive datasets to be combined on a computer screen using averaging, aliasing, and other methods (Keim, 1996; Keim and Kriegel, 1996). As the volume of spatial data available continues to grow, the need for methods for displaying much of it, perhaps simultaneously, will increase.

6) Displaying more pixels. We need to look for ways to use solutions that will provide desk-sized and wall-sized work areas with sufficient pixels to maintain the image quality we have become accustomed to on our desktop monitors. Although faster graphics engines are being developed, we continue to display information on less than two square feet of display area with one to three million pixels that subtend perhaps 25 degrees at normal viewing distance. Or we use a projection system and magnify each pixel onto a large screen that, when viewed from a distance, is the same 25-degree field of view. The human eye, however, has a field of view of greater than 180 degrees.

7) Displaying fewer pixels. At the same time that we advocate higher-resolution and larger display surfaces, we are increasingly working with smaller, lower-resolution displays found in PDAs, cell phones, car navigation systems, and other information appliances. "Consider PDAs and intelligent communicators on the palm-top, information appliances on the counter-top, automotive navigation systems on the dash-top, and messaging watches on the wrist-top. Making the best use of a limited number of pixels is in many ways more challenging than working with millions of pixels" (Foley, 2000).

8) User interfaces for three-dimensional creativity. Rather than doing creative work with traditional tools and then transferring the results into a computer graphics system for further work, we need to develop tools that allow the same subtle freedom as traditional tools in terms of greater expressiveness, more rapid development of prototypes, and sensory feedback. The challenge is to build interfaces and devices that allow the creative process of compiling a display to be developed with the computer as well.

11.7.3 Long-Term Visualization Research Challenges

1) Imaginative information visualization. There is endless opportunity for creativity in discovering new ways to present information. Advancements in this area will lie at the interface of computer graphics and graphic design. "Computer graphics empowers so many techniques concerning time variation and three dimensions and interactive navigation that knowledge of and enhancement of these computer graphics techniques is an inherent, essential part of the creative challenge, awaiting integration with graphic designers' awareness of aesthetics and perceptual issues" (Foley, 2000).

2) Automated creation of information and information visualizations. The underlying challenge is to automatically create informative and aesthetic visualizations that help the user understand the underlying data being presented. This requires that the user express what information is being sought and that an automated assistant then synthesize a visualization that conveys that information. The challenge is to have a system where the user state the problem, "Please show me the relationships between landscape conditions and in-stream habitats ..." and suddenly have an aesthetically pleasing and effective graphic created based on domain knowledge and graphic design knowledge. Work in this area has already been advanced to some level by Senay and Ignatius (1998) and Gahegan and O'Brien (1997).

3) Unified graphics. Geometry-based graphics contains points, lines, curves, planes, surfaces, and solid models. Image-based graphics contains image maps (also called bit maps), texture maps, environment maps, range maps, volume graphics (such as three-dimensional image maps), video, lumigraphs, and light fields, all representing point samples in two dimensions, three dimensions, or four dimensions. There is a need to unify geometry-based graphics, image-based graphics, and time-based graphics. Additionally, sound, touch, and even taste and smell all need to be integrated.

11.8 EXAMPLE PROJECTS

There are numerous excellent examples of cartographic visualization projects; most of them are already featured on the web and therefore available for demonstration. Prime examples are described below. Web site links are provided so that the reader can embark on a virtual tour of projects around the globe that is advancing the visualization of geospatial data.

1) Time-Series Animation Techniques for Visualizing Urban Growth, U.S. Geological Survey (USGS), USA. Acevedo and Masuoka, working with USGS's Urban Dynamics research program, have created a series of variants on animation sequences that show the historical growth of the Baltimore–Washington built-up area. They argue that before creating an animation, various issues which will affect the appearance of the animation should be considered, including the number of original data frames to use, the optimal animation display speed, the number of intermediate frames to create between the known frames, and the output media on which the animations will be displayed. Three-dimensional perspective animations were created by draping each image over digital elevation data prior to importing the frames to a movie file.
http://www.geom.unimelb.edu.au/envis/automap/Acevedo/ACEVEDO.HTM

2) TerraVision™ II, SRI International, USA. At SRI, a research group has pioneered the development of TerraVision, a distributed, interactive terrain visualization system which allows users to navigate, in real time, through a three-dimensional graphical representation of a real landscape created from elevation data and aerial images of that landscape. TerraVision is optimized to browse very large datasets, in the order of terabytes, distributed over multiple servers across the Web. Data are supported in three-dimensions VRML and GeoVRML. The software is web-enabled via a browser plug-in, and it is available as shareware. A visual demonstration of the system was used to advance the Digital Earth concept at the White House. *http://www.tvgeo.com/overview.shtml*

3) Floating Ring: A New Tool for Visualizing Distortion in Map Projections, Santa Cruz Laboratory for Visualization and Graphics, University of California at Santa Cruz, USA. At the University of California, Santa Cruz, research has produced an interactive tool for the visual exploration of the distortion associated with map projections. Called "the floating ring", the method involves a software tool that allows the specification of the size and coloration of a ring on the globe that is shown, with its distortions in three dimensions on the map.

Executables for several different platforms are available in an on-line paper that includes examples of the images and the types of interaction supported. *http://www.cse.ucsc.edu/research/slvg/map.html*

The following are described on the ICA "Projects" homepage (*http://www.geovista.psu.edu/icavis/projects.html*).

4) GeoVISTA Studio, Department of Geography, Penn State University, USA. GeoVISTA Studio is an open software development environment developed within the GeoVISTA center at Penn State. The main aim of the studio is to provide a programming-free environment within which to conduct both geo-computation and geographic visualization (*http://www.geovista.psu.edu/publications/geocomp2000/FinalStudioPaper.pdf*). Studio employs a visual programming interface, allowing users to quickly build their own applications using a workflow or dataflow paradigm. *http://www.geovista.psu.edu/products/studio/index.htm*

5) GeoVISTA Center–Geographic Visualization–Science, Technology & Applications Center, Department of Geography, Penn State University, USA. The Center is devoted to fundamental and applied scientific research on the visualization of georeferenced information, development of geographic visualization (GVis) technologies, and the application of both in science, industry, decision making, and education. The Center directs particular attention to research that links GVis with other components of geographic information science thus to the integration of many perspectives on geographic representation. *http://www.geovista.psu.edu/*

6) Commission collaboration with the Association for Computing Machinery Special Interest Group on Graphics (ACM-SIGGRAPH). In June 1996, a cross-organizational collaboration between the activities of the Association for Computing Machinery's Special Interest Group on Graphics (ACM SIGGRAPH) and the International Cartographic Association's (ICA) Commission on Visualization began. The "Carto Project" explores how viewpoints and techniques from the computer graphics and cartographic communities can be effectively integrated in the context of computer graphics applied to spatial data sets. *http://www.geovista.psu.edu/icavis/icavis/vis-acm.html*

7) Reliability Visualization (RVIS), Department of Geography, Penn State University, USA. This project focuses on development of exploratory spatial data analysis methods for depicting data and data reliability associated with changes in the health of the Chesapeake Bay. In addition, attention is given to design of interactive interfaces that make implementation of these methods possible. *http://www.geog.psu.edu/Howard/HowardHTML/RVIS.html*

8) Argus Project, Department of Geography, Leicester University, UK. The remit for Project Argus is to promote Visualization in the Spatial Sciences. The project members have identified a matrix of visualization techniques that can be applied to a series of data types. The application of technique to data type is being demonstrated by combining diverse data sets with existing visualization software, and developing new software. The data, software, and images and movies created from them constitute a Visualization Toolkit, which will provide a number of tools for visualizing spatial data for research and/or teaching purposes. *http://www.mimas.ac.uk/argus/*

9) Using Java to interact with geo-referenced VRML Department of Geography, Leicester University, UK. Virtual reality technology is providing earth scientists and cartographers with new, exciting, and interactive ways to model the world and real-life phenomena. One of the most important functions of traditional cartography is providing information about location: where the user is, where an object is located, and what is at a location. This is equally, if not more, important for navigating virtual worlds and referencing information from the real world. *http://www.geog.le.ac.uk/mek/usingjava.html*

10) MANET–Extensions to Interactive Statistical Graphics for Missing Data Department of Mathematics, Universität Augsberg, Germany. MANET is an object-oriented program designed for interactive graphical analysis of statistical data. Mainly, MANET focuses on missing values through use of interactive highlighting–how to tell something about the quality of a data set, if not knowing how many values (or which combinations) are relevant? *http://www.geovista.psu.edu/icavis/projects.html*

11) U.S. Environmental Protection Agency (USEPA) Scientific Visualization Center (SVC), USA. Information visualization is an important tool for environmental research. By representing numerical data in a visual format, information visualization allows environmental scientists and analysts to better understand the results of their research and to effectively convey those results to others. This site provides an overview of the many visualization research and application activities currently underway at the USEPA. *http://www.epa.gov/vislab/svc/index.html*

12) Examining Dynamically Linked Geographic Visualization, GeoVISTA Center, Department of Geography, Penn State University, USA. At the USEPA, efforts are underway to integrate the agency's geographic information systems (GIS), scientific visualization (SciVis), and World Wide Web (WWW) suite of tools for comprehensive environmental decision support. One result of these efforts is the USEPA Spatial Data Library System (ESDLS), which is currently accessible via the WWW. *http://www.geovista.psu.edu/sites/icavis/publications/featured.html*

13) The Apoala Project, GeoVISTA Center, Department of Geography, Penn State University, USA. Environmental risk assessments and decision support for ecosystem management requires the integration and analysis of large volumes of data derived from both models and direct measurement of the environment. The objective of the research being initiated is to design, implement, and assess a high performance computing prototype that can cope with spatiotemporal data and multiscale analysis, specifically, a prototype Temporal Geographic Information System (GIS) with integrated multivariate spatiotemporal visualization capabilities. *http://www.geovista.psu.edu/grants/apoala/tour1.htm*

   14) The ArcView/XGobi/XploRe Environment: Spatial Data Analysis in a Linked Software Environment, Iowa State University, Iowa, USA. The ArcView/XGobi/XploRe software environment closely couples three independent software packages: ArcView, a GIS, XGobi, a dynamic statistical graphics program, and XploRe, a statistical computing environment. Attribute data from point, polygon, and linear themes that are maintained within ArcView can be passed into XGobi and/or XploRe for a complete statistical analysis. One of the main features of the ArcView/XGobi/XploRe environment is linked brushing among the three software packages. Examples for the use of the ArcView/XGobi/XploRe environment include forest health data, precipitation data, and satellite imagery. *http://www.public.iastate.edu/~arcview-xgobi/*

11.9 CONCLUSIONS

Visualization engages, at a deeper level, the cognitive systems of the geographer. It also has the ability to present data in new ways, offering alternative perspectives, to stimulate the exploration of new hypotheses. It may be that the greatest contribution of visualization to the process of scientific thinking is liberating the brain from the fundamental activity of information retrieval and manipulation required to produce an image, thereby allowing the brain to devote its time and energy to higher levels of analysis and synthesis (McCormick *et al.*, 1987; Friedhoff and Benzon, 1989). To promote further advancements in geographic visualization, we need to coordinate and share perspectives among different communities of scientists and non-scientists. Finally, visualization can look cool—the appeal of a pleasing, intriguing, creative, or innovative display should not be underestimated. Visual appeal is captivating; it draws the eye and engages the brain. For some problems in Geographic Information Science, that may be the first step toward their solution!

11.10 REFERENCES

Andrienko, N. and Andrienko, G., 1999, Interactive maps for visual data exploration. *International Journal of GIS,* 13(4): 355–374.

Bertin, J., 1967, *Semiologie Graphique*, Mouton, Paris, 413 p.

Brodlie, K.W., Duce, D.A., Gallop, J.R. and Wood, J.D., 1998, Distributed cooperative visualization. In d. Sousa, A.A. and. Hopgood, F.R.A. (eds.), *State of the Art Reports at Eurographics98*, Eurographics Association, pp. 27–50.

Buckley, Aileen R., 1998, Visualization of multivariate geographic data for exploration. In Craglia, M. and Onsrud, H. (eds.), *Geographic Information Research: Bridging the Atlantic*, Volume 2, London: Taylor & Francis.

Buckley, A.R., 1997, *The Application of Spatial Data Analysis and Visualization in the Development of Landscape Indicators to Assess Stream Conditions*, Doctoral dissertation, Oregon State University, Department of Geography.

Buttenfield, B., 1996, Scientific Visualization for Environmental Modeling: Interactive and Proactive Graphics. *GIS and Environmental Modeling: Progress and Research Issues,* pp. 427–443. Colorado: GIS World Books.

Couclelis, H., 1998, Worlds of information: The geographic metaphor in the visualization of complex information. *Cartography and Geographic Information Science,* 25(4): 209–220.

DiBiase, D., 1990, Visualization in the Earth Sciences. Earth and Mineral Sciences, *Bulletin of the College of Earth and Mineral Sciences,* Penn State University, 59(2): 13–18.

DiBiase, D., MacEachren, A.M., Krygier, J.B. and Reeves, C., 1992, Animation and the role of map design in scientific visualization. *Cartography and Geographic Information Systems,* 19(4): 201–214, 265–266.

Fabrikant, S.I. and Buttenfield, B.P., 1997, *Envisioning user access to large data archive,* GIS/LIS '97, Cincinnati, Oct. 28–30, ASPRS/ACSM/AAG/URISA/ AM-FM/APWA, pp. 672–678.

Fabrikant, S.I., 2000, Spatialization browsing in large data archives. *Transactions in GIS,* 4(1): 65–78.

Foley, J., 2000, *Getting There: The Ten Top Problems Left, http://www.computer.org/ cga/articles/topten.htm*

Friedhoff, R. and Benzon, W., 1989, *Visualization: The Second Computer Revolution*, New York: Harry N. Abrams, Inc.

Funkhouser, Thomas and Kai Li, 2000, Guest editors' introduction: Large-format displays. *IEEE Computer Graphics & Applications,* 20(4).

Gahegan, M.N., 2000, Visualization as a geocomputational tool. In Openshaw, S. and Abrahart, B. (eds.), *GeoComputation,* London: Taylor & Francis, pp. 253–274.

Gahegan, M.N., 2000a, Visual exploration in geography: analysis with light. In Miller, H.J. and Han, J. (eds.), *Geographic Knowledge Discovery and Spatial Data Mining*, London: Taylor & Francis.

Gahegan, M.N. and O'Brien, D.L., 1997, A strategy and architecture for the visualization of complex geographical datasets. *International Journal of Pattern Recognition and Artificial Intelligence*, 11(2): 239–261.

Gilmartin, P. and Patton, J.C., 1984, Comparing the sexes on spatial abilities: Map-use skills. *Annals of the Association of American Geographers,* 74(4): 605–619.

Jacobson, R.D., 1998, Cognitive mapping without sight: Four preliminary studies of spatial learning. *Journal of Environmental Psychology,* 18: 289–305.

Keim, D. and Kriegel., H.-P., 1996, Visualization techniques for mining large databases: A comparison. *IEEE Transactions on Knowledge and Data Engineering,* (Special Issue on Data Mining).

Keim, D.A., 1996, Pixel-oriented visualization techniques for exploring very large databases. *Journal of Computational and Graphical Statistics*, 5(1): 58–77.

Kitchin, R.M., Jacobson, R.D., Golledge, R.G. and Blades, M., 1998, Belfast without sight: Exploring geographies of blindness. *Irish Geographer*, 31(1): 34–46.

Krygier, J., 1994, Sound and geographic visualization. *Visualization in Modern Cartography*, New York: Elsevier Science Ltd, pp. 149–166.

Kuhn, W., 1997, Handling data spatially: Spatializing user interfaces. *Proceedings of the Seventh International Symposium on Spatial Data Handling,* (Advances in GIS Research II), Kraak, M.-J., Molenaar, M. and Frendel, E., London: Taylor & Francis, pp. 877–893.

MacEachren, A. and the ICA Commission on Visualization, 1998, *Proceedings of the Polish Spatial Information Association Conference*, May, Warsaw, Poland. *http://www.geovista.psu.edu/icavis/draftAgenda.html*

MacEachren, A., 1994, Visualization in modern cartography: Setting the agenda. *Visualization in Modern Cartography,* New York: Elsevier Science Ltd, pp. 1–12.

MacEachren, A., 1995, *How Maps Work,* New York: The Guilford Press.

MacEachren, A., Buttenfield, B., Campbell, J., DiBiase, D. and Monmonier. M., 1992, Visualization. *Geography's Inner Worlds: Pervasive Themes in Contemporary American Geography*, New Jersey: Rutgers University Press, pp. 101–137.

MacEachren, A.M. and Kraak, M.-J., 2000, Overview: International Cartographic Association, *http://www.computer.org/cga/cg2000/g4toc.htm*

Macinlay, J., 1986, Automating the design of graphical presentation of relational information. *ACM Transactions on Graphics*, 5(2): 110–141.

McCormick, B., DeFanti, T. and Brown, M., 1987, Visualization. *Scientific Computing in Computer Graphics*, 21 i-E-8.

McGuinness, C., 1994, Expert/novice use of visualization tools. *Visualization in Modern Cartography,* New York: Elsevier Science Ltd, pp. 185–199.

Nyerges, T.L., 1993, How do people use geographical information systems? In Medyckyj-Scott, D. and Hearnshaw, H.M. (eds.), *Human Factors in Geographical Information Systems*, London: Belhaven Press, pp. 37–50.

Nyerges, T.L., Moore, T.J., Montejano, R. and Compton, M., 1998, Interaction coding systems for studying the use of groupware. *Journal of Human-Computer Interaction*, 13(2): 127–165.

Olsen, J.M. and Brewer, C.A., 1997, An evaluation of color selections to accommodate map users with color-vision impairments. *Annals of the Association of American Geographers*, 87(1): 103–134.

Oviatt, S., 1997, Multimodal interactive maps: Designing for human performance. *Human-Computer Interaction*, 12: 93–129.

Oviatt, S. and Cohen, P., 2000, Multimodal interfaces that process what comes naturally. *Communications of the ACM*, 43(3): 45–53.

Senay, H. and Ignatius, E., 1998, *Rules and principles of scientific data visualization*, URL: http://homer.cs.gsu.edu/classes/percept/visrules.htm.

Skupin, A. and Buttenfield, B.P., 1996, *Spatial metaphors for visualizing very large data archives,* GIS/LIS '96, Denver, Colorado, ACSM/ASPRS, pp. 607–617.

Skupin, A. and Buttenfield, B.P., 1997, *Spatial metaphors for visualizing information spaces.* AutoCarto 13, Seattle, Washington, ACSM/ASPRS, pp. 116–125.

Taylor, D.R.F., 1991, Geographic information systems. In Taylor, D.R.F. (ed.), *The Microcomputer and Modern Cartography*, Oxford, UK: Pergamon, pp. 1–20.

Vasconcellos, R., 1992, Knowing the Amazon through Tactile Graphics. *Proceedings of the 15th Conference of the International Cartographic Association*, Bournemouth, UK, 23 September–1 October 1993, International Cartographic Association, pp. 206–210, Germany.

CHAPTER TWELVE

Ontological Foundations for Geographic Information Science

David M. Mark, University at Buffalo
Barry Smith, University at Buffalo and University of Leipzig
Max J. Egenhofer, University of Maine
Stephen C. Hirtle, University of Pittsburgh

12.1 INTRODUCTION

Ontology has gained increased attention among researchers in geographic information science in recent years, and in the present paper we argue that ontology can play an important role in establishing robust theoretical foundations for geographic information science in the future. The growth of interest in the topic of geospatial ontology is documented by such activities as:

- A well-attended session on ontology at the 2000 Association of American Geographers (AAG) meeting in Pittsburgh, Pennsylvania, organized by Nadine Schuurman and David Mark, chaired by Max Egenhofer, and featuring presentations by Michael Curry, Greg Elmes, and Harvey Miller, as well as by Schuurman and Mark;

- A EuroConference on Ontology and Epistemology for Spatial Data Standards in LaLonde-sur-Mer, France, 2000, organized by Stephan Winter (*http://www.geoinfo.tuwien.ac.at/events/Euresco2000/gdgis.htm*);

- A special issue of the *International Journal of Geographical Information Science* (Winter, 2001) containing papers from the EuroConference by Frank (2001a), Kokla and Kavouras (2001), Kuhn (2001), Raubal (2001), Smith and Mark (2001), and Worboys (2001);

- Several sessions and papers on ontology at the first two International GIScience Conferences held in 2000 and 2002 (*http://www.giscience.org*); and

0-8493-2728-8/05/$0.00+$1.50

- A number of awards by the National Science Foundation and the National Imagery and Mapping Agency, including Kuipers (1995), Egenhofer (1999), Findler and Malyankar (1999), Mark and Smith (1999), and Gahegan *et al.* (2002).

12.2 THE MEANING OF "ONTOLOGY"

The term "ontology" has been used in philosophy and information science in a number of ways. We provide here a brief discussion of some of the issues surrounding the use of the term, drawing especially on Guarino and Giaretta (1995) and on Smith (2003). Three main senses of the term can be distinguished, corresponding to the three main types of ontological research:

1. Ontology as a branch of philosophy deals with "the nature and the organization of reality" (Guarino and Giaretta, 1995). Ontology in this sense seeks primarily to establish the types of entities that reality contains and to establish how these entities are related. The philosophical ontology of the geospatial domain would deal with the totality of geospatial objects, categories, relations, and processes—and with their interrelations at different resolutions.

2. Ontology in the information systems sense is defined as "a logical theory which gives an explicit, partial account of a conceptualization" (Guarino and Giaretta, 1995; compare Smith and Welty, 2001). Note that, in contrast to the philosophical sense of the term, where "ontology" is a mass noun that cannot be pluralized, in this information science sense one can speak of several ontologies.

3. Ontology, when conceived as a process of eliciting ontologies from different sorts of human subjects, is best understood as a branch of psychology. It involves the use of standard psychological methods in order to establish the conceptual systems that people use in relation to given domains of objects. This latter aspect of the ontology research domain has implications for geographic information science especially as concerns issues of usability.

For present purposes, the opposition between 1 and 2 is especially important. Ontology in the first sense can also be characterized as a conception of ontology as a *theory of reality*, by analogy with any other scientific theory, though specializing primarily in the preparation of *taxonomies* of the types of entities existing in a given domain (including, again, the types of relations which unify these entities together into complex wholes of different sorts). Ontology in the second sense can be characterized as a conception of ontology as a second-level theory relating to the first-level theories about reality or, more generally, to the various "conceptualizations" embraced by human beings for particular scientific and non-scientific purposes. Additionally, ontological research in sense 1 is pure or basic research carried out with the aim of increasing human knowledge, where ontological research in sense 2 forms theoretical foundations for the design of next-generation information systems and databases.

12.3 ONTOLOGICAL COMMITMENT

The philosopher Quine (1953) plays an especially important role in any consideration of the relations between the philosophical and information systems senses of the term "ontology," and it is almost certainly through the channel of Quine's influential paper "On What There Is" (1953) that the term first entered into the literature of artificial intelligence and then of information systems science.

For Quine, ontology studies the theories of the natural sciences, which he takes to be our best sources of knowledge as to what reality is like. Quine's aim is to find the ontology *in* science. Ontology for him comes down to the study of the *ontological commitments* of, or in other words of the ontological presuppositions underlying, different natural-scientific theories. Each natural science has its own repertoire of types of objects to the existence of which it is committed. For Quine this is defined by the repertoire of predicates used by the corresponding theory in its canonical formalization in first-order predicate logic (the logic of quantifiers). Quine's famous criterion of ontological commitment he formulated in the slogan: To be is to be the value of a bound variable. This might be interpreted in practical terms as follows: to determine what the ontological commitments of a scientific theory are it is necessary to determine the values of the quantified variables used in the canonical formalizations of its theories.

Information scientists have extended Quine's understanding of ontology, using the term to refer not just to the taxonomies of entities accepted by given theories of natural science, but also to the taxonomies associated with, for example, standardized lexica or databases, or with enterprise management systems. They seek to establish what the users of given information processing systems are *committed to* as a result of their underlying conceptualizations. Thus ontology in the information systems sense seeks canonical descriptions of knowledge domains and associated classificatory theories. In such fields as artificial intelligence and computer science, ontology typically refers to a vocabulary or classification system that describes the concepts operating in a given domain through definitions that are sufficiently detailed to capture the semantics of that domain. It has been applied almost exclusively not to the large and messy worlds of natural scientific investigation but rather to the relatively controlled worlds of databases and information systems, or geographic subsystems that are heavily regulated, such as ship navigation (Malyankar, 1999, 2001), or highway codes and driving regulations (Kuhn, 2001).

To see why the project of an ontology in either the traditional philosophical sense or the Quinean sense has been embraced by so many influential figures in the information systems community, it is useful to point to what we might call the Tower of Babel problem as this arises in the database domain. Each of the many different groups of database system designers has its own idiosyncratic terms and concepts by means of which it builds frameworks for information representation; methods must be found to resolve the terminological and conceptual incompatibilities, which then inevitably arise. While such incompatibilities were initially resolved on a case-by-case basis, it was recognized that the provision, once and for all, of a common terminological standard—a shared taxonomy of entities—could provide significant advantages over case-by-case resolution. The term "ontology" has come to be used by information scientists precisely to describe a canonical reference taxonomy of this sort.

12.4 OBJECTIVES

"Ontological Foundations for Geographic Information Science" has been recognized as an emerging research theme by the University Consortium for Geographic Information Science (UCGIS). Under this umbrella, we unify several interrelated research subfields, each of which deals with different perspectives on geospatial ontologies and their roles in geographic information science. While each of these subfields could be addressed separately, we believe it is important that ontological research be carried out in a unitary, systematic fashion. Three broad sets of foundational issues need to be resolved: (i) conceptual issues concerning what would be required to establish an exhaustive ontology of the geospatial domain, (ii) representational and logical issues relating to the choice of appropriate methods for formalizing ontologies, and (iii) issues of implementation regarding the ways in which ontology ought to influence the design of information systems.

An integrated approach is necessary, because there is a strong interdependency between the methods used to specify an ontology, and the conceptual richness, robustness, and tractability of the ontology itself. But such integration comes only at a price. For while the potential advantages of ontology for the purposes of information management are obvious, the task of providing a common reference ontology (a single consistent and stable set of category labels) which would be sufficiently rich to contain all the major taxonomical concepts used in all scientific disciplines is, even when we restrict ourselves to the spatial sciences, so enormous as to be unachievable without considerable compromises along the way.

We can understand the sorts of difficulties that can arise, if we consider that such an ontology, if it is to be widely accepted, must be neutral across different data communities. But there is, as experience has shown, a formidable trade-off between this constraint of neutrality and the requirement that an ontology be maximally wide-ranging and expressively powerful—that it should contain canonical definitions for the largest possible number of terms. One solution to this problem is the idea of a top-level ontology, which would confine itself to the specification of such highly general (domain-independent) categories as: time, space, inherence, instantiation, identity, measure, quantity, functional dependence, process, event, attribute, boundary, *etc.* (See for example *http://suo.ieee.org*)

The top-level ontology would then be designed to serve as a common, neutral backbone, which would then be supplemented by the work of ontologists working in more specialized domains on, for example, ontologies of geography, or climatology, or ecology.

Our choice as to the framework of formal representation of ontology will have implications when it comes to information system implementations, and the latter will themselves serve as testbeds for the purpose of establishing the correctness and completeness—and usability—of the ontologies developed. None of the current UCGIS research priorities provides such an integrative perspective, and the topic of "Ontological Foundations for Geographic Information Science" is unique in this respect.

12.5 THE UCGIS APPROACH

UCGIS will coordinate research in this area. Coordination is required because of the multiplicity of research communities involved. Work on geographic concepts, categories, relations, and processes from a theoretical perspective must be coordinated with geospatial data and software standards efforts on the one hand and with general (top-level) ontology projects on the other. The formal approaches in ontology will augment the process of laying down robust data and software standards in different sub-fields.

Ontology in the philosophical sense is an enterprise that cross-cuts all branches of science and information systems. This is because all sciences and information systems deal, in some way and at some level of generality, with reality. There then comes into play what we might call the ontologist's credo (Smith, 2003): *to create effective representations it is an advantage if one knows something about the things and processes one is trying to represent.*

Ontological *commitments*, on the other hand, underlie all forms of cognition. Hence, ontology in the information systems sense—ontology as the making precise of conceptualizations and commitments—is needed to support research in spatial information science insofar as the latter relates to how humans, both experts and non-experts, use and understand geospatial software and theories in spatial science. Ontology as the study of ontological commitments is thus close to the research issues dealt with by geographic information science under headings such as data modeling and representation.

Geospatial cognition and geographic information systems each relates in its own way to the spatial aspects of the real world, more specifically to the domain of objects and processes located in large spatial regions on or near the surface of the Earth. This latter domain exists independently of human cognition and concepts (even roads and cities, once built, enjoy an independent existence). The geospatial domain is in this respect contrasted, for example, to the domain targeted by banking information systems, where the objects and processes involved are themselves to a large degree constructs of the very systems humans build.

Spatial reality existed even before human beings entered the scene. The independence of the geospatial domain is made especially clear when we reflect that—again in contrast to the banking case—there are several independent branches of natural science—geology, geomorphology, pedology, climatology, oceanography, ecology, forestry, and perhaps even geography itself—that deal with the same reality as that which is targeted by geographic information systems. This independence implies, however, that the philosophical and the information systems conceptions of ontology are much more closely allied in the geographic domain than in some other domains.

The task of ontology building does not focus on the design of specific algorithms and data structures that would allow implementation and coding of geospatial information and processes. Rather it aims to build robust, comprehensive and usable taxonomies. On this basis it seeks appropriate representations for geospatial phenomena to match the underlying reality, especially insofar as this reality is salient to human beings. Ontology of the geospatial domain is not a subset of the original ten UCGIS research challenges, although several of them addressed ontological issues. The ontology topic certainly has close relations to the representations topic, the scale topic, and to the conceptual or semantic aspects of the interoperability topic.

12.6 THE IMPORTANCE OF THE EMERGING THEME TO NATIONAL RESEARCH NEEDS/BENEFITS

It is widely recognized that the semantics of geospatial information is critical for the development of interoperable geospatial data and software. It is also widely recognized that Geographic Information Systems (GIS) software and technology should be able to interoperate with other software and databases such as those involved in wireless applications, e-commerce, logistics, environmental health, and health care delivery. Such interoperability requires a common or shared ontology for the phenomena under consideration—any phenomena distributed over part or all of the Earth's surface. This means also that research in the ontology of geospatial phenomena should be coordinated with efforts designed to establish geospatial terminology standards.

12.6.1 Priority Research Areas

Serious research on the ontology of geographic phenomena has begun only recently, and thus far the work has been directed primarily toward the formal modeling of the geospatial world as this is experienced and conceptualized by non-experts (Smith and Mark, 2001). An exhaustive ontology of the geospatial domain would be vast, and given the complex interrelations between the different spatial sciences will likely never be finalized. More reachable objectives would be to develop a complete upper-level ontology for the geospatial domain, and to develop in stepwise fashion detailed ontologies for subdomains, ontologies that are consistent with this upper-level ontology. Subdomains of highest priority would be the principal domains of GIS application, together with areas of environmental and social science where GIS has been under-utilized due to ontological mismatches between the corresponding domains of objects and GIS software.

Since the geospatial sciences deal with phenomena across a variety of scales, a common ontology for the geospatial domain must relate to the entities addressed not only in common-sense spatial reasoning but also in scientific, engineering, and computational representations of geospatial phenomena. Above all, it must provide a framework within which all of these types of representation can be integrated.

The ontology of the geospatial domain will define taxonomies of the different types of geographic objects, fields, spatial relations, and processes. It will be accompanied by translation algorithms, mapping the ontology into the basic data models and representations necessary for scientific computing about geographic phenomena. The ontology will be formalized through axioms and definitions of classes, relations, and functions (Grenon and Smith, 2004).

A need for formalized ontological frameworks for data integration has been recognized by many disciplines that specialize in the gathering and exchange of information. However, this need has received much less attention from scientists themselves. This is because, within each discipline or field of study, a shared conceptual system is normally ensured through the education and training of the scientists involved. Where cross-disciplinary communication and collaboration is required, however, ontology provides the needed common integration platform. Different environmental and social sciences share in common the fact that they study phenomena that occur or act over geographic space. Yet very little integration across these disciplines has taken place. The ontology of geospatial phenomena will provide the tools to support such integration and it will correspondingly facilitate the interoperability of the geospatial information systems addressing phenomena in these different fields.

12.6.2 Kinds of Ontological Research

We have distinguished three types of ontology research, which must be regarded as complementary and mutually constraining, and which echo our distinction between three different senses of "ontology" above:

1. Research on ontology in the philosophical sense—ontology as theory of reality—which attempts to establish the types of objects, processes, and relations, at different levels of scale and granularity, from out of which the geospatial domain is constituted. The methods employed here should be maximally opportunistic, involving (1) interaction with scientists from domains such as geology, climatology, and many others, designed to establish the sorts of entities popu-

lating their respective domains, and (2) the development of formal methods for integrating these populations of entities, for example in terms of part-whole and granularity relations. At the same time, such research should be directed towards clarification of the relations between human knowledge, beliefs, and representations on the one hand, the models and representations embedded in our data systems on the other hand, and the real world of objects beyond.

2. Research on the methods and tools for describing, accessing, comparing, and integrating geo-ontologies. This area falls into the standard information science mode, which means specifying the conceptualizations underlying different types of GISystems software and associated datasets for purposes of interoperability and cross-system translation. Ontology-driven GISs (Fonseca *et al.*, 2002) and the Geospatial Semantic Web (Egenhofer, 2002) are domains that rely extensively on advances in this area.

3. Research in eliciting geo-ontologies from human subjects (both experts and non-experts) using standard psychological methods. This type of research is of importance in connection not only with usability issues but also with issues pertaining to observation error, data gathering, and data-formulation.

The three types of ontological research are mutually constraining in virtue of the fact that we want our information systems to relate to the same real-world domain of objects as is captured (at different scales) in scientific theories and in everyday action and perception: geo-ontologies in every case take the same real, spatial world as their object, a world of constant change, of multifarious causal processes at different levels of scale and granularity. The underlying complexity of geo-ontological research is thus higher than standard ontological research in information systems, which relates primarily to the types of closed world models specified in database design and characteristically involves simplifications motivated by specific short-term pragmatic goals (Smith, 2003).

12.6.3 Short-Term (2–3 Years)

All UCGIS ontological research will involve formalization, and progress will be maximized if common formal-ontological tools and concepts are employed. A key short-term priority for research in this area is thus to develop and distribute an upper-level ontology for geospatial phenomena that can be used as a common framework to ensure that independently developed subdomain ontologies are consistent and interoperable. An early agreement on a formal language for specifying the ontology will contribute substantially to the potential to achieve longer-term goals. Again, consistency and interoperability with broader ontology projects is highly desirable. Thus, researchers in geospatial ontology should form links with general ontology projects such as the IEEE Standard Upper Ontology (SUO) Study Group *(http://ltsc.ieee.org/suo/index.html)*, the work of the Laboratory for

Applied Ontology (LOA) in Trento under the direction of Nicola Guarino (*http://www.loa-cnr.it/*), and the Basic Formal Ontology (BFO) project at the University of Leipzig (*http://ifomis.de*). One medium-term project would be to study the family of formal mereotopological theories, establish their properties, and refine the best ones for use in ontologies of the geospatial domain.

12.6.4 Medium-Term (3–5 Years)

Ontology-based wayfinding systems and agents (Raubal *et al.*, 1997; Sorrows and Hirtle, 1999; Raubal, 2001) will provide a bridge between the data structures of a geographic information system and the users of the system. The conceptual hierarchy of a user is in many cases only partially related to structures used by an information system, which can result in abstruse or confusing directions. An important medium-term goal would be to provide a better understanding of the cognition of geographic concepts via the formalization of an ontology of naive geographical concepts (see Mark and Smith's recent work). We are making progress in this area, but need additional formalizations to develop further testable hypotheses.

Another important medium-term project would be research on the ontology of vagueness, following up on preliminary work on geographic objects with indeterminate boundaries reported in the book edited by Burrough and Frank (1996), and also in the various recent treatments of vagueness in geography by Varzi (2001), Bennett (2001), and Bittner and Smith (2001a).

An additional medium-term project would focus on the ontology of scale, and especially the issue of how to integrate spatial ontologies at different levels of granularity or resolution (Stell, 2000; Smith and Brogaard, 2002; Bittner and Smith, 2003). Scale is another of the original UCGIS research priorities or challenges. The interaction between map scale, map resolution, and the level of detail that can be depicted on graphic maps was well known in cartography, but there are important differences in how size, scale, and resolution interact in digital representations.

Specification of the ontology of change (Galton, 2001; Grenon and Smith, 2004) and geographic process would be another medium-term priority; geographic objects such as lakes, rivers, and storm fronts have very special dynamic properties not studied in standard ontologies.

Yet another project would attempt to answer questions regarding the degree to which folk geographic concepts, such as clouds and storms, fronts and air masses, which are conceptualized as objects in folk and common-sense weather models, are useful, or even indispensable, in scientific models of atmospheric behavior. The ontology of meteorology is itself an important research area complementary to that of the projects here described.

Finally, the discipline of social ontology would attempt to establish the nature of geospatial objects in the institutional realm, including political objects such as states and nations and legal and economic objects such as real estate parcels and boundaries of coastal waters (Smith, 1995; Coan and Egenhofer, 1996; Frank, 1997; Bittner, 2001; Smith and Zaibert, 2001).

These ontological studies will need to be complemented by the development of appropriate methods and tools to describe ontologies. The development of canonical languages for geo-ontologies is a medium-term project. Likewise, the development of computational methods to compare ontologies and to integrate them into web-based search engines is a medium-term research project.

12.6.5 Long-Term (10 Years and Beyond)

The long-term goal is to complete the description and formalization of the ontology of all phenomena at geographic scales, including those phenomena dealt with by other natural sciences such as climatology and oceanography. This needs to go hand-in-hand with the development of appropriate mechanisms that support the integration of geo-ontologies at different levels of explicitness, and the development of guidelines for the resolution of conflicts in geo-ontologies.

In addition, the development of geo-ontologies will complement other efforts underway, including the development of the semantic web (Hendler, 2001), particularly the Geospatial Semantic Web (Egenhofer, 2002). The long-term goal of the semantic web project (*http://www.semanticweb.org*) is to build an extension of the World Wide Web (WWW) in which meaning is encoded through ontologies. The geo-ontologies will be a critical piece of the larger project designed to provide greater intelligence to the web.

12.7 EXAMPLE RESEARCH PROJECTS

A good way to obtain examples of current research projects in geospatial ontology is to examine award abstracts from the National Science Foundation (NSF). A search of recent NSF awards relating to ontology on the NSF web site revealed 40 awards with some variant of "ontology" in the award title or abstract. The fact that half of these are from the last 3 years is evidence of the emergent nature of ontology as a topic in information science. Of the 40 NSF awards, four mention geographic, spatial, or geospatial themes explicitly. We also mention below some additional projects funded by other agencies.

The project directed by Kuipers (1995), entitled "An Ontological Hierarchy for Spatial Knowledge", formalized the Spatial Semantic Hierarchy (SSH), a model for representations of spatial knowledge (Kuipers, 2000). This project, which ended in 1998, did not explicitly address spatial knowledge at geographic scales, and was mainly intended to support simulated and physical robots. The other three spatial ontology projects funded by NSF all began in 1999 and will extend from one to three years; all are focused on the geospatial domain. The fact that these funded projects were all initiated in 1999 is clear evidence that the topic is in an emerging phase, at least under the name "ontology."

Egenhofer's (1999) ontology project, entitled "NSF-CNPq Collaborative Research on Integrating Geospatial Information", involves collaboration with the Brazilian National Institute for Space Research (INPE), and focuses on semantic interoperability of spatial and geographic databases (Fonseca *et al.*, 2002). Under the National Imagery and Mapping Agency (NIMA) NIMA University Research Initiative (NURI) project "Similarity Assessments Based on Spatial Relations and Attributes" (*http://www.spatial.maine.edu/~max/nima.html*), this group is also designing computational methods to determine the similarity of different ontologies.

At the Pennsylvania State University, Mark Gahegan is working on "Enabling Collaboration and Improving Understanding: The Management of Semantics for Geospatial Information", funded by a recent NIMA NURI award.

Findler and Malyankar (1999) were funded by NSF's "Digital Government" program under the title "Digital Government: Representation and Distribution of Geospatial Knowledge", a project to determine an ontology for coastal entities such as shorelines and tide tables, in partnership with the U.S. Coast Guard and National Oceanic and Atmospheric Administration (NOAA; Malyankar, 1999, 2001).

Mark and Smith (1999) recently began a 3-year NSF-funded project entitled "Geographic Categories: An Ontological Investigation", designed to determine the ontology of geographic objects and associated cognitive categories; the context of Mark and Smith's project is general common-sense or naive geography, and the project emphasizes human subjects testing in a variety of languages. The results of the project are intended to contribute to spatial data transfer and semantic interoperability of general-purpose geographic software and data, and the work thus far has led to the clarification of a range of foundational and representational issues in geospatial ontology.

12.8 POSSIBLE SHOWCASE DEMONSTRATIONS

The use of ontologies to facilitate retrieval from spatial databases can provide a visual demonstration of the possible outcomes of the research here described. The recent work of Egenhofer and his colleagues on sketched-based interfaces and the work of Sara Fabrikant on interfaces for spatial data suggest high-impact, portable demonstrations that can highlight the benefits of an ontological approach. Other demonstration projects may include the use of geospatial lexicons, for instance in intelligent web geo-services and advanced spatial similarity search engines. The application of methods for comparing computationally different ontologies will also lead to tools for comparing geospatial standards.

12.9 REFERENCES

Bennett, B., 2001, What Is a Forest? On the Vagueness of Certain Geographic Concepts. *Topoi,* 20: 189–201.

Bittner, T., 2001, The Qualitative Structure of Built Environments. *Fundamenta Informaticae*, 46: 97–126.

Bittner, T. and Smith, B., 2001a., Vagueness, Granularity and Partitions. In Welty, C. and Smith, B. (eds.), *Formal Ontology and Information Systems*, New York: ACM Press.

Bittner, T. and Smith, B., 2001b., A Taxonomy of Granular Partitions. In Montello, D. (ed.), *Spatial Information Theory, Proceedings of COSIT 2001*, Berlin/New York: Springer.

Bittner, T. and Smith, B., 2003, A Theory of Granular Partitions. In Duckham, M., Goodchild, M.F. and Worboys, M.F., *Foundations of Geographic Information Science*, London: Taylor & Francis Books.

Burkhardt, H. and Smith, B. (eds.), 1991, *Handbook of Metaphysics and Ontology*, two vols., Munich/Philadelphia/Vienna: Philosophia.

Burrough, P.A. and Frank, A.U. (eds.), 1996, *Geographic Objects with Indeterminate Boundaries*, London and Bristol, PA: Taylor & Francis.

Casati, R. and Varzi, A.C., 1995, *Holes and Other Superficialities*, Cambridge: M.I.T. Press.

Casati, R. and Varzi, A C., 1999, *Parts and Places*, Cambridge: M.I.T. Press.

Coan, M. and Egenhofer, M., 1996, The Ontology of Land Boundaries under Natural Change: Erosion and Accretion of Sandy Shoals in Nantucket Sound. *Proceedings, ACSM/ASPRS Conference*, Baltimore, MD.

Degen, W., Heller, B., Herre, H. and Smith, B., 2001, GOL: A General Ontological Language. In Welty, C. and Smith, B. (eds.), *Formal Ontology and Information Systems*, New York: ACM Press, pp. 36–46.

Egenhofer, M.J., 1999, *NSF-CNPq Collaborative Research on Integrating Geospatial Information*, National Science Foundation Award IIS 9970123.

Egenhofer, M.J., 2002, Toward the Semantic Geospatial Web. *ACM–GIS 2002— 10th ACM International Symposium on Advances in Geographic Information Systems*, McLean, VA, pp. 1–4.

Egenhofer, M.J. and Mark, D.M., 1995, Naive geography. In Frank, A.U. and Kuhn, W. (eds.), *Spatial Information Theory. A Theoretical Basis for GIS*, Lecture Notes in Computer Sciences No. 988. Berlin: Springer-Verlag, pp. 1–15.

Fabrikant, S.I., 2000, The Ontology of Semantic Information Spaces. In *Geographical Domain and Geographical Information Systems—EuroConference on Ontology and Epistemology for Spatial Data Standards*, September.

Farquhar, A., Fikes, R., Pratt, W. and Rice, J., 1995, *Collaborative Ontology Construction for Information Integration*, Technical Report KSL-95-10, Stanford, California: Knowledge Systems Laboratory, Stanford University.

Findler, N.V. and Malyankar, R.M., 1999, *Digital Government: An Ontology for Geospatial Knowledge*, National Science Foundation Award EIA 9876604.

Fonseca, F., Egenhofer, M., Agouris, P. and Câmara, G., 2002, Using Ontologies for Integrated Geographic Information Systems. *Transactions in GIS*, 6(3): 231–257.

Frank, A.U., 1997, Spatial Ontology: A Geographical Point of View. In Stock, O. (ed.), *Spatial and Temporal Reasoning*, Dordrecht: Kluwer Academic Publishers, pp. 135–153.

Frank, A.U., 2001a, Tiers of Ontology and Consistency Constraints in Geographic Information Systems. *International Journal of Geographical Information Science*, 15(7): 667–678.

Frank, A.U., 2001b, The Rationality of Epistemology and the Rationality of Ontology. In Smith, B. and Brogaard, B., (eds.), *Rationality and Irrationality, Proceedings of the 23rd International Ludwig Wittgenstein Symposium*, Kirchberg am Wechsel, August 2000, Vienna: öbv&hpt, pp. 110–120.

Gahegan, M.N., Post, E.S., Barron, E.J. and Fonseca, F., 2002, *ITR: Ontologies In Action: An Information Architecture to Support Investigation of Linked Health–Environment Interactions*. Nation Science Foundation Award BCS-0219025.

Galton, A., 2001, *Qualitative Spatial Change*, Oxford: Oxford University Press.

Grenon, P. and Smith, B., 2004, SNAP and SPAN: Towards a Dynamic Spatial Ontology. *Spatial Cognition and Computation*, 4(1): 69–103.

Guarino, N., and Giaretta P., 1995, Ontologies and Knowledge Bases: Towards a Terminological Clarification. In Mars, N.J.I. (ed.), *Towards Very Large Knowledge Bases*, IOS Press, pp 25–32.

Guarino, N. (ed.), 1998, *Formal Ontology in Information Systems*, Amsterdam, Berlin, Oxford: IOS Press. Tokyo, Washington, DC: IOS Press (Frontiers in Artificial Intelligence and Applications).

Hayes, P., 1985a, The Second Naive Physics Manifesto. In Hobbs, J. and Moore, R. (eds.), *Formal Theories of the Commonsense World*, Norwood, NJ: Ablex, pp. 1–36.

Hayes, P., 1985b, Naive Physics I: Ontology of Liquids. In Hobbs, J. and Moore, R. (eds.), *Formal Theories of the Commonsense World*, Norwood, NJ: Ablex, pp. 71–108.

Hendler, J. and Lassila, O., 2001, Agents and the Semantic Web. *IEEE Intelligent Systems*, 16(2), 30–37.

Johansson, I., 1989, *Ontological Investigations. An Inquiry into the Categories of Nature, Man and Society*, New York and London: Routledge.

Kokla, M. and Kavouras, M., 2001, Fusion of Top-Level and Geographic Domain Ontologies Based on Context Formation and Complementarity. *International Journal of Geographical Information Science*, 15(7): 679-687.

Kuhn, W., 2001, Ontologies in Support of Activities. *International Journal of Geographical Information Science*, 15(7): 613-631.

Kuipers, B.J., 1978, Modeling Spatial Knowledge. *Cognitive Science*, 2: 129–153.

Kuipers, B.J., 1995, *An Ontological Hierarchy for Spatial Knowledge*, National Science Foundation Award IIS 9504138.

Kuipers, B., 2000, The Spatial Semantic Hierarchy. *Artificial Intelligence*, 119: 191–233.

Malyankar, R.M., 1999, *Creating a Navigation Ontology*, Workshop on Ontology Management, AAAI-99, Orlando, FL. In Tech. Rep. WS-99-13, AAAI, Menlo Park, CA.

Malyankar, R.M., 2001, *Acquisition of Ontological Knowledge from Canonical Documents*, IJCAI 2001 Workshop on Ontology Learning.

Mark, D.M. and Smith, B., 1999, *Geographic Categories: An Ontological Investigation*. National Science Foundation Award BCS 9975557.

Mark, D.M., Egenhofer, M.J., and Hornsby, K., 1997, *Formal Models of Commonsense Geographic Worlds: Report on the Specialist Meeting of Research Initiative 21*. Report 97-2. Santa Barbara, CA: National Center for Geographic Information and Analysis.

Mark, D.M., Skupin, A. and Smith, B., 2001, Features, Objects, and Other Things: Ontological Distinctions in the Geographic Domain. In Montello, D. (ed.), *Spatial Information Theory, Proceedings of COSIT 2001*, Berlin/New York: Springer, pp. 488–502.

Mark, D.M., Smith, B. and Tversky, B., 1999, Ontology and Geographic Objects: An Empirical Study of Cognitive Categorization. In Freksa, C. and Mark, D. M. (eds.), *Spatial Information Theory: A Theoretical Basis for GIS*, Lecture Notes in Computer Sciences. Berlin: Springer-Verlag, pp. 283–298.

Peuquet, D.J., Smith, B., and Brogaard-Pederson, B., 1999, *Ontology of Fields. Varenius Project Specialist Meeting Report*, Santa Barbara, CA: National Center for Geographic Information and Analysis.

Quine, W.V.O., 1953, On What There Is. As reprinted in *From a Logical Point of View*, New York: Harper & Row.

Raubal, M., 2001, Ontology and Epistemology for Agent-Based Way-Finding Simulation. *International Journal of Geographical Information Science*, 15(7): 653–665.

Raubal, M., Egenhofer, M.J., Pfoser, D. and Tryfona, N., 1997, Structuring Space with Image Schemata: Wayfinding in Airports as a Case Study. In Hirtle, S.C. and Frank, A.U. (eds.), *Spatial Information Theory*. Heidelberg: Springer-Verlag, pp. 85–102.

Smith, B., 1994, Fiat Objects. In Guarino, N., Vieu, L. and Pribbenow, S. (eds.), *Parts and Wholes: Conceptual Part-Whole Relations and Formal Mereology, 11th European Conference on Artificial Intelligence, Amsterdam, 8 August 1994*, Amsterdam: European Coordinating Committee for Artificial Intelligence, pp. 15–23. Revised and expanded version as: Fiat Objects, *Topoi*, 20:2, September 2001, pp. 131–148.

Smith, B., 1995, On Drawing Lines on a Map. In Frank, A.U. and Kuhn, W. (eds.), *Spatial Information Theory, Proceedings of COSIT '95*, Berlin: Springer-Verlag, pp. 475–484.

Smith, B., 2003, Ontology. In Luciano F. (ed.), *Blackwell Guide to Philosophy, Information and Computers*, Oxford: Blackwell, pp. 155–166.

Smith, B. and Brogaard, B., 2002, Quantum Mereotopology. *Annals of Mathematics and Artificial Intelligence*, 35(1–2): 153–175.

Smith, B. and Mark, D.M., 1998, Ontology and Geographic Kinds. In Poiker, T. K. and Chrisman, N. (eds.), *Proceedings. 8th International Symposium on Spatial Data Handling (SDH'98), Vancouver*, International Geographical Union, pp. 308–320.

Smith, B. and Mark, D.M., 1999, Ontology with Human Subjects Testing. *American Journal of Economics and Sociology*, 58(2): 245–272.

Smith, B. and Mark, D.M., 2001, Geographic Categories: An Ontological Investigation. *International Journal of Geographical Information Science*, 15(7): 51–62.

Smith, B. and Welty, C., 2001, Ontology: Towards a New Synthesis, editors' introduction to: Welty, C. and Smith, B. (eds.), *Formal Ontology and Information Systems*, New York: ACM Press, pp, iii–ix.

Smith, B. and Zaibert, L., 2001, The Metaphysics of Real Estate. *Topoi*, 20(2): 161–172.

Sorrows, M.E. and Hirtle, S.C., 1999, The Nature of Landmarks for Real and Electronic Spaces. In Freksa, C. and Mark, D.M. (eds.), *Spatial Information Theory*, Heidelberg: Springer-Verlag, pp. 37–50.

Stell, J.G., 2000, The Representation of Discrete Multi-Resolution Spatial Knowledge, *Proceedings of Seventh International Conference on Principles of Knowledge Representation and Reasoning* (KR2000), San Francisco: Morgan Kaufmann Publishers, pp. 38–49.

Uschold, M. and Grüninger, M., 1996, Ontologies: Principles, Methods and Applications. *Knowledge Engineering Review*, 11(2): 93–155.

Varzi, A.C., 2001, Vagueness in Geography. *Philosophy and Geography,* 4(2): 49–65.

Winter, S. (ed.), 2001, Ontology in the Geographic Domain. *Special Issue of the International Journal of Geographical Information Science*, 15(7): 587–590.

Worboys, Michael, 2001, Nearness Relations in Environmental Space. *International Journal of Geographical Information Science*, 15(7): 633–651.

CHAPTER THIRTEEN

Remotely Acquired Data and Information in GIScience

George F. Hepner, University of Utah and
American Society for Photogrammetry and Remote Sensing
Dawn J. Wright, Oregon State University
Carolyn J. Merry, The Ohio State University
Sharolyn J. Anderson, Texas State University–San Marcos
Stephen D. DeGloria, Cornell University

13.1 INTRODUCTION

The science and technology of remotely acquired data and information, often termed remote sensing, is not a new, emerging research theme *per se*, but rather a theme that offers new potential for research and development in many innovative directions. In the last 5 years, this area has undergone a true revolution in the number and type of sensors, data availability, potential new applications, and governmental and commercialization activities. There has been a massive increase in our ability to acquire radiometrically sensitive, geospatially-referenced sensor data from aircraft, satellite and undersea instrument platforms. Furthermore, new innovations in sensor systems are being developed to exploit various types of acoustic and electromagnetic (EM) data, including interferometric radar, infrared detector arrays, thermal, LIDAR (LIght Detection And Ranging), and other laser illumination techniques. Many new sensors are being developed that acquire EM data in novel ways, such as the acquisition of hundreds of narrow band spectra, termed hyperspectral remote sensing.

The wide range of uses for remotely acquired data is changing as well. In the past, most research efforts were on terrestrial landscapes, but the structure and composition of the atmosphere and hydrosphere of the Earth are being recognized as increasingly important to the quality of life and survival of humans. The submerged portions of our planet (71 percent of the Earth's total surface) are the

0-8493-2728-8/05/$0.00+$1.50
© 2005 by CRC Press LLC

focus of recent development of sophisticated sensors for ocean data collection and management (Garver *et al.*, 1994; Siegel and Michaels, 1996; Fornari *et al.*, 1997; Eleveld, 1999). These sensors hold tremendous potential for mapping and interpreting the ocean environment in unprecedented detail.

Use of remotely acquired data is entering the everyday life of the typical person. Weather radar, on-board automobile navigation, transportation management, law enforcement, recreational wayfinding, and many other citizen uses depend on advancements in remotely acquired data and information. The new frontiers in sensors, modes of data acquisition, and domains of investigation require research into these new and, in some cases, radical conceptual approaches to remote sensing.

As in many areas, technology development has outpaced our programs for user education. Issues of appropriateness of certain sensors to applications, the cost effectiveness of remote sensing data use, and informed procurement of new sensors, systems, and data need to be addressed very rapidly.

The increased capabilities of sensing from remote aerial and satellite platforms have increased the potential for surveillance that impinges on the constitutional rights of individual privacy. In a broader context, the transparency of remotely acquired data across national borders has major policy implications that need to be recognized and analyzed.

13.2 RESEARCH OBJECTIVES

The following objectives represent research arenas for UCGIS researchers. These have been defined using perspectives related to scientific merit, funding potential, and national policy implications.

1) Encourage UCGIS research into the methods by which advances in sensor systems, data sources, and analysis procedures can be utilized with the other capabilities of geographic information science: global positioning systems (GPS), measuring systems, visualization, data mining, real-time geographic information systems (GIS), and other geospatial scientific advancements.

2) Develop a UCGIS-based research capacity as an unbiased evaluator of new sensor technologies and comparative data analyses. This is needed to insure that informed use and procurement keep pace with the technological advancements.

3) Provide a structure for UCGIS members to increase their involvement in research associated with new sensors, digital sensor calibration, and sensor data systems.

4) Assess the social, legal, and policy implications of new sensor surveillance capabilities as they relate to individual rights of privacy, social adjustments and global transparency.

5) Encourage the increased use of remote sensing to study our land, oceans, lakes, rivers, and the atmosphere of the Earth.

6) Foster the conversion of military remote sensing assets to civilian uses in the environmental, agricultural, natural hazards, and other domains.

7) Investigate methods of using one or more existing data sources of modern ancillary geospatial data to leverage the massive amounts of historical remote sensing data for research on global environmental change.

13.3 THE UCGIS APPROACH

The UCGIS was formed to focus university interdisciplinary research and educational efforts on major national issues of GIScience. The basic and applied research activities necessary to meet the above objectives mesh well with the founding rationale for the UCGIS. In fact, the UCGIS ability to pool interdisciplinary talent is essential to meeting the objectives, which will foster the efficient and valid utilization of remote sensing data in the integrated GIScience enterprise.

This proposed research theme will focus on remotely-sensed phenomena, the measurement of these phenomena, integration of these sensor-derived data into information, and expansion of the array of analytical products that can be derived from this information.

UCGIS desires to have a role in the formulation of science and technology policy for the nation. A more directed focus on the field of remote sensing will foster internal UCGIS research coordination efforts, while improving the comprehension of UCGIS capabilities by the outside community. The proposed theme aligns very well with the recent initiatives in the major GIScience and policy communities in the U.S. For example, declassification and resultant adaptation of military remote sensing assets to the civilian community, the Open Skies Policy, newly, expanded governmental and private sector programs in remote sensing beyond the Landsat program, and the efforts in global environmental change detection and monitoring could and should involve focused UCGIS participation.

These research initiatives are a large portion of the future GIScience programs of the U.S. Congress, National Aeronautical and Space Administration (NASA), the National Geospatial Intelligence Agency (NGA—formerly the National Imagery and Mapping Agency, NIMA), the National Oceanic and Atmospheric Administration (NOAA), the U.S. Department of Transportation (USDOT), the National Security Agency (NSA), and the National Science Foundation (NSF). UCGIS can use this proposed theme to better focus our intellectual resources, and to provide credibility to our efforts in assisting these agencies and the private sector in addressing the use of remote sensing in these new research programs.

In addition, the involvement of the UCGIS, as an innovative and credible organization, is important to elevate the new era in remote sensing as an increasingly important component of GIScience. In the last 40 years, science and technologies that allowed travel to outer space have provided the satellite systems and massive computer capabilities that allow us to better study and understand the

Earth through remote sensing. In the past, university researchers in remote sensing and GIScience were forced to spend most of their resources dealing with hardware, software, data acquisition, and archival issues. Recent advancements in these areas, as well as lower costs, now allow multidisciplinary teams of university-based researchers to share capabilities to address fundamental science questions and their associated applications.

In the 1998 National Research Council publication, *People and Pixels: Linking Remote Sensing and Social Science,* social scientists Ronald Rindfuss and Paul Stern cite 1) building a community of scholars, 2) training future scholars, and 3) providing the necessary data as the critical institutional needs for linking remote sensing and social science to address important scientific and public policy questions (Liverman *et al.*, 1998). This proposed research theme will contribute significantly to meeting these recommendations through collaborative research between various disciplines. The UCGIS is the natural home for major efforts in this research area.

13.4 THE IMPORTANCE OF THE EMERGING THEME TO NATIONAL RESEACH NEEDS/BENEFITS

Recent interrelated, fundamental changes in remote sensing policy, basic science, technology transfer, and the private-public mixture of investment and control in remote sensing represents a nexus of activities that has heightened the importance of remote sensing to the national research and economic agendas. This proposed theme is a means for UCGIS to convey directly to the scientific and policy communities that we recognize these changes, and are attempting to address them as they relate to national research needs and benefits. These recent changes include:

1) Governmental controls on the spatial and spectral resolution of sensors and data have been relaxed. Until recently, the remote sensing community in the U.S.A. was constrained to broad spectral band sensors and a spatial resolution of 30 meters. Foreign competition and a changing view of the future have resulted in the easing of constraints on the spatial and spectral resolution for the U.S.A. In addition, it is likely that even more classified remotely acquired data and information will be released for public use for economic development, global environmental research, and other scientific purposes. The proposed research theme will place the UCGIS in the forefront of policy and application as these changes continue to occur in the future.

2) There has been a significant advance in the scientific development and technology transfer of usable sensor systems producing reliable, cost-effective data. The newest developments have occurred in high spatial resolution satellite data in the visible and near infrared spectrum, radar and interferometric radar, hyperspectral data, and laser illumination sensors (LIDAR). In the oceans, new remotely-operated vehicles (ROVs) carry out mapping tasks at unprecedented m- to cm-scales, both on the ocean floor and in the water column (Psuty *et al.*, 2004).

This is far beyond the resolution available from satellite altimetry or from systems towed or mounted on research vessels (Wright, 1999). University investigators have analyzed Landsat data in every conceivable manner for years. Suddenly, we are confronted with new possibilities in remotely acquired data and information to support both basic and applied science never before available to the university community.

3) The last few years has seen a major growth in the private sector commercialization of remote sensors, data and information products. A recent study by the American Society for Photogrammetry and Remote Sensing (ASPRS) and NASA has estimated that the remote sensing industry will grow at 10–15 percent per year over the next 5 years (see *http://www.asprs.org*). The impact of commercialization is taking esoteric science and moving it to the public consumer level. This has led to increased availability and reduced costs of remote sensing data, which provides great opportunities for university-based science (Mondello, 2004). Also, the commercial entities are in dire need of university research in sensor calibration, algorithms for analysis, and enhanced applications to bolster their link to the expanding global markets for imagery products.

4) NASA, Department of Defense (DOD), NOAA, USDOT and U.S. Geological Survey (USGS) are major GIScience agencies in the U.S. government. While they have major initiatives in many areas, the leading GIS-related ones are in new sensor systems, space-based sensors, imagery analysis, and accelerating the supply of remotely acquired data for geographic information systems.

The rapid growth and changing nature of remote sensing has prompted NASA, in collaboration with ASPRS, to undertake a dynamic 10-year forecast project for the commercial, academic and governmental sectors. This study is a basis for NASA's future responses to meet the needs for relevant systems and data and to assist in the workforce development necessary for the future.

NASA's newest satellite systems indicate the trends of the future. They include the Terra and EO-1 satellites. Sensors on Terra include the MODerate Resolution Imaging Spectroradiometer (MODIS—a 36-channel instrument at a 250-, 500-, or 1,000-meter resolution) and Advanced Spaceborne Thermal Emission and Reflectance Radiometer (ASTER—a visible and thermal channel instrument at 15- and 30-meter resolutions). The EO-1 satellite, that was recently launched, includes the Hyperion, which is a hyperspectral scanner (220 bands at 30-meter resolution). In cooperation with NASA, the USDOT is supporting a major new program for university research in the application of remote sensing to transportation infrastructure, transportation flow, disaster assessment, safety and hazards, and the environment (see *http://www.ncrst.org*).

The High Performance Computing and Communications initiative of NOAA has been funding projects that incorporate satellite Advanced Very High Resolution Radiometer (AVHRR) and ocean color imagery, color aerial photographs, and LIDAR imagery into GIS and digital libraries (see *http://www.hpcc.noaa.gov*).

Also for the oceans, the USGS has recently invested in several new systems for characterizing the U.S. continental shelf. These include acoustic sidescan-sonar imaging and multibeam swath-imaging systems that produce both sidescan-sonar and bathymetric data (see *http://walrus.wr.usgs.gov/infotech/*).

The DOD lists laser, infrared, radar, and acoustic sensors as militarily critical remote sensing technologies for spaceborne and airborne sensing of terrestrial, atmospheric, and ocean environments (see *http://www.dtic.mil/mctl*). The Shuttle Radar Topographic Mission (SRTM) for the acquisition of interferometric radar data and the Navy's Warfighter hyperspectral satellite are on-going programs that will support both military and civilian science research and applications.

5)   A major change is the recognition of the broader potential for remote sensing in the atmospheric and oceanographic domains. This trend has been stimulated by the recognition that the oceans and atmosphere are critical components in the analysis of global environmental change. In addition to acoustic sensing, investigation of thermal and microwave sensors for sea level measurement, wave (wind) speed and direction, and surface temperatures are areas of needed research for global weather, climate change, and biological resources management. Shallow water chemical and biological analysis can utilize hyperspectral sensor data for biological productivity and pollution studies.

6)   The advent of high-resolution remote sensing in all portions of the EM spectrum has given rise to the potential for unmatched surveillance of nations and individual citizens (Baker *et al.*, 2001). These capabilities are great advancements for national security, environmental monitoring, and law enforcement. However, as with other technological advancements in our society, use of these remote sensing capabilities has outpaced needed legal and social infrastructure (Julie, 2000). Our focus should be to anticipate these needs with preemptive research and policy rather than react to crisis situations.

These changes in the posture of remote sensing science and policy frame a major component of the national needs and benefits in the GIScience area. Along with its member universities and the ASPRS, it is imperative that the UCGIS coordinate a response to these changes in the national research and policy agendas.

13.5 EXAMPLE RESEARCH PROJECTS / SHOWCASE DEMONSTRATIONS

1)   **Coral Reef Mapping and Marine Sanctuary Management.** Coral reefs are among the most diverse and valuable ecosystems on Earth, rivaling tropical rainforests in their productivity. On a global scale, these regions contribute greatly to the production of seafood, new medicines for treating everything from colds to cancer, recreation, protection of shorelines from the impact of waves and storms, and billions of dollars in revenues each year to local communities and

national economies (Jones *et al.*, 1999; Wolanski, 2001). Unfortunately, coral reefs are also recognized as being among the most threatened marine ecosystems on the planet, having been seriously degraded by pollution, over-fishing, boat groundings, aquarium collection techniques, incorrect snorkeling and diving practices, and bleaching. (Bryant *et al.*, 1998). Remote sensing techniques (digital aerial photography, and airborne LIDAR and hyperspectral imaging) are increasingly being used to map and manage endangered coral reefs, particularly within the U.S. National Marine Sanctuary System (NMSS). Using state-of-the-art acoustic multibeam swath mapping technology, scientists from Oregon State University and the University of South Florida have been funded by the NSF and the NMSS to obtain complete bathymetric coverage of the Fagatele Bay National Marine Sanctuary (FBNMS) in American Samoa. FBNMS is the remotest and least explored of all the national marine sanctuaries, and the only true tropical coral reef in the sanctuary system. Until recently it was largely unexplored below depths of ~30 meters, with no comprehensive documentation of the plants, animals, and submarine topography. Indeed, virtually nothing is known of shelf-edge (50–120 meters deep) coral reef habitats throughout the world, and no inventory of benthic-associated species exists (Koenig *et al.*, 2000). Multibeam surveys are helping to answer key research and management questions for the FBNMS, and to establish future survey, sampling, monitoring, and management protocols (Wright *et al.*, in press). Surveys are also being ground-truthed and calibrated by information collected by underwater digital videography during SCUBA and rebreather dives.

 2) Urban Transportation. In 2000, the USDOT Research and Special Programs Administration (RSPA) and NASA created a new initiative to establish the National Consortia on Remote Sensing in Transportation (NCRST), with funding for four years under the Transportation Equity Act for the 21st century, Section 5113. There are four university consortia that are pursuing research in applying remote sensing and spatial information technologies in four focus areas of transportation: environmental assessment (NCRST-E), infrastructure management NCRST-I), traffic flow (NCRST-F), and disaster assessment, safety and hazards ((NCRST-DASH). State Departments of Transportation (DOTs) have used GIS technology over the past 20 years. Recently, commercial satellite and aircraft systems have been developed that have the high resolution needed for many transportation applications, particularly in the engineering area. LIDAR, hyperspectral data, unmanned autonomous vehicles (UAVs), and the 1-meter satellite data all have strong possibilities for use in demonstration projects of value to the engineering transportation community. Several potential application areas where remote sensing could be used in transportation demonstration projects are outlined in Morrison *et al.* (1999) (also see *http://www.cfm.ohio-state.edu/info/summary.html*).

 3) Urban Remote Sensing. In general, our understanding of urban theory, structure, and process is insufficient to explain and predict how urban change will be manifested in land use patterns and environmental alterations in urban and suburban areas. As cities continue to grow at unprecedented rates, understanding the relation between changes to land cover, local climate, and transportation systems becomes increasingly more important. Although relatively small in areal extent, cities have major impacts on the surrounding regional and the global environment. For cities located in developing areas of the world, there is virtually no geographic data system infrastructure to provide essential information on transportation, three-dimensional urban structure, land use, and surface materials (International Geosphere-Biosphere Committee, 1995). To help compensate for this deficiency, this demonstration project would serve to highlight the potential of the use of a remotely-sensed information systems approach to delineate, measure, and characterize urban landscapes. The project will demonstrate the use of satellite and airborne interferometric radar, hyperspectral sensor systems, and photographic products within a geographic information systems framework. These methods for inventory and monitoring of urban areas provide the necessary data and model validation information for the analyses of land use change, urban heat island formation and mitigation, non-point source pollution modeling, flood hazard potential, and emergency response to disasters (Hepner *et al.*, 1998). These analyses based on remotely acquired information assets are especially crucial in the developing regions of the world.

4) Precision Technologies for Managing Agronomic and Environmental Resources. Remotely acquired data and information in the GIScience context are used to discriminate agronomic and environmental features at variable scales of space, time, and complexity. Once discriminated, patterns of these features are mapped, then linked to biophysical processes that influence the sustainability of agronomic and environmental systems. Improved hyperspectral characterization of biophysical variables at field, farm, and watershed scales and at higher levels of taxonomic, temporal, and radiometric resolution improve estimates of crop performance and quality. This characterization also improves our abilities to better manage farmland related to crop varieties, crop quality, soils, fertilizers, pesticides, and invasive species. Databases of edaphic factors critical to crop production (*e.g.*, soil and climate) are integrated with remotely acquired data and information to implement and validate spatially-explicit models of crop growth and quality, nutrient and pesticide fate and transport, and spatial variability of soil characteristics and weather on crop yield and quality.

Of increasing importance and utility are location-based services where on-site, real-time assessments and management recommendations are provided to stakeholders. Such services are enhanced web-based spatial database queries and analyses for GPS-referenced field sites. In principle, the coordinate location of a field site is transmitted using cellular technology to a centralized server containing biophysical and socio-economic databases. Depending on the query invoked, the set of agronomic and environmental conditions of the geo-reference field site are transferred to resource-specific look-up tables for the generation and real-time transmission of management recommendations to the field site for implementation or evaluation.

Increased emphasis must be placed on integrating remote sensing science in GIScience curricula, strengthening the argument that remotely acquired data form the basis of most spatially-explicit geographic information used by society. This will ensure that scientific goals and stakeholder needs drive remote sensing technology development, application, and evaluation.

5) Coastal Landscape and Aquatic Habitat Analysis and Modeling. Landslides in forested areas are of great interest because they may pose significant threats to human safety, human structures, and other resources within these regions of the U.S.A. (*e.g.*, Lancaster *et al.*, 1999; Wing, 2000). In addition, salmon habitats in forested streams are critical to the economies of many states, particularly those in the Pacific Northwest. The State of Oregon and the USDA Forest Service have funded a large interdisciplinary study to predict forested patterns and aquatic habitats in 5-year increments. The study uses Landsat satellite imagery, land ownership data, and stream habitat survey data with advanced GIS techniques. In addition, knowledge of current land management is used to develop forest stand simulation models for deriving future conditions, which can then be incorporated into future resource management policies (Garman *et al.*, 1999). The project will be critical to the effective implementation of the Northwest Forest Plan and the Governor of Oregon's Salmon Initiative.

13.6 PRIORITY AREAS OF RESEARCH

13.6.1 Short Term (2–3 Years)

1) Fundamental research in the spectral signatures provided in imaging spectroscopy from airborne and satellite platforms. Hyperspectral data provide new opportunities for study of the Earth. Signatures of actual earth surface materials differ from laboratory-derived spectra of these materials, which also vary from airborne and satellite imagery data. These variations are due to many issues, including calibration, atmospheric correction, intimate mixtures, and other confounding factors in the real world environment. Research is needed in all of these areas to fully exploit the capabilities of this new remote sensing approach. Libraries of context-appropriate hyperspectral signatures should be developed and placed into easily accessible databases for researchers. These signature libraries provide a needed database for comparison with satellite sensor data. Appropriate tools to manage, display, and process the signatures should also be developed.

2) Calibration of digital cameras and sensors for radiometric characteristics and for evaluation of absolute geo-referencing accuracy. With the advent of digital cameras and sensors, additional research is necessary to understand and interpret the performance parameters of these devices in different settings and for varied uses. The establishment of additional geo-referencing field test sites with diverse geometries, complex surface roughness, and a variety of surface materials is essential. This would include the creation of a standardized set of performance parameters for every camera and sensor. This effort would involve coordination with government agencies and sensor manufacturers.

3) LIDAR-interferometric radar (IFSAR). Several important applications require high spatial resolution and three-dimensional measurement of the bare earth. In addition to the bare earth, the three-dimensional geometries of urban structures and systems, snow and ice structures, and surface reconstruction after earthquakes, warfare, and subsidence can be obtained using LIDAR and IFSAR data. Forestry applications using LIDAR to measure forest structure is another promising research area. New missions, such as IceSAT (Ice, Cloud and land Elevation SATellite), will carry a satellite laser altimeter that is designed to measure along-track topography over land and water for use in measuring temporal changes in the cryosphere. These emerging areas require significant research in image correlation, noise suppression, radiometric calibration, layover effects, feature measurement, and applications development.

4) Oceanographic multi-platform sensing. Bathymetric data from a swath acoustic mapping system located underneath a ship need to be geo-referenced to underwater video images or sidescan sonar data. These data are integrated with sample site observations, temperature measurements, and/or earthquake data obtained from an ocean bottom seismometer. Analytical methods need to be devised to handle the different dimensionalities, resolutions, and accuracies that will be produced by these different sensors. As transmission rates at sea of up to several gigabytes per day become more and more commonplace, the ability to assess and geocode ocean floor data collected at these different scales, in varying formats, and in relation to data from other disciplines has become crucial.

The introduction of these sophisticated tools has necessitated the development of reliable, spatially-referenced data management systems for the various data streams. The primary goals are to use these data integration systems to advance the analysis of ocean dynamics, optimize scientific interpretations, and to facilitate the rethinking or reformulating of hypotheses for this largely unknown component of the earth system.

5) Legal and social implications of increased surveillance capabilities. Research is required on two primary social and legal issues involving increased surveillance capabilities. This research is vital for formulation of policy concurrent with the technological advancements. At the local and national levels, increased surveillance capabilities by government and private firms may infringe on constitutional rights against unreasonable search and an expectation of privacy. At the national and global level, the foreign policy implications of increased abilities to "view" foreign nations with satellite sensors (global transparency) need to be better understood.

6) Exploitation of high spatial resolution. Spatial data of 1-meter resolution are available from commercial satellites. These high resolution data along with the repetitive, large regional coverage provide opportunities for analysis of urban dynamics, natural hazards, and agriculture to name a few. These data are further enhanced by leveraging the image data with GIS, GPS, and ground data. The efficient and effective use of information products derived from these data is a primary area of research.

7) Research in the adaptation of imagery data to new citizen/consumer applications, such as intelligent vehicle navigation, recreation, and economic development. The price, utility, and availability of remote-acquired data and information are such that their use in the consumer market is becoming a reality. University research can play a critical role in transferring these technologies to the average citizen. Also, a major role lies in the education of the broader public to the intelligent, responsible use of these data, devices, and technology.

13.6.2 Medium Term (3–5 Years)

1) Use of new sensor data to resurrect legacy remote sensing data. A primary constituent of the assessment of global environmental change is estimation of environmental baseline conditions. Only when a baseline is determined can relative change be measured. An archive of remote sensing data is available dating back to the early 1970s. These data sets should be resurrected and processed into compatible standard formats for use in merging with the new sensor data. The archive provides an excellent opportunity to monitor and map changes in the global environment. The resource and value of these data sets will help us to develop a better understanding of how the earth system has been changing over the past 30 years.

2) Development of autonomous control and on-board data processing systems to enhance remote sensing data collection and automated classification. Improvements in autonomous, on-board systems design, performance, and consistency are needed for automated collection and analysis/classification of data for a variety of sensors. The massive amounts of data collected can be effectively reduced and transformed to information using these systems, thereby, eliminating the need for real-time raw data transmission to processing stations. Approaches include expert systems, artificial intelligence, neural networks, and agent-based information extraction strategies. The applications include satellite and aircraft on-board feature detection and image classification, hyperspectral signature extraction, and DEM creation from LIDAR and IFSAR data acquisition. In the ocean, these systems would improve classification of sea-surface reflected sonar pulses, extraction of surface wave slopes from polarimetric synthetic aperture radar images, and mapping of mesoscale wind fields using Radarsat-1 ScanSAR images.

3) Development of artificial intelligence/expert systems to enhance remote sensing data collection and automated classification of features. Improvements in expert system design, performance, and consistency are needed for automated collection and analysis/classification of data for a variety of sensors. The massive amounts of land cover and environmental data cannot be analyzed directly in a timely and productive fashion. Research in artificial intelligence combined with spatial decision support systems is needed to automate the identification of imagery features and classes and define meaningful spatial and temporal change. In the oceanographic domain, needed research includes the automated analysis of directional spectra for ocean surface waves, extraction of surface wave slopes from polarimetric synthetic aperture radar images, improved mapping of mesoscale wind fields, and the automated classification of sediment types on the ocean floor.

13.6.3 Long Term (10 Years and Beyond)

1) Research in analysis of time series information from remote sensing data. Most current geographic information systems have only rudimentary support for time series analysis, a weakness still rooted in imperfect conceptual understandings of temporal data and, more specifically, of the spatial dynamics of many terrestrial, atmospheric, marine, and coastal environments (Wright and Bartlett, 2000). The greatest problem facing all means of change detection and analysis is simply the lack of consistent, and often continuous, data over long time spans. Temporally formatted remote sensing products can form the basis for research into this problem that can have a major impact on understanding earth processes and patterns through time and space.

2) Fundamental research in environmental acoustics for imaging and mapping. Imaging and mapping the water column and sea floor must be done primarily with the aid of sound, as sound waves are transmitted both farther and faster through seawater than electromagnetic energy. Therefore, to correctly interpret and map this environment research is needed in several areas including the effects of sound speed perturbations on the performance of various acoustic sensors, and the effects of horizontal refraction of broadband acoustic signals by moving internal solitary wave packets in shallow water.

13.7 CONCLUSIONS

The remote sensing area is an increasingly significant portion of GIScience. The authors firmly believe that the proposed research theme fulfills an important need within the UCGIS research agenda. The proposed theme is a credible addition to the UCGIS research agenda on scientific merit alone. However, given the trends in the governmental and commercial sectors, the need to establish the UCGIS as a leader in remote sensing research has important implications to the prominence of the UCGIS in the funding and policy areas. Given the challenges and opportunities presented by recent changes in remote sensing and allied areas it is critical for the UCGIS to respond in an effective manner by creation and support of this theme as a new and separate research initiative.

13.8 REFERENCES

Baker, J.C., O'Connell, K.M. and Williamson, R.A., (eds.), 2001, *Commercial Observation Satellites: at the Leading Edge of Global Transparency*, Rand Corporation, Arlington, Virginia and ASPRS, Bethesda, Maryland, 643 pp.

Bryant, D., Burke, L., McManus, J. and Spalding, M., 1998, *Reefs at Risk: A Map-Based Indicator of Potential Threats to the World's Coral Reefs*, World Resources Institute, Washington, D.C., 60 pp.

Eleveld, M.A., 1999, *Exploring Coastal Morphodynamics of Ameland (The Netherlands) with Remote Sensing Monitoring Techniques and Dynamic Modelling in GIS*, Ph.D. Thesis, University of Amsterdam, Amsterdam, The Netherlands, 225 pp.

Fornari, D.J., Humphris, S.E. and Perfit, M.R., 1997, Deep submergence science takes a new approach. EOS, *Transactions, American Geophysical Union*, 78(38): 402–408.

Garman, S.L., Swanson, F.J. and Spies. T.A., 1999, Past, present, future landscape patterns in the Douglas-fir region of the Pacific Northwest. In Rochelle, J.A., Lehmann, L.A. and Wisniewski, J. (eds.), *Forest Fragmentation: Wildlife and Management Implications*, Leiden, The Netherlands: Brill Academic Publishing, pp. 61–86.

Garver, S., Siegel, D. and Mitchell, B., 1994. Variability in near surface particulate absorption spectra: What can a satellite ocean color imager see. *Limnology and Oceanography*, 39: 1349–1367.

Hepner, G.F., Houshmand, B., Kolikov, I. and Bryant, N., 1998, Investigation of the integration of AVIRIS and IFSAR for urban analysis. *Photogrammetric Engineering and Remote Sensing*, 64-8: 813–820.

International Geosphere-Biosphere Committee, 1995, *Land Use and Land Cover Change-Science Research Plan*, IGBP Report No. 35 IGBP, Stockholm.

Jones, G.P., Milicich, M.J., *et al.*, 1999, Self-recruitment in a coral reef fish population. *Nature*, 402: 802–804.

Julie, R.S., 2000, High-tech surveillance tools and the fourth amendment: Reasonable expectations of privacy in the technological age. *American Criminal Law Review*, Winter, 37: 127–135.

Koenig, C.C., Coleman, C.F., Grimes, C.B., Fitzhugh, G.R., Scanlan, K.M., Gledhill, C.T. and Grace, M., 2000, Protection of fish spawning habitat for the conservation of warm-temperate reef fisheries of shelf-edge reefs of Florida. *Bulletin of Marine Science*, 66, 3: 593–616.

Lancaster, S.T., Hayes, S.K. and Grant, G.E., 1999, The interaction between trees and the landscape through debris flows. EOS, *Transactions, American Geophysical Union*, 80(46), suppl., p. F425.

Liverman, D., Moran, E., Rindfuss, R. and Stern, P. (eds.), 1998, *People and Pixels; Linking Remote Sensing to Social Science*. National Academy Press, Washington, D.C., 256 pp.

Mondello, C., Hepner, G.F. and Williamson, R.A., 2004, Ten-year forecast of the remote sensing industry, *Photogrammetric Engineering and Remote Sensing*, 70:1, pp 5–58, *http://www.asprs.org*

Morrison, J.M., Merry, C.J., McCord, M.R. and Goel, P.K., 1999, *Workshop on Applications of Remotely Sensed Data to Transportation, Final Report to Federal Highway Administration*, The Ohio State University, Research Foundation: Columbus, Ohio, December, 17 p.

Psuty, N.P., Steinberg, P. and Wright, D.J., 2004, Coastal and marine geography. In Gaile, G.L. and Willmott, C.J. (eds.), *Geography in America at the Dawn of the 21st Century*, New York: Oxford University Press.

Siegel, D. and Michaels, A., 1996, On non-chlorophyll light attenuation in the open ocean: Implications for biogeochemistry and remote sensing. *Deep-Sea Research*, 43: 321–345.

Wing, M., 2000, Landslide and debris flow influences on aquatic habitat conditions. *Proceedings of the 4th International Conference on Integrating GIS and Environmental Modeling, (GIS/EM4): Problems, Prospects and Research Needs*. Banff, Alberta, Canada.

Wolanski, E. (ed.), 2001, *Oceanographic Processes of Coral Reefs: Physical and Biological Links in the Great Barrier Reef*, Boca Raton, Florida: CRC Press, 356 pp.

Wright, D.J., 1999, Getting to the bottom of it: Tools, techniques, and discoveries of deep ocean geography. *The Professional Geographer*, 51, 3: 426–439.

Wright, D.J. and Bartlett, D.J., eds. 2000, *Marine and Coastal Geographical Information Systems*, London: Taylor & Francis.

Wright, D.J., Donahue, B.T. and Naar, D.F., 2002, Seafloor mapping and GIS coordination at America's remotest marine sanctuary (American Samoa). In Wright, D.J. (ed.), *Undersea with GIS*. Redlands, California: ESRI Press, 240 pp.

CHAPTER FOURTEEN

Geospatial Data Mining and Knowledge Discovery

May Yuan, University of Oklahoma
Barbara P. Buttenfield, University of Colorado
M. N. Gahegan, Pennsylvania State University
Harvey Miller, University of Utah

14.1 INTRODUCTION

The advent of remote sensing and survey technologies over the last decade has dramatically enhanced our capabilities to collect terabytes of geographic data on a daily basis. However, the wealth of geographic data cannot be fully valued when information implicit in data is difficult to discern. This confronts geographic information (GI) scientists with an urgent need for new methods and tools that can intelligently and automatically transform geographic data into information and, furthermore, synthesize geographic knowledge. It calls for new approaches in geographic representation, query processing, spatial analysis, and data visualization (Yuan, 1998; Miller and Han, 2000; Gahegan, 2000a). Information scientists face the same challenge as a result of the digital revolution that expedites the production of mountains of data from credit card transactions, medical examinations, telephone calls, stock values, and other numerous human activities. Collaborative efforts in artificial intelligence, statistics, and databases communities have been developing technologies of knowledge discovery in databases (KDD) to extract useful information from massive amounts of data in support of decision-making (Gardner, 1996; Hedberg, 1996; Bhandari *et al.*, 1997).

Knowledge discovery technology has emerged as an empowering tool in the development of the next generation database and information systems through its abilities to extract new, insightful information embedded within large heterogeneous databases and to formulate knowledge. A KDD process includes *"data warehousing, target data selection, cleaning, preprocessing, transformation and reduction, data mining, model selection (or combination), evaluation and inter-pretation, and finally consolidation and use of the extracted knowledge"* (Fayyad, 1997, p5). Specifically, data mining (DM) aims to discover something new from the facts recorded in a database. It prescribes the steps toward efficient development of knowledge discovery applications. Hitherto, data mining tools mostly adopt tech-niques from statistics (Glymour *et al.*, 1996), neural networks (Lu *et al.*, 1996), and visualization (Lee and Ong, 1996) to classify data and extract patterns. But ulti-mately, KDD aims to enable an information system that transforms the information to knowledge through hypothesis testing and theory formation. It sets new challenges for database technology: new concepts and methods are needed for basic operations, query languages, and query processing strategies (Lmielinski and Mannila, 1996).

This chapter provides a research frame for GI scientists to study the inte-gration of geospatial data mining and knowledge discovery. Following an exami-nation of the current state of DM and KDD technology, we identify special needs for geospatial DM and KDD, and outline research challenges and the significance to national research needs. We outline research frontiers in geographic knowledge discovery briefly and propose a research agenda to highlight short-term, mid-term, and long-term objectives.

14.2 BACKGROUND: AN OVERVIEW OF THE STATE-OF-ART IN DATA MINING AND KNOWLEDGE DISCOVERY

There is currently a good deal of interest in geospatial data as a rich source of structure and pattern, making it ideal for data mining research (*e.g.*, Koperski and Han, 1995; Ester *et al.*, 1996, 1998; Knorr and Ng, 1996; Koperski *et al*, 1999; Roddick and Spiliopoulou, 1999). Many of the very large consumer, medical, and financial transaction databases now being constructed contain spatial and tem-poral attributes and hence offer the possibility of discovering or confirming geo-graphical knowledge (Miller and Han, 2001). For decision makers this knowledge represents improved decision power. While DM and KDD quickly become popular among GI scientists, misconception appears not uncommon. Through discussions of the essence of DM/KDD and academic heritage, this section aims to draw a clear picture of the involved science and technology.

14.2.1 What DM/KDD Is, and Is Not

A generally accepted definition of DM and KDD is given by Fayyad *et al.* (1996) as: *"…the non-trivial process of identifying valid, novel, potentially useful and ultimately understandable patterns in data."* From this definition we can see the following.

1) DM is not straightforward analysis or machine learning. It is non-trivial, usually in the sense that the dataset under consideration is massive. If this were

not so, then an exhaustive statistical analysis should be possible, and is usually preferable since it is more rigorous. Data mining methods contain a degree of non-determinism to enable them to scale to massive datasets. Smythe (2000) provides some clarification in later work where he challenges the somewhat prevalent view that applying an established inductive machine learning technique to data (such as a decision tree) qualifies as mining.

2)   Some aspect is unknown at the start of the DM process and must be found. The term DM does not apply in cases where the outcome is already known, *i.e.* deterministic or deductive problems. Perhaps ideally, DM should be an abductive task (originally named by Peirce [1878] as hypothesis), simultaneously uncovering some structure within the data and a hypothesis to explain it. However, this would require sophisticated conceptual structures by which hypotheses might be represented within a machine. Currently, the focus of KDD seems to be on inductive learning methods, where the aim is to construct a model for the intension of some category from identified training examples. Because the structure is largely known (by way of these training examples) this is not mining *per se*, but rather a form of knowledge generalization. One exception to this is where the training examples themselves are a hypothesis only, generated from the data rather than given *a priori*, in an effort to establish classes with which to represent the data.Tools such as AutoClass (Cheeseman and Stutz, 1996) function in this manner.

3)   The uncovered structure needs to be valid; *i.e.* shown to be a significant or reliable inference with some level of confidence. Reliability metrics are required to support the hypotheses presented and to differentiate the significant from the marginal.

4)   The findings should be novel, that is, unknown at the outset. Obviously, the machine has no concept of what is known by experts, so has no means by which to map novelty to the application domain of discourse. However, it is possible to post-process results so that similar inferences are grouped together in generalized forms. Bradsil and Konolige (1990) refer to this as meta-learning. Each discovery is thus assured of being distinct from its peers.

5)   The uncovered structure needs to be useful, *i.e.*, be explainable and applicable in a manner that makes sense within the context of the current application domain. Large datasets may contain a great deal of structure that is not in itself useful. Focusing effort on those parts that are interesting is problematic because they are by definition unknown at the outset.

14.2.2 Academic Heritage

Successful applications of data mining are not common, despite the vast literature now accumulating on the subject. The reason is that, although it is relatively straight-forward to find pattern or structure in data, establishing its relevance and explaining its cause are both very difficult problems. Furthermore, much of what can be "discovered" may well already be common knowledge to the expert. Addressing these problematic issues requires the synthesis of underlying theory from the database, statistics, machine learning, and visualization communities. The issues relevant to DM from each of these disciplines, including database, statistics, and artificial intelligence, are described below (after Smythe, 2000).

14.2.2.1. Database

The database community draws much of its motivation from the vast digital datasets now available online and the computational problems involved in analyzing them. Almost without exception, current databases and database management systems are designed without thought to KDD, so the access methods and query languages they provide are often inefficient or unsuitable for mining tasks (Rainsford and Roddick, 1999).

In geographical analysis, Openshaw's Geographical Analysis Machine (Openshaw *et al.*, 1990) is an example of a more-or-less exhaustive DM tool. However, such brute force approaches do not scale well. As pointed out above, DM usually begins from the assumption that the dataset is massive, and accordingly the analysis tools must be designed so that computational performance is given the utmost priority. Approaches to improving performance can take the following forms.

- **Optimization of existing methods**—Many analysis techniques scale somewhere between $O(n^3)$ and $O(n\log(n))$ in terms of computational complexity (Martin, 1991), with the majority falling somewhere around $O(n^2)$. For smaller datasets this causes no problems, but where the number of features (attributes) is large, or the number of records is large, or both, such scaling renders existing techniques unusable. Many breakthroughs have been reported in the last few years to improve computational complexity so that it approaches $O(n)$. Techniques are typically based around optimistic optimization of hierarchical methods, such as decision trees, and include RIPPER (Cohen, 1995) and BOAT (Gehrke *et al.*, 1999).

- **Approximation of existing methods**—Algorithms attempt to encapsulate all the important structure contained in the original data, so that information loss is minimal and mining algorithms can function more efficiently. The premise here is that the functionality of some existing analysis methods can be approximated either by (a) sampling the data or (b) re-expressing the data in a simpler form. Sampling strategies must try to avoid bias, which is difficult if the target and its explanation are unknown. Data reduction approaches must attempt to "squash" the data into some lower dimensional form, similar in concept to a principal component transformation or a self-organizing (Kohonen) map.

- **New methods for DM**—Smythe (2000) points out that a variety of new approaches to DM have been created, which can function well using standard query interfaces and languages. He cites association rules (Agrawal *et al.*, 1993) as the most established example, but goes on to caution that they are rather impoverished in the analytic sense as they need further processing before they can represent the statistical significance of findings. However, one of the few documented successes of DM so far has been in analyzing consumer behavior by applying association rules to databases of purchases (*e.g.*, Berry and Lino, 1997). Such rules can be used to uncover likelihood of one type of purchase, given a set of others. They form the basis of some on-line, consumer analysis applications, too.

14.2.2.2. Statistics

The algorithmic basis of many DM methods can be traced back to multivariate statistical principles such as maximum likelihood, linear discriminant, and k-means functions. Good accounts of multivariate analysis in statistics are given by Dunteman (1984) and Mardia *et al.* (1979). These parametric approaches to analysis are complemented by clustering methods, as exemplified by the works of Anderberg (1973), Devijver and Kittler (1982), and Kaufman and Rousseeuw (1990). More recently, further progress has been made with the development of specifically spatial clustering techniques (see Murray and Estivill–Castro, 1997, for a useful summary). From a statistical perspective, the challenges posed by DM are fundamental, forcing the development of new types of inferential analysis techniques focused on discovering and evaluating local patterns within the data (*e.g.*, Anselin, 1995, 1996) rather than validating or refuting established global models.

- **Validating the findings**—Many of the techniques used to uncover local structure are not statistically rigorous and the challenge is to make them so (Elder and Pregibon, 1996). Data mining techniques, such as association rule construction, are less rigorous than existing statistical methods and do not confirm to significance testing using established statistical theory (Glymour *et al.*, 1997; Smythe, 2000). In a predictive sense this makes reliability assessment problematic. Furthermore, DM proceeds by constructing many (millions or even billions) of local hypotheses; even using a very high significance value we might reasonably expect a very large, even massive, number of "false positives". This causes two distinct problems. Firstly, how might more reliable measures of significance be constructed and secondly, how can false positives be differentiated from truly significant findings?

14.2.2.3. Artificial Intelligence (AI)

From an AI perspective, difficult problems of a representational nature present themselves. A variety of machine learning methods can be used to perform some of the generalization and inductive learning tasks associated with knowledge construction, including: case-based reasoning, neural networks, decision trees, rule induction, Bayesian belief networks, genetic algorithms, fuzzy and rough sets theory. See Mitchell (1997) for details of the workings of these methods.

- **Explaining the findings**—As noted above, to be truly abductive, structure must be simultaneously discovered and explained by a hypothesis of some sort. Ideally, this hypothesis would be constructed in the domain of the expert, *i.e.*, a high-level or abstract reason that makes sense within a specific problem context. But more realistically, hypotheses are given in the lower-level language of the data and clustering tools (*e.g.*, an induction rule hierarchy), making them difficult to interpret by the human expert. The need here is for more complex models of geography (or other application domains) to be represented within the computer, which would provide the structure required for a higher (more abstract) form of abduction to take place (*e.g.*, Sowa, 1999, Ch. 7). That is not to say the existing methods are not useful, since any clues to structure in data may well help trigger abductive reasoning by the expert, mapping the low level hypothesis into the application domain.

- **Representing the findings**—If new objects or categories are being uncovered, then they will also need to be represented in some manner. This topic also involves a significant database component. If findings are to be worked back into the database schema, then this schema must be capable of dynamic update (Drew and Ying, 1998). Furthermore, the semantics of the schema will need to be rich enough to encode the discovered relationships, or again capable of evolving the required relationship-types (*e.g.*, Luger and Stubblefield, 1998). This latter requirement is more difficult because it involves extending the semantic richness of a data model, rather than simply adding in new tables and populating them. Within the geographic sphere, this requirement causes particular difficulty, because implementations of conceptual models vary widely in terms of functionality and level of abstraction. Furthermore, there are only a handful of academic models that might be able to represent discovered spatio-temporal relationships, and none in commercial production at this time.

- **Reporting the findings**—Related to the above, discovered or uncovered knowledge must be reported to the expert (Gains, 1996), especially since it is unlikely that it can be directly represented in the system (see above). Textual reporting can produce an overwhelming amount of data in an indigestible form. Visual approaches to DM and knowledge discovery are therefore becoming popular and form part of a growing arsenal of visualization methods by which complex data may be depicted and explored (see below).

- **Visualization**—Visual approaches that might support DM and knowledge discovery have arisen independently in the statistics and database communities as well as within many other branches of science (Gahegan *et al.*, 2001 provide a more detailed overview). However, the terms used to describe these approaches differ by community. Within the database community, the phrase "visual data mining" is used to describe vast datasets rendered in some summarized form (*e.g.*, Keim and Kriegel, 1996; Card *et al.*, 1998; Ribarsky *et al.*, 1999). Statisticians, on the other hand, use the term "exploratory data analysis", but this also includes statistical techniques as well as graph-based visual methods (*e.g.*, Tukey, 1977; Asimov, 1985; Haslett *et al.*, 1991; Mihalisin *et al.*, 1991). These strands are largely convergent, aiming to capitalize on the pattern recognition of human experts. But perhaps even more important are the rich cognitive structures and mental models that human experts can apply to provide hypotheses and theories to help explain outcomes of computational knowledge discovery (Valdez–Perez, 1999).

Some visualization tools have recently been developed to directly support spatial DM and knowledge construction activities, such as the selection of useful data dimensions and the search for structure or pattern (*e.g.*, Lee and Ong, 1996; MacEachren *et al.*, 1999; Gahegan *et al.*, 2000). Useful overviews of visual DM are provided by Wong (1999) and Hinneburg *et al.* (1999).

14.2.3 Summary of techniques and approaches

A summary of the intersection of academic communities and the knowledge discovery tasks of: *finding* structure, *reporting* and *representing* the findings, *validating* their significance, and *optimizing* computational performance are given in Table 14.1. Though not exhaustive, this table indicates some of the key research initiatives and directions. The high level of interest in knowledge discovery is likely to lead to many additional techniques in the near future. But as is often the case with newer academic areas, there is little research evaluating and comparing techniques as yet, so it is difficult to judge their relative merits for a given application.

Table 14.1 A summary of DM/KDD techniques and approaches

	Databases	**Statistics**	**Artificial intelligence**	**Visualization**
Finding	Association rules	Local pattern analysis and global inferential tests	Neural networks, decision trees	Exploratory visualization Visual data mining
Reporting	Rule lists	Significance and power	Likelihood estimation, information gain	A stimulus within the visual domain
Representing	Schema update, metadata	Fitted statistical models, local or global	Conceptual graphs, meta models	Shared between the scene and the observer
Validating	Weak significance testing	Significance tests	Learning followed by verification	Human subjects testing
Optimizing	Reducing computational complexity	Data reduction and stratified sampling strategies	Stochastic search, gradient ascent methods	Hierarchical and adaptive methods, grand tours

14.3 MAJOR RESEARCH PERSPECTIVES

The increasing ability to capture, store, and process digital data and information is not unique to geographic information science: similar information revolutions are occurring in diverse fields such as marketing, biology, astronomy, and meteorology, just to name a few. Is there anything special about geographic data that requires unique tools and presents unique research challenges? Will progress in geographic knowledge discovery create broader impacts, leading to a better geographic information science?

In this section of the chapter, we identify the unique aspects of geographic knowledge discovery and its potential impacts on geographic information science and geographic research more broadly. These challenges and impacts can be classified into three main areas: namely, geographic information in knowledge discovery, geographic knowledge discovery in geographic information science, and geographic knowledge discovery in geographic research. This section summarizes discussion in Miller and Han (2001); see the original source for more detail and references.

14.3.1 Geographic Information in Knowledge Discovery

Geographic data has unique properties that require special consideration and techniques. First, geographic information exists within highly dimensioned geographic measurement frameworks. While other KDD applications involve highly dimensioned information spaces, geographic data is unique since up to four dimensions of the information space are interrelated and provide the measurement framework for the remaining dimensions. The most commonly adopted measurement framework is the topology and geometry associated with Euclidean space. However, some geographic phenomena have properties that are non-Euclidean; examples include travel times within urban areas, mental images of geographic space, and disease propagation over space and time (see Cliff and Haggett, 1998; Miller, 2000). Projecting geographic data into alternative, more appropriate measurement frameworks can aid the search for patterns in geographic DM. The information inherent in the geographic measurement framework is often ignored in induction and machine learning tools (Gahegan, 2000b).

Measured geographic attributes often exhibit the properties of spatial dependency and spatial heterogeneity. The former refers to the tendency of attributes at some locations in space to be related; typically, these are proximal locations. The latter refers to the non-stationarity of most geographic processes, meaning that global parameters do not reflect well the process occurring at a particular location. While these properties have been traditionally treated as nuisances, contemporary research fueled by advances in geographic information technology provides tools that can exploit these properties for new insights into geographic phenomena (*e.g.*, Getis and Ord, 1992, 1996; Anselin, 1995; Brunsdon, Fotheringham, and Charlton, 1996; Fotheringham, Charlton, and Brunsdon, 1997). Some preliminary research in geographic knowledge discovery suggests that ignoring these properties affects the patterns derived from DM techniques (Chawla *et al.*, 2001). More research is required on scalable techniques for capturing spatial dependency and heterogeneity in geographic knowledge discovery.

A third unique aspect of geographic information in knowledge discovery is the complexity of spatiotemporal objects and patterns. In most non-geographic domains, data objects can be meaningfully represented as points within the information space without losing important properties. This is often not the case with geographic objects: size, shape, and boundaries can affect geographic processes, meaning that geographic objects cannot necessarily be reduced to points without information loss. Relationships such as distance, direction, and connectivity are also more complex with dimensional objects (see Egenhofer and Herring, 1994; Okabe and Miller, 1996; Peuquet and Zhang, 1987). Transformations among these objects over time are complex but information-bearing (see Hornsby and Egenhofer, 2000). The scales and granularities for measuring time can also be complex, preventing a simple "dimensioning up" of space to include time (Roddick and Lees, 2001). Developing scalable tools for extracting patterns from collections of diverse spatiotemporal objects is a critical research challenge. Also, since the complexity of derived spatiotemporal patterns and rules can be daunting, a related challenge is making sense of these derived patterns, perhaps through "meta-mining" of the derived rules and patterns (Roddick and Lees, 2001).

The range and diversity of geographic data formats also presents unique challenges. The digital geographic data revolution is creating new types of data formats beyond the traditional "vector" and "raster" formats. Geographic data repositories increasingly include ill-structured data such as imagery and georeferenced multi-media (see Câmara and Raper, 1999). Discovering geographic knowledge from geo-referenced multimedia data is a more complex sibling of the problem of knowledge discovery from multimedia data (see Zaïne *et al.*, 1998).

14.3.2 Geographic Knowledge Discovery in Geographic Information Science (GIS)

There are unique needs and challenges for building discovered geographic knowledge in geographic information science. Most digital geographic databases are at best a very simple representation of geographic knowledge at the level of basic geometric, topological, and measurement constraints. Knowledge-based GIS attempts to build higher-level geographic knowledge into digital geographic databases for analyzing complex phenomena (see Srinivasan and Richards, 1993; Yuan, 1997, 2001). Geographic knowledge discovery is a potentially rich source for knowledge-based GIS and intelligent spatial analysis. A critical research challenge is developing representations of discovered geographic knowledge that are effective for knowledge-based GIS and spatial analysis.

14.3.3 Geographic Knowledge Discovery in Geographic Research

Geographic information has always been a central commodity of geographic research. For much of history, geographic research has occurred within a data-poor environment. Many of the revolutions in geographic research can be tied to improved technologies for georeferencing, capturing, storing, and processing geographic data; examples include sailing ships, satellites, clocks, the map, and GIS. The current explosion in digital geographic data may be the most dramatic shift in the environment for geographic research in the history of science. This leads to perhaps one of the most important "meta-questions" for geographic research in the 21st century, namely, what are the questions that we could not answer in the past?

14.4 THE UCGIS APPROACH

The UCGIS seeks to facilitate a multidisciplinary research effort on the development of geospatial DM/KDD science and technology. As discussed above, the development of DM and KDD technology has opened new avenues in information science research. It also plays an important role in any research endeavor based upon geospatial information. The ability to mine data pre-supposes that data delivery mechanisms and access mechanisms are in place. While data delivery services are becoming available in local and distributed computing environments, many impediments remain. In addition to research perspectives discussed above, one emphasis for this UCGIS research theme must address infrastructure support for DM and knowledge discovery. What mechanisms exist are not designed to handle problems specific to geospatial information. Yet, a robust data foundation is critical to promoting a multidisciplinary involvement in the research arena.

Three characteristics of geospatial data create special challenges to development of a robust data foundation. The characteristics that make geospatial data "special" as a computing problem have been acknowledged in many other writings, of course. Moreover, development of a data infrastructure needed to support GIScience in general forms a focus in another UCGIS initiative (spatial data infrastructure). Let us point out that the focus here is not on developing the spatial data infrastructure *per se* but on developing DM within the emerging infrastructure. As argued below, the research problems solved by generating a solid data foundation can be shown to create the need for new developments in DM and knowledge discovery. Many UCGIS researchers have the expertise with geospatial data coupled with an understanding of the limitations of existing and emerging infrastructures and pursue a research agenda addressing the DM/KDD topics in a geospatial context.

The first characteristic relevant to DM/KDD is that geospatial data repositories tend to be very large. Data volume was a primary factor in the transition at many federal agencies from delivering public domain data via physical mechanisms (CD–ROM, for example) to electronic mechanisms (National Research Council Data Foundations report). Moreover, existing GIS datasets are often splintered into feature and attribute components that are conventionally archived in hybrid data management systems. Algorithmic requirements differ substantially for relational (attribute) data management and for topological (feature) data management (Healey, 1991). Computational procedures from knowledge discovery must also be diversified if they are to become fully operational within a geospatial computing environment, and this forms an important component of this research theme. Even with deployment of newer integrated GIS data models (such as the Environmental Systems Research Institute's [ESRI, Inc.] *geodatabase* data model and Smallworld's object-oriented data model), the hybrid data model will be preserved. In practice, knowledge integration will begin to span not only disparate data models in a single archive, but disparate archives in disparate database management systems.

A second characteristic of geospatial data is the extremely short phase of data, which are collected cyclically. Data discovery must accommodate collection cycles that may be unknown (as for example identifying the cycles of shifts in major geological faults) or that may shift from cycle to cycle in both time and space (for example the dispersion patterns of a health epidemic, or of toxic waste). Integration of information from multiple data models (point source field data with multispectral raster data) is acknowledged as a focus in an established UCGIS research theme (Spatial Data Acquisition and Integration), and does not need to become a focus of attention here. Instead, knowledge discovery researchers can turn attention to problems of reasoning and modeling on very short temporal cycles. For example, geospatial knowledge discovery could support real-time tornado tracking, or avalanche prediction, or other localized weather events. Infrastructure issues that need to be researched include (for example) the development of real-time DM, and to utilize knowledge discovery tools to guide correlation of discovered data patterns across time, determination of temporal drift, validation of data trends across temporal discontinuities, and so forth. Because we understand much less about the nature of time than of space, methodologies for archiving data to facilitate cyclic spatial searching remain crude. The extent to which one can identify data patterns will be determined in whole or in part by the organization of the data in an archive. Research on how best to structure data or to reorder data for specific knowledge discovery tasks is not covered in other UCGIS research themes, and must be addressed before a robust geospatial data foundation can developed.

A third characteristic of relevance to DM/KDD applies to a characteristic of the data foundation rather than of the data. Emergence of the Internet has supported development of data clearinghouses, digital libraries, and online repositories wherein one does not access data, but pointers to data. It is paradoxical that as increasing amounts of digital data become available via the Internet, they become increasingly difficult to locate, retrieve, and analyze. This is due in large part to the fact that the Internet lacks a comprehensive catalog or index (Buttenfield, 1998). Without a coordinating infrastructure, many data sources and services available today remain essentially inaccessible. Currently, over three million Websites are online; and yet even the best search engines can locate only one third of the accessible pages (National Public Radio, 1998; National Research Council, 1995). Data mining tools need to be established to locate environmental data sources in the "gray literature" areas of the Internet. Such data sources include but are not limited to field data collected in developing countries, very localized community data sites such as inner city neighborhood and community activist sites, and similar data sources not known to or known by conventional doorways into the geospatial data infrastructure. This type of knowledge discovery treats the entire Internet as a very large, decentralized data repository, and provides a venue for contributions to a global information infrastructure.

The decentralization of data delivery via ftp and the Internet revokes many assumptions of what can be known in advance about an archive about to be explored. In addition to the format and data model issues described above, one must consider the semantic issues. Infrastructure support for DM/KDD must facilitate thesaurus support. Data definitions are acknowledged to vary widely from agency to agency within a single country. Witness for example the difference between the definition of "address" by 911 Dispatchers (the location of a front door) and by the U.S. Post Office (the location of a mailbox). In urban areas, these two items (front door and mailbox) may be co-located. In rural areas, however, the locations may differ by half a kilometer or more (example from Jack Estes, 1993, personal communication). Knowledge discovery tools working across data sets must be embedded with functions to discriminate semantic differences from errors; and in a decentralized DM environment, linkages between data and thesauri may not be explicit. Herein lies another important research area for this UCGIS theme.

To summarize, the development of DM and knowledge discovery tools must be supported by a solid geographic foundation. The emergence of a geo-spatial data infrastructure has been *ad hoc*. Contributed data has not been coupled with contributed tools for data analysis and modeling. Data mining, knowledge discovery methods have not been implemented to deal effectively with geospatial data, whose sensitivities are known widely to geographers. As our understanding of the nature of geographic information and its sensitivities to spatial, temporal, and spectral resolution improve, it is probable that refinement of DM algorithms will prove insufficient; and design of new procedures and knowledge validation procedures will begin to emerge. We view the acceptance of the need for new DM/KDD designs as one of the primary indicators of the success of the research agenda we propose.

14.5 IMPORTANCE TO NATIONAL RESEARCH NEEDS

In the information age, geospatial data are collected from diverse sources at a rate that exceeds state-of-the-art capabilities for data management. In a historical con-text, the amount of information generated in 1999 alone is estimated to be more than 12 percent of the total volume generated by humankind in all of recorded history (Bradley *et al.*, 2001). Recent studies on the increasing amounts of data collected and stored digitally suggest a widening gap between the volumes of stored data and human's ability to process the data. Moore's Law states that computing power doubles every 18 months. Gray and Shenoy (2000) claim that the cost of storage media drops in half every 9 months. Given this, it is not surprising that data archives will continue to grow and we can expect that the vast majority of archived items will not be accessible by human searches, which are based upon existence of a catalog.

Many human activities could benefit from advances in DM and KDD research. In some cases, timely access to archived information could minimize loss of life, harm to geographic or societal groups, and costs to society as a whole. This is particularly compelling in the context of mitigating natural hazards.

New developments in technologies, such as Internet 2, will extend Internet protocols and increase bandwidths for data transmission. It will become possible for people to search larger archives which are globally distributed, according to their own particular needs. As individuals are able to extract data from electronic repositories more effectively, they become more empowered to participate in their own governance. Additionally, they become alert to inconsistencies in archived data. Particularly for data of local geographic significance, this opens the door for grass-roots contributed data updates and correction notices. "*Developing the tech-nical and institutional means to support incorporation of local knowledge into networked repositories presents a novel challenge*" (National Research Council, 1999: 2).

The actual benefits to the nation of improved access to information are difficult to predict. Many are intangible; others do not have a fixed cost. However, the value of an informed and participatory citizenry to environment and society makes a clear and compelling argument for pursuit of research in DM and KDD, and points also to the importance of teaching the nation's youth in utilizing such tools to tap into the vast archives of geospatial information that are and will continue to be archived.

14.6 EXAMPLE PROJECTS

We are still at a very early juncture in the history of geospatial knowledge discovery (GKD). Many attempts are being made to develop new GKD tools and applications. Below, we outline a list of geographic KDD applications in geographic information science and broader geographic research.

14.6.1 Map Interpretation and Information Extraction

Malebra *et al.* (2001) demonstrate the use of inductive machine learning tools within a GIS environment. Their system can extract and interpret complex human and physical features from topographic maps for input into a GIS and for analysis.

14.6.2 Information Extraction from Remotely Sensed Imagery

The increasing detailed spatial, temporal and spectral resolutions provided by advances in remote sensing technologies are creating massive imagery databases. These databases are overwhelming the ability of researchers to analyze and understand the information implicit within these data. Gopal, Liu, and Woodcock (2001) use artificial neural networks combined with visualization techniques to interpret and understand the patterns extracted from remotely sensed images.

14.6.3 Mapping Environmental Features

Many geographic phenomena have complex, multidimensional attributes that are difficult to summarize and integrate using traditional analytical methods. Eklund, Kirkby, and Salim (1998) use inductive learning techniques and artificial neural networks to classify and map soil types. Lees and Ritman (1991) use decision tree induction methods for mapping vegetation types in areas where terrain and unusual disturbances (*e.g.*, fire) confound traditional remote sensing classification methods.

14.6.4 Extracting Spatiotemporal Patterns

Identifying unusual patterns in massive spatiotemporal databases can be difficult since the number of possible patterns can be very large. Mesrobian *et al.* (1996) develop the Open Architecture Scientific Information System (OASIS) for querying, exploring and visualizing geophysical phenomena from large, heterogeneous and distributed databases. The Conquest Scientific Query Processing System, a component of OASIS, identifies cyclonic activity from weather and climate data by extracting unusual patterns in air pressure and winds over time. In another domain, Openshaw and colleagues (Openshaw *et al.*, 1987; Openshaw, 1994) develop exploratory techniques based on simple querying and artificial life methods for spotting spatial-temporal clusters in crime data.

14.6.5 Interaction, Flow and Movement

Spatial interaction, flow, and movement in geographic space can provide insights into the spatial structure of physical and human geographic systems. Spatial structure and spatial interaction are intimately related: location influences interaction patterns while interaction patterns influence the location of entities and activities. For tractability purposes, traditional spatial and network analyses make strong assumptions about influences among flow, interaction, movement, and location, essentially only capturing direct and proximal effects in space and time. More complex *n*-th order influences may be buried in the massive interaction, flow, and movement databases are being captured by real-time monitoring systems, intelligent transportation systems and "position-aware" devices such as cellular telephones and wireless internet clients. Marble *et al.* (1997) describe visualization methods for exploring massive interaction matrices. Smyth (2001) explores the possibilities for geographic KDD from the space-time trajectories of mobile devices.

14.7 RESEARCH FRONTIERS AND PRIORITY AREAS FOR RESEARCH

There are several critical research challenges in geographic KDD and DM. Miller and Han (2000) offer the following list of emerging research topics in the field.

14.7.1 Developing and Supporting Geographic Data Warehouses

To date, a true geographic data warehouse (GDW) does not exist. Spatial properties are often reduced to simple aspatial attributes in mainstream data warehouses. Creating an integrated GDW requires solving issues in spatial and temporal data interoperability, including differences in semantics, referencing systems, geometry, accuracy, and position.

14.7.2 Better Spatiotemporal Representations in Geographic Knowledge Discovery

Current GKD techniques generally use very simple representations of geographic objects and spatial relationships. Geographic DM techniques should recognize more complex geographic objects (lines and polygons) and relationships (non-Euclidean distances, direction, connectivity, and interaction through attributed geographic space such as terrain). Time needs to be more fully integrated into these geographic representations and relationships.

14.7.3 Geographic Knowledge Discovery Using Diverse Data Types

GKD techniques should be developed that can handle diverse data types beyond the traditional raster and vector models, including imagery and geo-referenced multimedia.

14.7.4 User Interfaces for Geographic Knowledge Discover

GKD needs to move beyond technically oriented researchers to the broader GIScience and other research communities. This requires interfaces and tools that can aid diverse researchers in applying these techniques to substantive questions.

14.7.5 Proof of concepts and benchmarking

As in other KDD and DM domains, there needs to be some definitive test cases or benchmarks to illustrate the power and usefulness of GKD to discover unexpected and surprising geographic knowledge. A related issue is benchmarking to determine the effects of varying data quality on discovered geographic knowledge.

14.7.6 Building Discovered Geographic Knowledge into GIS and Spatial Analysis

We require effective representations of discovered geographic knowledge that are suitable for GIS and spatial analysis. This may include inductive geographic databases, online analytical processing (OLAP)-based GIS interfaces and intelligent tools for guiding spatial analysis.

Expanding upon the above list, we propose the following research agenda in geospatial DM and KDD.

14.7.7 Short-Term Objectives

- Apply DM and KDD techniques to the new generations of geospatial data models and identify analytical and visualization needs for geospatial DM and KDD;

- Survey the existing spatial analysis methods, evaluate their potential for large data sets, and, when appropriate, extend their computational abilities in large data sets; and

- Apply data warehousing techniques and models to the geographic context for the development of a strong geospatial data foundation and examine methodologies for distributed databases and distributed processing that accommodate the spatial nature of both the data and potential retrieval queries.

14.7.8 Medium-Term Objectives

- Develop a taxonomy of geographic knowledge and categorize models (methods) for geographic information computing;

- Develop a system for geographic knowledge acquisition and synthesis;

- Develop robust spatial and temporal representations and develop algorithms to automate complex geographic queries in large, distributed, heterogeneous, and dynamic databases;

- Develop robust spatial and temporal reasoning and analytical models to support geographic knowledge formulation through interactive and recursive query processes; and

- Develop multi-dimensional, interactive visualization techniques with dynamic links to distributed GIS databases to greatly enhance user's capabilities to detect hidden patterns and inspect potential correlations among geographic variables.

14.7.9 Long-Term Objectives

- Develop an integrated theory for geographic information representation, processing, analysis, and visualization. The theory will suggest the best geographic representation, analytical methods, and visualization techniques to extract the highest level of geographic information and knowledge in a GIS database; and

- Enable a full implementation of geographic KDD across distributed databases that allow the general public to inspect climate patterns and regional demographic dynamics, for example, on the Internet.

14.8 ACKNOWLEDGEMENTS

We would like to thank the 40 participants of the one-day workshop on geospatial DM and KDD at UCGIS Summer Assembly in Buffalo June 20, 2001, in particular, Dr. Robert Edsall, Dr. Kathleen Hornsby, Dr. Lin Lou, and Dr. Shashi Shekhar. Comments and suggestions from the 40 participants have strengthened the manuscript. Yuan's effort is in part supported by the National Science Foundation under Grant No. 0074620.

14.9 REFERENCES

Agrawal, R., Imielinski, T. and Swami, A., 1993, Mining association rules between sets of items in large databases. *ACM SIGMOD*, pp. 207–216.

Anderberg, M.R., 1973, *Cluster Analysis for Applications*, Boston, USA, Academic Press.

Anselin, L., 1995, Local indicators of spatial association—LISA. *Geographical Analysis*, 27, 93–115.

Anselin L., 1996, The Moran scatterplot as an ESDA tool to assess local instability in spatial association. In Fischer, M., Scholten, H. and Unwin, D. (eds.), *Spatial Analytical Perspectives on GIS*, London: Taylor & Francis, pp. 111–125.

Asimov, D., 1985, The grand tour: A tool for viewing multidimensional data. *SIAM Journal of Science and Statistical Computing*, Vol. 6, pp. 128–143.

Berry, M.J.A. and Lino, G., 1997, *Data Mining Techniques For Marketing, Sales, and Customer Support*, New York, NY: John Wiley & Sons.

Bhandari, E., Colet, E., Parker, J., Pines, Z, Pratap, R., Pratap, R. and Ramanujam, K., 1997, Advanced scout: Data mining and knowledge discovery in NBA data. *Data Mining and Knowledge Discovery*, 1, 121–125.

Bradley, P.S., Fayyad, U., Sarawagi, S. and Shim, K. 2001, Editorial. *SIGKDD Explorations*, 2, 2, i–iii.

Bradsil, P.B. and Konolige, K. (eds.), 1990, *Meta-Learning, Meta-Reasoning and Logics*, Boston, MA: Kluwer Academic Press.

Brunsdon, C., Fotheringham, A S. and Charlton, M.E., 1996, Geographically weighted regression: A method for exploring spatial nonstationarity. *Geographical Analysis*, 28, 281–298.

Butterfield, D.P., 1998, Looking Forward: Geographic Information Services and libraries in the future. *Cartography and GIS*, 25(3): 161–171.

Câmara, A.S. and Raper, J. (eds.), 1999, *Spatial Multimedia and Virtual Reality*, London: Taylor & Francis.

Card, S., Mackinlay, J. and Shneiderman, B., 1998, Information visualization. In Card, S., Mackinlay, J. and Shneiderman, B. (eds.), *Information Visualization*, San Francisco: Morgan-Kaufmann, pp. 1–34.

Chawla, S., Shekhar, S., Wu, W.L. and Ozesmi, U., 2001, Modeling spatial dependencies for mining geospatial data: An introduction. In Miller, H.J. and Han, J. (eds.), *Geographic Data Mining and Knowledge Discovery*, London: Taylor & Francis.

Cheeseman, P. and Stutz, J., 1996, Bayesian classification: Theory and results. In Fayyad, U., Piatetsky-Shapiro, G, Smyth, P. and Uthurusamy, R. (eds.), *Advances in Knowledge Discovery and Data Mining,* Cambridge, MA: AAAI/MIT Press, pp. 153–189.

Cliff, A.D. and Haggett, P., 1998, On complex geographical space: Computing frameworks for spatial diffusion processes. In Longley, P.A., Brooks, S.M., McDonnell, R. and MacMillan, B. (eds.), *Geocomputation: A Primer*, Chichester, U.K.: John Wiley & Sons, pp. 231–256.

Cohen, W.W., 1995, Fast, effective rule induction. *Proceedings 12th International Conference on Machine Learning*, San Francisco, CA: Morgan-Kaufmann, pp. 115–123.

Devijver, P.A. and Kittler, J., 1982, *Pattern Recognition: A Statistical Approach*, London: Prentice Hall International.

Drew, P. and Ying, J., 1998, Metadata management for geographic information discovery and exchange. In Sheth, A. and Klas, W. (eds.), *Multimedia Data Management: Using Metadata to Integrate and Apply Digital Media*, New York: McGraw-Hill, pp. 89–121.

Dunteman, G.H., 1984, *Introduction to Multivariate Analysis*, Beverly Hills, CA: Sage.

Egenhofer, M.J. and Herring, J.R., 1994, Categorizing binary topological relations between regions, lines and points in geographic databases. In Egenhofer, M., Mark, D.M. and Herring, J.R. (eds.), *The 9-intersection: Formalism and its Use for Natural-language Spatial Predicates*, National Center for Geographic Information and Analysis Technical Report 94-1, pp. 1–28.

Eklund, P.W., Kirkby, S.D. and Salim, A., 1998, Data mining and soil salinity analysis. *International Journal of Geographical Information Science*, 12, 247–268.

Elder, J.F. and Pregibon, D., 1996, A statistical perspective on knowledge discovery in databases. In Fayyad, U., Piatetsky-Shapiro, G, Smyth, P. and Uthurusamy, R. (eds.), *Advances in Knowledge Discovery and Data Mining,* Cambridge, MA: AAAI/MIT Press, pp. 83–113.

Ester, M., Kriegel, H.-P., Sander, J. and Xu, X., 1996, A Density Based Algorithm for Discovering Clusters in Large Spatial Databases with Noise. *Proceedings 2nd International Conference on Knowledge Discovery and Data Mining (KDD-96)*, pp. 226–231.

Ester, M., Kriegel, H.-P. and Sander, J., 1998, Algorithms for characterization and trend detection in spatial databases. *Proceedings 4th International Conference on Knowledge Discovery and Data Mining (KDD'98)*, New York, USA, pp. 44–50.

Fayyad, U., 1997, Editorial. *Data Mining and Knowledge Discovery,* 1, 5–10.

Fayyad, U., Piatetsky-Shapiro, G. and Smyth, P., 1996, From data mining to knowledge discovery in databases. *AI Magazine*, Fall, pp. 37–54.

Fotheringham, A.S., Charlton, M. and Brunsdon, C., 1997, Two techniques for exploring non-stationarity in geographical data. *Geographical Systems*, 4, 59–82.

Gahegan, M., 2000a, The case for inductive and visual techniques in the analysis of spatial data. *Geographical Systems*, 7(2): 77–83.

Gahegan, M., 2000b, On the application of inductive machine learning tools to geographical analysis. *Geographical Analysis*, 32, 113–139.

Gahegan, M., Takatsuka, M., Wheeler, M. and Hardisty, F., 2000, GeoVISTA Studio: A Geocomputational Workbench. *Proceedings 4th Annual Conference on GeoComputation*, UK, August 2000. URL: *http://www.ashville.demon.co.uk/gc2000/*

Gahegan, M., Wachowicz, M., Harrower, M. and Rhyne, T.-M., 2001, The integration of geographic visualization with knowledge discovery in databases and geocomputation. *Cartography and Geographic Information Systems* (special issue on the ICA research agenda).

Gains, B.R., 1996, Transforming Rules and Trees into Comprehensible Knowledge Structures. In Fayyad, U., Piatetsky–Shapiro, G., Smyth, P. and Uthurusamy, R. (eds.), *Advances in Knowledge Discovery and Data Mining* Cambridge, MA: AAAI/MIT Press.

Gardner, C., 1996, *IBM Data Mining Technology*, IBM Corporation, Stamford, Connecticut.

Gehrke, J., Ganti, V., Ramrakrishnan, R. and Loh, W.-Y., 1999, BOAT—Optimistic decision tree construction. *Proceedings SIGMOD 1999*, ACM Press: New York, pp. 169–180.

Getis, A. and Ord, J.K., 1992. The analysis of spatial association by use of distance statistics. *Geographical Analysis,* 24, 189–206.

Getis, A. and Ord, J.K., 1996, Local spatial statistics: An overview. In Longley, P. and Batty, M. (eds.), *Spatial Analysis: Modelling in a GIS Environment*, Cambridge, UK: GeoInformation International, pp. 261–277.

Glymour, C., Madigan, D., Pregibon, D. and Smyth, P., 1996, Statistical inference and data mining *Communications of the ACM*, 39(11): 35–41.

Glymour, C., Madigan, D., Pregibon, D. and Smyth P., 1997, Statistical themes and lessons for data mining. *Journal of Data Mining and Knowledge Discovery*, Vol. 1, pp. 11–28.

Goodchild, M.F., Buttenfield, B.P., Adler, P., Krygiel, A., Onsrud, H. and Kahn, R., 1999, *Distributed Geolibraries*. National Research Council Monograph. Washington, D.C.: National Academy Press.

Gopal, S., Liu, W. and Woodcock, C., 2001, Visualization based on fuzzy ARTMAP neural network for mining remotely sensed data. In Miller, H.J. and Han, J. (eds.), *Geographic Data Mining and Knowledge Discovery*, London: Taylor & Francis.

Gray, J. and Shenoy, P., 2000, Rules of thumb in data engineering. *Proceedings of 16th International Conference on Data Engineering, IEEE*, pp. 3–13.

Haslett, J., Bradley, R., Craig, P., Unwin, A. and Wills, G., 1991, Dynamic graphics for exploring spatial data with application to locating global and local anomalies. *The American Statistician*, Vol. 45, No. 3, pp. 234–242.

Healey, R., 1991, Database management systems. In Maguire, D., Goodchild, M.F. and Rhind, D. (eds.), *Geographic Information Systems: Principles and Applications*. London: Longman.

Hedberg, S.R., 1996, Search for the mother lode: Tales of the first data miners. *IEEE Expert*, 11(5): 4–7.

Hinneburg, A., Keim, D. and Wawryniuk, M., 1999, HD-Eye: Visual mining of high dimensional data. *IEEE Computer Graphics and Applications*, September/October 1999, pp. 22–31.

Hornsby, K. and Egenhofer, M.J., 2000, Identity-based change: A foundation for spatio-temporal knowledge representation. *International Journal of Geographical Information Science*, 14, 207–224.

Kaufman, L. and Rousseeuw, P.J., 1990, *Finding Groups in Data: An Introduction to Cluster Analysis*, New York: Wiley.

Keim, D. and Kriegel, H.-P., 1996, Visualization techniques for mining large databases: A comparison. *IEEE Transactions on Knowledge and Data Engineering* (Special Issue on Data Mining).

Knorr, E.M. and Ng, R.T., 1996, Finding aggregate proximity relationships and commonalities in spatial data mining. *IEEE Transactions on Knowledge and Data Engineering*, Vol. 8, No. 6, pp. 884–897.

Koperski, K. and Han, J., 1995, Discovery of Spatial Association Rules in Geographic Information Databases. *Proceedings 4th International Symposium on Large Spatial Databases* (SSD95), Maine, pp. 47–66.

Koperski, K. Han, J. and Adhikary, J., 1999, Mining knowledge in geographic data. *Communications of the ACM*, URL: *http://db.cs.sfu.ca/sections/publication/kdd/kdd.html*

Lee, H. and Ong, H., 1996, Visualization support for data mining. *IEEE Expert*, 11(5):69–75.

Lees, B.G. and Ritman, K., 1991, Decision-tree and rule-induction approach to integration of remotely sensed and GIS data in mapping vegetation in disturbed or hilly environments. *Environmental Management*, 15, 823–831.

Lmielinski, T. and Mannila, H., 1996, A database perspective on knowledge discovery. *Communications of the ACM*, 39(11): 58–64.

Lu, H., Setiono, R. and Liu, H., 1996, Effective data mining using neural networks. *IEEE Transactions on Knowledge and Data Engineering*, 8(6): 957–961.

Luger, G.F. and Stubblefield, W.A., 1998, *Artificial Intelligence: Structures and Strategies for Complex Problem Solving*, Reading, MA: Addison-Wesley.

MacDougall, E.B., 1992, Exploratory analysis, dynamic statistical visualization and geographic information systems. *Cartography and Geographical Information Systems*, Vol. 19, No. 4, pp. 237–246.

MacEachren, A.M., Wachowicz, M., Edsall, R., Haug, D. and Masters, R., 1999, Constructing knowledge from multivariate spatio-temporal data: integrating geographical visualization with knowledge discovery in database methods. *International Journal of Geographic Information Science*, Vol. 13, No. 4, pp. 311–334.

Malerba, D., Esposito, F., Lanza, A. and Lisi, F.A., 2001, Machine learning for information extraction from topographic maps. In Miller, H. J. and Han, J. (eds.), *Geographic Data Mining and Knowledge Discovery*, London: Taylor & Francis.

Marble, D.F., Gou, Z., Liu, L. and Saunders, J., 1997, Recent advances in the exploratory analysis of interregional flows in space and time. In Kemp, Z. (ed.), *Innovations in GIS 4*, London: Taylor & Francis.

Mardia, K.V., Kent, T. and Bibby, J.M., 1979, *Multivariate Analysis*, London: Academic Press.

Martin, J.C., 1991, *Introduction to Languages and the Theory of Computation*, New York: McGraw-Hill.

Mesrobian, E., Muntz, R., Shek, E., Nittel, S., La Rouche, M., Kriguer, M., Mechoso, C., Farrara, J., Stolorz, P. and Nakamura, H., 1996, Mining geophysical data for knowledge. *IEEE Expert*, 11(5): 34–44.

Mihalisin, T., Timlin, J. and Schwegler, J., 1991, Visualizing multivariate functions, data and distributions. *IEEE Computer Graphics and Applications*, Vol. 19, No. 13, pp. 28–35.

Miller, H.J., 2000, Geographic representation in spatial analysis. *Journal of Geographical Systems*, 2, 55–60.

Miller, H.J. and Han, J., 2000, Discovering geographic knowledge in data rich environments: A report on a specialist meeting. *SIGKDD Explorations: Newsletter of the (Association for Computing Machinery) Special Interest Group on Knowledge Discovery and Data Mining*, 1(2): 105–108; available at *http://www.acm.org/sigs/sigkdd/explorations/*

Miller, H.J. and Han, J. (eds.), 2001, *Geographic Data Mining and Knowledge Discovery*, London: Taylor & Francis.

Miller, H.J. and Jiawei, H., 2001, Geographic data mining and knowledge discovery: An overview. In Miller, H. J. and Han, J. (eds.), *Geographic Data Mining and Knowledge Discovery*, London: Taylor & Francis.

Mitchell, T.M., 1997, *Machine Learning*, New York: McGraw-Hill.

Murray, A.T. and Estivill-Castro, V., 1997, Cluster discovery techniques for exploratory spatial data analysis. *International Journal of Geographical Information Science*, 12(5): 431–443.

National Public Radio Morning Edition, 1998, *Report on emergence of commercial Internet search engines such as Yahoo and other "dot-coms"*, 3 April, 1998.

National Research Council, 1995, *Data Foundation for the National Spatial Data Infrastructure*, Mapping Science Committee, Sugarbaker, L.A., Chair. Washington, D.C.: National Academy Press.

National Research Council, 1999, *Distributed Geolibraries: Spatial Information Resources. Panel on Distributed Geolibraries,* Goodchild, M.F., Chair, Washington, D.C.: National Academy Press.

Okabe, A. and Miller, H.J., 1996, Exact computational methods for calculating distances between objects in a cartographic database. *Cartography and Geographic Information Systems*, 23, 180–195.

Openshaw, S., Cross, A. and Charlton, M., 1990, Building a prototype geographical correlates machine. *International Journal of Geographical Information Systems*, Vol. 4, No. 4, pp. 297–312.

Openshaw, S., 1994, Two exploratory space-time-attribute pattern analysers relevant to GIS. In Fotheringham, A.S. and Rogerson, P.A. (eds.), *Spatial Analysis and GIS*, London: Taylor & Francis, 83–104.

Openshaw, S., Charlton, M., Wymer, C. and Craft, A., 1987, A mark 1 geographical analysis machine for automated analysis of point data sets. *International Journal of Geographical Information Systems*, 1, 335–358.

Peuquet, D.J. and Zhang, C.-X., 1987, An algorithm to determine the directional relationship between arbitrarily-shaped polygons in the plane. *Pattern Recognition*, 20, 65–74.

Pierce, C.S., 1878, Deduction, induction and hypothesis. *Popular Science Monthly*, 13, 470–482.

Rainsford, C.P. and Roddick, J.F., 1999, Database issues in knowledge discovery and data mining. *Australian Journal of Information Systems*, Vol. 6, No. 2, pp. 101–128.

Ribarsky, W., Katz, J. and Holland, A., 1999, Discovery visualization using fast clustering. *IEEE Computer Graphics and Applications*, September/October 1999, pp. 32–39.

Roddick, J.F. and Spiliopoulou, M., 1999, A bibliography of temporal, spatial and spatio-temporal data mining research. *SIGKDD Explorations*, Vol. 1, No. 1, URL: *http://www.cis.unisa.edu.au/~cisjfr/STDMPapers/*

Roddick , J.F. and Lees, B., 2001, Paradigms for spatial and spatio-temporal data mining. In Miller, H.J. and Han, J. (eds.), *Geographic Data Mining and Knowledge Discovery*, London: Taylor & Francis.

Smyth, C.S. 2001, Mining mobile trajectories. In Miller, H J. and Han, J. (eds), *Geographic Data Mining and Knowledge Discovery*, New York: Taylor and Francis, pp: 337–362.

Smythe, P., 2000, Data mining: Data analysis on a grand scale? *Statistical Methods in Medical Research*, September, 2000.

Sowa, J.F., 1999, *Knowledge Representation: Logical, Philosophical, and Computational Foundations*, Pacific Grove, CA: Brooks/Cole.

Srinivasan, A. and Richards, J.A., 1993, Analysis of GIS spatial data using knowledge-based methods. *International Journal of Geographical Information Systems*, 7, 479–500.

Tukey, J.W., 1977, *Exploratory Data Analysis*. Reading, MA: Addison-Wesley.

Valdez-Perez, R.E., 1999, Principles of human computer collaboration for knowledge discovery in science. *Artificial Intelligence*, Vol. 107, No. 2, pp. 335–346.

Wong, P.C., 1999, Visual data mining. *IEEE Computer Graphics and Applications*, Vol. 19, No. 5, pp. 20–21.

Yuan, M., 1997, Use of knowledge acquisition to build wildfire representation in geographic information systems. *International Journal of Geographical Information Systems,* 11, 723–745.

Yuan, M., 1998, Representing Spatiotemporal Processes to Support Knowledge Discovery in GIS databases. *Proceedings 8th International Symposium on Spatial Data Handling Spatial Data Handling*, Poiker, T.K. and Chrisman, N. (eds.), pp. 431–440.

Yuan, M., 2001, Representing complex geographic phenomena with both object- and field-like properties. *Cartography and Geographic Information Science,* 28(2). 83–96.

Zaïane, O.R., Han, J., Li, Z.-N. and Hou, J., 1998, Mining Multimedia Data. *Proceedings, CASCON'98*: Meeting of Minds, Toronto, Canada, November 1998; available at: *http://db.cs.sfu.ca/sections/publication/smmd/smmdb.html*

CHAPTER FIFTEEN

Postscript on the UCGIS and Research

David M. Mark, University at Buffalo
E. Lynn Usery, University of Georgia and U.S. Geological Survey
Robert B. McMaster, University of Minnesota

At the meeting in Boulder, Colorado, in December 1994, that led to the formation of the University Consortium for Geographic Information Science UCGIS), the delegates identified the promotion and coordination of GIScience research as a major goal for the fledgling organization. Over the following 18 months, this general objective was instantiated in Columbus through the process described in Chapter 1 of this book. By the end of the Columbus meeting, many of us were exhausted—but UCGIS had 10 Research Priorities, with writing teams identified to flesh out the topics into White Papers. Great, but what next?

Right there in Columbus, we began to develop plans for promoting the Priorities, especially to the government and private sectors. And, under urging to "Think Big", someone came up with the idea of holding a Congressional Breakfast. Why not? John Bossler and his staff at the Center for Mapping at Ohio State contacted the office of Senator John Glenn of Ohio for sponsorship, and in January 1997, UCGIS hosted its first Congressional Breakfast on Capitol Hill, with a program structured around five of the Research Priorities. More than a dozen congressional staff members attended, but it seemed that cutting-edge priorities for basic research are of little interest to politicians and their staff members—such people are motivated by practical considerations of how implemented information systems can save lives or money or both, make the nation and the world safer and better through efficient, effective, and equitable technology. UCGIS learned from this experience, and in subsequent congressional breakfasts and Capitol Hill briefings, we emphasized geographic information system (GIS) applications, showing GIS success stories and pointing out how improved systems and data could lead to even more effective applications of the technology.

What then of the Research Priorities? First, as an organization including many research universities, dedicated to fostering GIScience research, education, and practice, it was very important that we had established a set of priorities or research challenges—perhaps their detailed content is not of interest to the legislative side of the government, but their existence showed that we were serious and that we were organized. Second, the Priorities and their associated white papers continue to be an important part of our interactions with Federal agencies and with the GIS industry, organizations that must make the connections between theory and practice. Third, the UCGIS Research White Papers, most of which have matured into chapters in this book, have provided foundations for graduate education and research training in GIScience. In 1997, the 10 topics and their white papers formed the main topics for the first UCGIS Virtual Seminar, and to this day we use them in our courses, either as sources for lecture material or as readings for our students.

The UCGIS agenda for GIScience research provides a basis for funding agencies, research scientists, students, and others to determine a focus for science. The agenda spans the range of topics in broad areas that can be summarized as follows. Problems of data acquisition and the integration of various types of data are discussed and referenced in several of the chapters of this book, especially Chapter 2—Spatial Data Acquisition and Integration and Chapter 13—Remotely Acquired Data and Information in GIScience. The fundamental basis of cognition and representation of geographic data and information are discussed and referenced in Chapter 3—Cognition of Geographic Information, Chapter—4 Scale, Chapter 5—Extensions to Geographic Representations, and Chapter 12—Ontological Foundations for Geographic Information Science. The problems of using geographic data and information for analysis and modeling are examined in Chapter 6—Spatial Analysis and Modeling in a GIS Environment, Chapter 7—Research Issues on Uncertainty in Geographic Data and GIS-Based Analysis, and Chapter 14—Geospatial Data Mining and Knowledge Discovery. The display and visualization of geographic data for both presentation and data exploration are discussed in Chapter 11—Geographic Visualization. Societal and institutional issues and geographical information engineering are examined in Chapter 8—The Future of the Spatial Information Infrastructure, Chapter 9—Distributed and Mobile Computing, and Chapter 10—GIS and Society: Interrelation, Integration, and Transformation. The range of topics covered in these chapters is sufficient to incorporate most scientific work in the field, while the specific detail and referenced literature provide in-depth treatment of major research issues.

Thus we have witnessed both the birth of a national-level GIS organization and the development of a comprehensive research agenda over a decade-long period. In addition to the research agenda—the challenges—documented in this book, it should also be noted that other activities within the organization have produced an education agenda, and a set of application challenges. Each of these activities is reported in the literature. A notable difference in the development of these "agenda creating" activities (for research, education, and applications) is the iterative nature of the research activity. The process, which started at the first UCGIS Assembly in Columbus (1996), was revisited at the 1998 Park City, Utah, Assembly (where expert panels on each of the 10 topics were brought together), the 2000 (Oregon Assembly) where a series of new challenges were proposed, and the 2002 Athens, Georgia, Assembly where a final list of 10 plus 4 long-term challenges was confirmed. It has been a carefully thought-out, time-consuming, and, at times, debatable process.

The final set of challenges developed by the UCGIS and its membership is certainly dependent on the model that was used, and specific decisions made. Other models undoubtedly would have led to different outcomes. In terms of the original model—that of soliciting topics from UCGIS member institutions—only those ideas from a limited number of institutions were included. The actual process of clustering topics into larger themes before the first assembly was also a subjective yet necessary activity that resulted in certain "biases". Furthermore, at the first assembly a decision was made to "de-emphasize" applications, thus focusing the agenda on basic research. A decision at the 1998 assembly in Park City was to fix the 10 then-called Priorities (later changed to challenges), and not allow for an expansion. Later the set was expanded, but a topic was only allowed if a set of institutions, not just one, supported it. Clearly, the model and methods used by the UCGIS have helped to determine the final set just as much as the intrinsic value of the topic itself. For instance, one can point to the missing topic of cartography and visualization at the first assembly, simply since other priorities dominated (and no individual, or set of individuals, pushed for this).

It would be interesting to see what other models might have produced in a national research agenda. What if researchers had arrived at the first assembly without topics submitted beforehand, and the process had started from ground zero at the event? What if the UCGIS Board had derived the first pass for discussion? What if the first assembly had been dominated by national research laboratories, not academic researchers? And, as discussed below, what if foreign researchers had been allowed to participate in the process (initially, the UCGIS was only open to U.S. based institutions)?

The UCGIS research agenda was developed as a national agenda in the United States by members of the GIScience research community primarily in North America. This agenda, while broad and inclusive, is not comprehensive and has different emphases than GIScience research in other parts of the world. For example, the Association for Geographic Information Laboratories in Europe (AGILE) hosts a conference each year addressing GIScience research. AGILE conference themes in 2003 included geographic information and policy, e-government, disaster and risk management, decision support systems, semantic interoperability, real-time GIS, and other topics, reflecting key issues in the European context. While the common areas of research interests with UCGIS are broad, there are differences in emphasis and approach with other organizations. The Online Resource for Managing Ecological Data and Information (*http://www.ecoinformatics.org*) has projects examining a knowledge network for biocomplexity, metadata entry for environmental data, a science environment for ecological knowledge, and a vegetation databank. Geoinformatics, which includes content similar to GIScience, has several conferences worldwide, among them those sponsored by the Chinese

Professionals in GIScience (CPGIS). The CPGIS 2004 conference themes include: location based services and mobile GIS, GIS and planning support systems, spatial decision support systems, information extraction from remote sensing data, data fusion and digital mapping, multi-scale and multi-media representation, spatio-temporal modeling/analysis, geovisualization and virtual environments, spatio-temporal databases, geocomputational modeling, uncertainty and spatial data quality control, and other topics.

The UCGIS agenda is distinctive with its themes for GIScience research from other parts of the world and other organizations in its focus on basic research rather than applications of the science. It also differs with its evolution to a two-tiered approach of long-term research challenges and shorter-term high profile research topics or priorities. The long-term challenges represent more basic research issues that will not yield solutions with a few focused research projects. These challenges are continual and result from the basic characteristics, *e.g.*, locational basis, cognitive dependencies, spatial autocorrelation, subjectivity of areal units, scale, that make geographic data unique. The short-term research areas may be directly addressed and require advancements in current concepts and methodologies to provide solutions.

A high-priority goal for UCGIS when it was established was to work to increase the nation's investment in GIScience research. The National Science Foundation (NSF) is the primary U.S. agency for the support of non-medical basic research, and so it was appropriate when NSF sponsored and hosted a "Workshop on Geographic Information Science and Geospatial Activities at NSF" in January 1999. Although not officially a UCGIS event, most of the participants were from UCGIS institutions. The resulting report (Mark, 1999) was circulated widely at NSF, and reprinted in the *URISA Journal* (Mark, 2000). Whereas no funding program specifically dedicated to GIScience research has emerged from NSF to date, the new "Human and Social Dynamics" competition (HSD; NSF 04-537) reflects a new prominence for a spatial perspective on research at NSF. The next decade will tell whether the persistent efforts of UCGIS to promote the importance of GIScience research pay off in enhanced research funding, increased research progress, and advances in Geographic Information Systems and Technology.

REFERENCES

Mark, D.M., (ed), 1999, *Geographic Information Science: Critical Issues in an Emerging Cross-Disciplinary Research Domain*. Report on a workshop held at the National Science Foundation, January 14–15, 1999, *http://www.geog. buffalo.edu/ncgia/workshopreport.html*

Mark, D.M., (ed), 2000, Geographic Information Science: Critical Issues in an Emerging Cross-Disciplinary Research Domain. *Journal of the Urban and Regional Information Systems Association*, 12(1): 45–54.

INDEX